EAI/Springer Innovations in Communication and Computing

Series Editor
Imrich Chlamtac, CreateNet, Trento, Italy

Editor's Note

The impact of information technologies is creating a new world yet not fully understood. The extent and speed of economic, life style and social changes already perceived in everyday life is hard to estimate without understanding the technological driving forces behind it. This series presents contributed volumes featuring the latest research and development in the various information engineering technologies that play a key role in this process.

The range of topics, focusing primarily on communications and computing engineering include, but hardly limited to, wireless networks; mobile communication; design and learning; gaming; interaction; e-health and pervasive healthcare; energy management; smart grids; internet of things; cognitive radio networks; computation; cloud computing; ubiquitous connectivity, and in mode general smart living, smart cities, Internet of Things and more. The series publishes a combination of expanded papers selected from hosted and sponsored European Alliance for Innovation (EAI) conferences that present cutting edge, global research as well as provide new perspectives on traditional related engineering fields. This content, complemented with open calls for contribution of book titles and individual chapters, together maintain Springer's and EAI's high standards of academic excellence. The audience for the books consists of researchers, industry professionals, advanced level students as well as practitioners in related fields of activity include information and communication specialists, security experts, economists, urban planners, doctors, and in general representatives in all those walks of life affected ad contributing to the information revolution.

About EAI

EAI is a grassroots member organization initiated through cooperation between businesses, public, private and government organizations to address the global challenges of Europe's future competitiveness and link the European Research community with its counterparts around the globe. EAI tens of thousands of members on all continents together with its institutional members base consisting of some of the largest companies in the world, government organizations, educational institutions, strive to provide a research and innovation platform which recognizes excellence and links top ideas with markets through its innovation programs.

Throughs its open free membership model EAI promotes a new research and innovation culture based on collaboration, connectivity and excellent recognition by community.

More information about this series at http://www.springer.com/series/15427

Mubashir Husain Rehmani • Riadh Dhaou

Editors

Cognitive Radio, Mobile Communications and Wireless Networks

 Springer

RESEARCH MEETS INNOVATION

Editors
Mubashir Husain Rehmani
Telecommunications Software
and Systems Group
Waterford Institute of Technology
Waterford, Ireland

Riadh Dhaou
National Polytechnic Institute of Toulouse
Toulouse, France

ISSN 2522-8595 ISSN 2522-8609 (electronic)
EAI/Springer Innovations in Communication and Computing
ISBN 978-3-030-08154-6 ISBN 978-3-319-91002-4 (eBook)
https://doi.org/10.1007/978-3-319-91002-4

This Springer imprint is published by the registered company Springer Nature Switzerland AG.
The registered company address is: Gewerbestrasse 11, 6330 Cham, Switzerland

Contents

Author Biography

Mubashir Husain Rehmani (M'14-SM'15) received the B.Eng. degree in computer systems engineering from Mehran University of Engineering and Technology, Jamshoro, Pakistan, in 2004, the M.S. degree from the University of Paris XI, Paris, France, in 2008, and the Ph.D. degree from the University Pierre and Marie Curie, Paris, in 2011. He is currently working at the Telecommunications Software and Systems Group (TSSG), Waterford Institute of Technology (WIT), Waterford, Ireland. He served for 5 years as an Assistant Professor at COMSATS Institute of Information Technology, Wah Cantt., Pakistan. He is currently an Area Editor of the *IEEE Communications Surveys and Tutorials.* He served for 3 years (from 2015 to 2017) as an Associate Editor of the *IEEE Communications Surveys and Tutorials.* Currently, he serves as Associate Editor of *IEEE Communications Magazine*, Elsevier *Journal of Network and Computer Applications (JNCA)*, and the *Journal of Communications and Networks (JCN)*. He is also serving as a Guest Editor of Elsevier *Ad Hoc Networks* journal, Elsevier *Future Generation Computer Systems* journal, the *IEEE Transactions on Industrial Informatics,* and Elsevier *Pervasive and Mobile Computing* journal. He has authored/edited two books published by IGI Global, USA, one book published by CRC Press, USA, and one book with Wiley, UK. He received "Best Researcher of the Year 2015 of COMSATS Wah" award in 2015. He received the certificate of appreciation, *"Exemplary Editor of the IEEE Communications Surveys and Tutorials for the year 2015"* from the IEEE Communications Society. He received Best Paper Award from IEEE ComSoc Technical Committee on Communications Systems Integration and Modeling (CSIM), in IEEE ICC 2017. He consecutively received research productivity award in 2016–2017 and also ranked # 1 in all Engineering disciplines from Pakistan Council for Science and Technology (PCST), Government of Pakistan. He also received Best Paper Award in 2017 from Higher Education Commission (HEC), Government of Pakistan.

Dr. Riadh Dhaou Associate Professor with the Toulouse INP (Institut National Polytechnique de Toulouse). He is attached to the Telecom and Networking Department of the ENSEEIHT. He is member, since 2003, of the IRT team of the IRIT

(Institut de Recherche en Informatique de Toulouse) Laboratory. He received the Engineering degree in computer science (Diplome d'Ingénieur Concepteur en Informatique) from the ENSI (Ecole Nationale des Sciences de l'Informatique), University of Tunis II in 1997, and the D.E.A. (Diplome d'Etudes Approfondies) in Computer Systems from the Université Pierre et Marie Curie in Paris (Paris VI), in 1998. He was awarded, respectively, a Ph.D. degree in Computer Systems, Telecommunication and Electronic by the University of Paris VI (in November 2002) and the HDR (Habilitation à Diriger des Recherches) by the Toulouse INP (in November 2017). His research interests include statistical characterization and modelling of mobility, mobile and space communications, cross layer schemes modelling and optimization, performance analysis of wireless networks, autonomous multi-hop/cooperative communications systems, capacity and outage analysis of multi-user heterogeneous wireless systems, resource allocation, design and performance evaluation of wireless sensor networks and energy consumption optimization. Since 2003, he is scientific chief project with the cooperative laboratory TéSA, a non-profit association, leading research studies and PhDs in Telecommunications for Space and Aeronautics. Since November 2017, he is the carrier of the satellite theme within the IRT team. He jointly supervised 14 Ph.D. theses (9 were defended) and 3 master-degree theses. He published about 78 papers (7 journals and 5 book chapters) and achieved 35 research grants in satellite and sensor networks (CNES, Thales-Alenia Space, Airbus D&S). He has been technical leader to 7 research grants in satellite networks domain and participated in several industrial and academic grants. He was involved in the Technical Program Committee of 7 International Conferences. He was General Chair of PSATS'2013 and was member of one Organization Committee of two other International Conferences. He is, since 2013, part of the Editorial Board of WINET (The springer Wireless Networks journal). He participated in 11 Ph.D. thesis committees. He participated in several European and National projects: CAPES-COFECUB Project MMAPS (Management, Mobility, Security, Architecture and Protocols for the Future Internet of Things) – ANR Project CAPTEURS – RNRT Project DILAN – ESPRIT Project BISANTE (Broadband Integrated Satellite Network Traffic Evaluation) – RNRT Project CONSTELLATIONS. He also participated in the Network of Excellence NoE Euro-NGI, particularly on the evolution of IP networks.

Chapter 1
Towards Spectrum Sharing in Virtualized Networks: A Survey and an Outlook

Furqan Ahmed, Adrian Kliks, Leonardo Goratti, and Shah Nawaz Khan

1.1 Introduction

The 5G vision entails a diverse set of requirements including $1000\times$ increase in aggregate data rates, to be achieved by the year 2020 [1]. To this end, apart from physical and medium access control layer innovations, new spectrum sharing paradigms have to be defined in conjunction with efficient network virtualization techniques [2]. In fact, flexible yet effective spectrum management is considered as one of the key enablers for the success of 5G networks. The benefits achievable by the dense deployments of heterogeneous mobile networks (HMNs) in 5G networks can be further enhanced by employing advanced spectrum sharing techniques. Given that the spectrum is to be allocated to multiple operators, with potentially high deployment densities, classical coexistence solutions within the licensed as well as unlicensed spectrum may not provide required isolation and efficiency. In particular, approaches based on exclusive spectrum access in licensed bands may be inefficient in small-cell scenarios with non-uniform user density, variable area coverage of operators and variable load. In this regard, alternatives to licensed-only techniques, such as licensed shared access (LSA), licensed assisted access (LAA), license exempt, co-primary shared access (CSA), and pluralistic licensing, are of paramount

F. Ahmed (✉)
KTH Royal Institute of Technology, Stockholm, Sweden
e-mail: furqanah@kth.se

A. Kliks
Poznan University of Technology, Poznań, Poland
e-mail: adrian.kliks@put.poznan.pl

L. Goratti · S. N. Khan
CREATE-NET, Trento, Italy
e-mail: leonardo.goratti@create-net.org; s.khan@create-net.org

© Springer International Publishing AG, part of Springer Nature 2019
M. H. Rehmani, R. Dhaou (eds.), *Cognitive Radio, Mobile Communications and Wireless Networks*, EAI/Springer Innovations in Communication and Computing,
https://doi.org/10.1007/978-3-319-91002-4_1

importance and constitute the emerging vision for spectrum utilization. In order to enable this in an efficient manner in 5G radio access network (RAN), a virtualization-based control and coordination framework is required [3].

In the context of abstraction of RAN and its virtualization, exclusive use seems to be one of the simplest strategies, as static spectrum allocation to operators and services simply eliminates the spectrum aspects from the virtualization process. Furthermore, network operators will be able to provide reliable statistics on various systems available on certain geographical areas, such as the amount of spectrum, allowed data rates, feasible traffic density, maximum transmit power, etc. These statistics will be stable (i.e. already averaged over the long observation time) and thus could be efficiently used in the network abstraction process. However, the usage of frequency resources will require permanent monitoring in order to assess the amount of remaining resources. This approach could provide benefits for both intra- and inter-operator resource sharing schemes. The main issue with the static assign-ment is that it reduces the degrees of freedom, which can be utilized for more efficient system virtualization [4]. Starting with a concise overview of notable spectrum sharing schemes and related challenges relevant to 5G, we focus on virtualization and discuss an architecture for enabling an efficient management of spectrum.

The main contributions of this chapter can be summarized as follows:

- It gives a concise overview of spectrum sharing schemes relevant to 5G networks. This includes key spectrum sharing techniques, regulations, and recent trials.
- Challenges for spectrum sharing in virtualized networks are discussed.
- Finally, a spectrum management framework for enabling virtualized spectrum sharing is discussed.

The rest of the chapter is organized as follows. Section 1.2 gives an overview of different spectrum sharing schemes. Sections 1.3 and 1.4 discuss the regulatory aspects and recent trials, respectively. In Sect. 1.5, we discuss virtualization in conjunction with a spectrum management architecture for virtualized networks. In Sect. 1.6, the focus is on main challenges for spectrum sharing in virtualized networks. Finally, conclusions are given in Sect. 1.7.

1.2   Spectrum Sharing for 5G: An Overview

It is highly likely that in 5G networks, two traditional models of spectrum manage-ment and licensing schemes, namely, exclusive use and license exempt, will be replaced by their more flexible versions [5, 6]. Moreover, new spectrum sharing and aggregation schemes are also emerging rapidly. For example, a small-scale spectrum aggregation and sharing scheme is proposed in [7]. A pluralistic licensing scheme is discussed in [19], and in [60], authors present a geolocation database-inspired spectrum sharing scheme with network-wide optimality. Spectrum sharing schemes based on public-private partnerships are particularly important from a 5G perspective [8]. Although the classical static solutions are easily implementable in practice, flexibility is essential for accommodating the expected traffic growth and diverse use cases of 5G. Various virtualization approaches towards resource sharing (e.g. network, spectrum, and infrastructure) have been foreseen as a solution to the problem of spectrum scarcity and underutilization [4].

Mobile operators have traditionally avoided sharing the spectrum, as well as the infrastructure, mainly due to the competition. Likewise, the incumbents usually do not prefer to share their respective networks since it may lead to the loss of some customers in favour of the rival operators. In recent years, a steady increase has been observed in sharing activities between the mobile operators. However, this is mainly limited to the infrastructure. The motivation behind sharing infrastructure or spectrum stems from the fact that average revenue per user has decreased, while capacity demands have increased substantially [9]. In principle, the peak traffic at mobile sites has evolved from a few tens of Mbps to hundreds or in some cases to several Gbps. Other factors that have contributed to making infrastructure sharing favourable include a multitude of costs associated with deployment and maintenance of the network [10]. Thus, the main advantage of infrastructure sharing is that it leads to a decrease in the capital expenditure (CAPEX), especially in remote areas where the traffic is low. Moreover, multi-operator spectrum sharing in cooperative settings is beneficial, especially with dynamic and asymmetric traffic [11]. The main issue is to establish a trustworthy relation between operators. Therefore, the complicating aspect in sharing is fairness, data transparency, and service quality agreements among the operators.

From a 5G spectrum sharing perspective, the role of dedicated databases and the application of cognitive radio technologies are of considerable importance. In such approaches, the information of available spectrum at certain location is stored in a dedicated database which can be accessed by any authorized player (e.g. operator, regulator, policy-maker) or user/device (such as mobile terminals, base stations). These databases can contain various types of data, ranging from the definition of operational policies and legal regulations (delivered by the regulators) to the limitations identified on the basis of transmit technology (defined by, e.g. operators) and the allowed interference level or transmit powers. A number of solutions based on databases have been proposed in the literature [12–15]. In [12], spectrum sensing augmented with databases is proposed for secondary access to television white space (TVWS) spectrum. This leads to an improved protection of primary users, which include incumbents such as digital terrestrial television and program making and special events users. Dynamic spectrum access-based secondary access to other licensed spectrum bands between 30 MHz and 6 GHz is discussed in [13]. In order to support traditional spectrum databases, an open access spectrum database based on measurements optimized by big data processing is considered in [14]. Likewise, in [15], the authors discuss a radio environment map construction based on heterogeneous sensing, for supporting the centralized database.

One of the key aspects associated with the use of databases is their need of periodic, continuous, and accurate updates. This can be enabled on the basis of channel measurements carried out by devices (e.g. mobile terminals or dedicated sensors). Cooperative spectrum sensing techniques constitute an important tool for updating the databases. Moreover, the application of databases and spectrum sensing can be treated either as a distinct spectrum sharing approach or merely a tool for realization of a given spectrum sharing scheme. In the former case, the interested party queries the database seeking permission for starting data transmission, with the specified parameters (e.g. preferred frequency band, transmit power, expected connection duration, expected data rate, etc.). It is worth noting that in such approach, the detailed spectrum sharing rules stored in the databases will be defined by the regulator, operators, and other key players. Consequently, a dedicated entity will make use of the database (augmented by artificial intelligence or machine learning, etc.) to allocate the spectrum and monitor and enforce execution of its decisions and application of necessary changes in spectrum assignment, if needed. This approach differs from the standard spectrum sharing schemes and is closely related to the original concept of cognitive radio where all spectrum utilization constraints can be relaxed and handled in an adaptive manner.

The latter case entails the use of the database as a specific tool required for application of spectrum sharing scheme. In this case, the database will need to be queried to obtain the necessary information about spectrum sharing rules (e.g. LSA sharing rules), applicable at a given location and/or time. Thus, databases seem to be the natural and necessary solution for practical realization of any spectrum sharing schemes. In fact, realization of most spectrum sharing schemes would rely on some sort of databases, the key difference will be in their level of complexity, size, update frequency, and flexibility. Furthermore, the use of databases paves the way for application of advanced algorithms for the management and processing of the massive data stored in them, thereby maximizing the efficiency of spectrum sharing. In what follows, we review the most promising spectrum sharing strategies briefly, highlighting their advantages and drawbacks.

1.2.1 Exclusive Use of Spectrum (Individual Licenses)

In the traditional case, the whole radio spectrum is partitioned and statically allocated to multiple operators for license exempt use. Such spectrum division is made usually at the national level. However, the pertinent international agreements (such as those made at World Radio Communication Conference level) have to be respected. Following these decisions, each country creates its own frequency allocation plan and defines the rules governing the use of frequency bands (e.g. the maximum transmit power is defined, types of services that can be delivered to the end-user, etc.). This is an efficient scheme for the management of interference between various systems as each frequency spectrum band has its own strict transmission rules, which are ensured by the service provider. However, such static spectrum assignment often leads to significant underutilization of spectrum as the frequency bands which are not in use at a given time and geographical location cannot be used by any other operator or service provider [16].

Together with careful system design (i.e. proper definitions of spectrum masks and right measurements of signal power leaking to neighbouring systems), this approach guarantees minimization of the mutual interference. The operating parameters are known (as these are defined by the standards and legal regulations), thus the management system and spectrum monitoring strategies are rather simple. From an operator point of view, exclusive use of dedicated spectrum allows for more stable investment plans and definition of developing strategies. Therefore, static assignment of frequency bands to services and licensed operators is feasible only in the case when the number of different services is relatively low. As the traffic demands are growing rapidly, exclusive use is not feasible anymore and will lead to severe underutilization of spectrum. With exclusive access, the operators cannot maximize revenue generation from the radio spectrum, when it is not completely used by their own network and end-users.

1.2.2   License-Exempt Rules (Unlicensed or Commons)

This scheme is essentially the opposite of exclusive access. Each user is allowed to access the spectrum band with no additional licenses or permissions. Notable examples include the ISM bands, in which various systems can coexist in the same geographical area (e.g. WiFi and Bluetooth operating in 2.4 GHz band). A key benefit of this approach is that a user can start utilizing the spectrum at any time and/or location. Consequently, the license-exempt spectrum bands are interference limited. In particular, due to absence of proper interference control mechanisms, unexpected performance degradation is inevitable. Various solutions to this problem have been proposed. However, in future, the unlicensed spectrum bands would eventually be facing the problem of tragedy of commons, where a given unlicensed band will not be usable at all due to collisions in accessing the spectrum or high in-band interference.

The main advantage of unlicensed spectrum is that it relaxes the requirement of obtaining a license. A direct consequence is the reduction in costs, along with an opportunity to use the spectrum freely. Open access spectrum also provides incentives for new networks and services to be deployed at low cost. A major downside is the difficulty in providing quality-of-service guarantees. It is also difficult to maintain fair use of spectrum in an open access scenario.

1.2.3   Licensed Shared Access (LSA) and Authorized Shared Access (ASA)

Licensed Shared Access (LSA) concept, originally introduced by the European Commission (EC) to respond to the industry interests, is a sharing model that aims at introducing additional licensed users on spectrum bands currently used by other incumbent systems [17]. The LSA concept has gained growing interest in regulation, standardization, and research, particularly for the deployment of mobile communication systems in the bands that are allocated to mobile but currently involve other types of incumbent usage. It enables the introduction of a limited number of new radio systems based on an individual licensing scheme with quality of service (QoS) guarantees for both the new entrant and the incumbent. The LSA concept is based on voluntariness and thus requires acceptance from the involved stakeholders and agreement on the terms and conditions for sharing. The rights to access the band are given to the entrant LSA licensee by the national regulatory authority (NRA) according to the agreement. The first application area under consideration for LSA in Europe is the 2.3–2.4 GHz band. It is allocated to the mobile service but is currently being used for other types of incumbents such as the programme making and special events (PMSE) service depending on the national situation. Regulatory efforts in European Conference of Postal and Telecommunications Administrations (CEPT) have developed an initial sharing framework for LSA. Other studies on the subject have developed harmonized technical conditions, cross-border coordination, guidelines for the LSA sharing framework, incumbent usage, implementation examples, and technical sharing solutions specifically between the mobile broadband and incumbent PMSE service in the 2.3–2.4 GHz band. As a result, the regulatory framework for LSA in the 2.3–2.4 GHz is ready for national adoption.

From the architecture and implementation perspective, the LSA concept is envisaged to be realized with two new additional components on top of existing cellular architecture: LSA Repository and LSA Controller. These have key roles in the management of interference along with the varying LSA band availability. Consequently, incumbents do not suffer from interference. The LSA Repository has a role in storing and updating the information about the LSA spectrum band availability and its usage conditions. It acts as the middle point between the incumbent and LSA licensee domains and collects the spectrum usage information from the incumbent. The LSA Controller is located in the mobile network side, and its role is to ensure the protection of the incumbent user and mobile network by calculating the protection areas based on the information received from the LSA Repository and the information on the mobile network layout. These two additional building blocks can be integrated into the existing cellular architecture in a straightforward manner and is currently being considered in the standardization activities.

The LSA concept is a sharing model that aims at providing attractive operational conditions for both the incumbent and the entrant systems, in terms of protection from harmful interference. The approach of mobile network operators (MNOs) for accessing new bands with LSA is based on building a limited number of new functionalities on top of the existing cellular architecture, thereby making the adoption easier. However, the challenge is the voluntary adoption of LSA. In particular, the incumbents need to be convinced to open up their deployed bands for sharing, which is not easy without true incentives.

1.2.4   Citizen Broadband Radio Service with Spectrum Access System

The Citizens Broadband Radio Service (CBRS) sharing model is a three-tier sharing model introduced by the Federal Communications Commission (FCC) in the USA for a 3550–3700 MHz band [18]. It enables additional usage on a band with existing incumbent usage on both licensed and license-exempt basis, while protecting the incumbent's rights. It introduces additional licensed users (Priority Access Licenses (PALs)), which have operational certainty similar to the LSA licensees, as well as additional license-exempt General Authorized Access (GAA) users. These GAA users need to be registered as CBRS users so there will be a finite number of GAA users and their operations are not protected from other CBRS users as PALs. The key component for the management of interference in this concept is the Spectrum Access System (SAS). It coordinates the spectrum usage of the CBRSs to protect the incumbents and PALs from other CBRS users. SAS is a combination of control functions and database for the coordination of interference. Additionally, the CBRS concept has adopted environmental sensing capability (ESC) for monitoring the incumbent activity which would detect the appearance of specific incumbents.

The US three-tier CBRS model is a more complex sharing model than the LSA concept as it introduces a third tier of opportunistic access which is not present in the LSA. The CBRS model provides a rich ecosystem by paving the way for new players to access the market. It enables more dynamic operations and improves the spectrum usage efficiency. However, the complexity of the concept is high. PAL licenses are issued on a census track level, which may result in very small areas for the licenses and complicated interference scenarios.

1.2.5   Pluralistic Licensing

The main idea of pluralistic licensing was proposed as an innovative approach to spectrum sharing between primary and secondary players (operators) [17]. It can be understood as the award of licenses under the assumption that opportunistic secondary spectrum access will be allowed and that interference may be caused to the primary with parameters and rules that are known to the primary at the point of obtaining the license [17]. Here we assume that the primary operator will choose from a range of offered pluralistic licenses, each with a different fee structure, and each specifying alternative opportunistic access rules that can be mapped to associated interference characteristics [6, 19]. As the core license is still granted to the primary operator, the main control mechanisms are kept by the primary. The primary might trade-off the form and degree of opportunistic access for a various licensing fees or another incentive. The decisions under pluralistic licensing can be made on short- or long-term, as well as on the geographical basis. It is worth noting that there might be some control mechanisms applied between co-primary operators at the same locations. Once the decisions are made and the offers are available to interested secondary operators, a cognitive mechanism is used by the operators to access the band. The details of the access mechanism (i.e. if the use of sensing or databases will be required, etc.), as well as its radio characteristics, depend on the context within which the band is chosen to operate. This context (agreed by the primary operator) defines the extent to which the secondary must avoid interfering with the primary.

Like any dynamic solution, the concept of pluralistic licensing creates new degrees of freedom in the system design and management. The spectrum holders (operators) can create their own rules for spectrum sharing depending on their respective priorities associated with, e.g. company development strategy, financial status, or geographical conditions (such as expected traffic). On the other hand, a natural consequence of increasing system flexibility is the increase in the complexity of its management.

1.2.6   Licensed Assisted Access (LAA)

The use of unlicensed frequency band located around 5 GHz has been garnering significant attention recently. In particular, the use of the long-term evolution (LTE) technology in the unlicensed bands has led to the development of LTE-Unlicensed (LTE-U) solution. The use of unlicensed band requires coexistence of WiFi and LTE-U users. One way to enable this is to dynamically select clear channels, which would lead to avoidance of WiFi interference. However, the presence of vacant WiFi channels is not always guaranteed. Thus, clear channel assessment (CCA) or listen-before-talk (LBT) approach is needed for such unlicensed spectrum use. Further-more, LAA introduced in Rel. 13 of the LTE standard is based on the assumption that unlicensed band can be used with the licensed one. Different scenarios can be considered relevant here, for example, transmitting control data using licensed band and user data transmission using aggregation of licensed and unlicensed bands. This is under the assumption that the unlicensed bands will be vacated in the case where traffic will be low enough to be supported by using licensed bands. Usually, LAA is considered for supporting downlink traffic by the use of unlicensed spectrum bands. This issue has been discussed extensively in a number of recent studies [20–25].

1.2.7   Co-primary Shared Access

Co-primary shared access (CSA) is a relatively recent concept, which assumes that operators (that are peers) decide to jointly utilize a certain portion of their licensed spectrum. For details see [5, 26]. Specifically, two CSA cases can be defined: mutual renting, where operators keep their respective licenses, but they can mutually rent part of the spectrum under certain conditions; and spectrum pooling, in which case the dedicated chunk will be in common use, based on the group licenses. It is important to note that mutual renting can be regarded as a specific form of LSA among a limited set of operators. On the other hand, different cooperation-based approaches can be considered in the case of spectrum pooling.

1.3   Legal Regulations for Spectrum Sharing

The proliferation of networks, services, and subsequently the ever-increasing demands for more radio spectrum has pushed the network operators, researchers, and regulators to look beyond simple licensed and unlicensed spectrum use and adopt more efficient sharing and management concepts in future network architectures. The level of attention a particular technology receives from these main players is usually a good indicator of its true potential. Spectrum sharing and its different application models have received particular attention from networks researchers as can be assessed by the amount of research literature available on these topics. In recent history, the regulators have also followed suit and looked at different ways to facilitate inter-operator/networks spectrum sharing in many spectrum bands. Theoretically, a simpler approach towards meeting the demands of mobile networks for more radio spectrum is to rearrange the existing allocation. A simple rearrangement of spectrum allocation can free up more radio spectrum for licensed and unlicensed use and keep the existing primitive method of spectrum sharing, i.e. division/ partitioning into different spectrum bands for different mobile networks. Practically, however, this approach incurs many costs associated with upgrading the infrastructure and devices of the network operators and users together with many administrative challenges. For example, in the USA, vacating 1710–1755 MHz band for commercial use cost an estimated 1.5 billion US dollars and took 6 years to clean up the spectrum. Additionally, this rearrangement approach does not solve the spectrum underutilization problem. Therefore, spectrum sharing must be incorporated in future network architectures not only through technology development but also through well-defined rules and regulations.

Legal regulations pertaining to spectrum use in different parts of the world are determined by country-/region-specific regulatory bodies. In the recent past, many initiatives have been taken by different regulators across the world to adopt some of the above-mentioned spectrum sharing models. The main drivers for these initiatives have been the demands for more spectrum and the potential new revenue streams which could be generated from not only basic spectrum sharing but also through sophisticated business models among the network operators. This section summarizes some of the recent regulatory efforts for facilitating spectrum sharing in current and future networks scenarios.

1.3.1 Spectrum Sharing Regulations in the USA

Determining spectrum access regulations in the USA is the responsibility of the FCC and National Telecommunications and Information Administration (NTIA) who jointly formulate the rules for spectrum access in the country. They regulate the RF spectrum from 3 KHz to 300 GHz and jointly decide upon Federal (Fed) and Nonfederal (NFed) use. A cursory look at these regulations shows that a stronger focus has recently been laid on its adoption and facilitation in future networks landscape. As the list of spectrum sharing initiatives by FCC and NTIA in different spectrum bands is long, this section only provides a concise summary. In 2013, based on a US presidential memorandum, the FCC and NTIA initiated a report on their activities for facilitating spectrum sharing in the USA. Subsequently, 13 different initiatives were taken to support both conventional and new models of spectrum sharing in different spectrum bands ranging from 50 MHz to 100 GHz. These initiatives focus on supporting secondary use (licensed and unlicensed) in presence of existing Fed, NFed, and broadcast (e.g. TV) incumbent users. The coordination for spectrum access between primary and secondary users in these initiatives is based on manual coordination, TV white space databases, dynamic spectrum selection, spectrum access system, and directional transmissions with emission bounds. In addition, many federal agencies in the USA are exerting significant R&D efforts on spectrum sharing driven by $100 million funding through the National Science Foundation, NTIA, DARPA, and the National Institute of Standards and Technology (NIST). Under this umbrella, several research groups including Wireless Spectrum Research and Development (WSRD), Enhancing Access to the Radio Spectrum (EARS), National Spectrum Consortium (NSC), etc. have been formed to primarily focus on RF spectrum utilization. A complete list and description of these initiatives together with technical specifications can be found in [27].

1.3.2 Spectrum Sharing Regulations in Europe

Regulators in Europe have also identified spectrum sharing as a key technology to support mobile broadband applications and have taken several steps for its realization in future networks. As the formulation of actual regulations on spectrum sharing can only follow a detailed research and analysis, the EC has initiated several works to assess the costs, benefits, and challenges of the different spectrum sharing models. In 2012, a special Radio Spectrum Policy Program (RSPP) was approved by EC with the mandate to harmonize spectrum use across Europe and support efficient use and management. In this respect, collective use of spectrum (CUS) and LSA have been identified as two broad classes of spectrum sharing. The CUS essentially looks to offer more RF spectrum for license-free access to many networks and is therefore hindered by spectrum re-farming and capacity issues. Nevertheless, several trials under the RSPP have demonstrated successful CUS implementation in

863–870 MHz frequency range, sharing spectrum among a large number of wireless microphones, sensors, and tracking devices. The LSA concept on the other hand is based on shared spectrum rights. It provides a more predictable level of QoS over the shared spectrum guaranteed by the spectrum regulators. The LSA approach and its different variants are considered within Europe as the enabler for MBB within the 2.3–2.4 GHz frequency range. Another main activity for supporting shared spectrum access in Europe is based around the TV white space utilization for secondary access, where a white space is defined as a time-/location-specific spectrum access opportunity. A preliminary assessment was carried out for white space devices (WSDs) in the UHF frequency band between 470 and 790 MHz for the adoption of a geolocation database and other technical requirements to enable the operation of WSDs. The development of telecommunication standards within EU falls under the responsibility of ETSI. Recently, ETSI Reconfigurable Radio Systems (RRS) group has produced server technical reports on spectrum sharing and have explored different scenarios including spectrum on demand [28], and the applicability of cognitive radio concepts [29]. In addition, the group has exerted efforts on mapping high-level functions required for LSA [30] and the realization of WSDs through geolocation database. The advancements in mobile communication networks in Europe are guided by 3GPP through different releases. The concept of Multi-Operator Core Network was formulated in release-6 to not only support core network but also spectrum sharing among different network operators. The more recent releases have expanded the concept to other related topics such as traffic offloading to unlicensed bands and LAA. Spectrum sharing has been identified to play a key role in realizing the objectives of 5G networks, and many EU projects are focusing on exploring its true potential under the EU 2020 vision such as 5G-XHaul, COHERENT, METIS-II, etc.

1.3.3   Spectrum Sharing Regulations Elsewhere

The Australian Communications and Media Authority (ACMA) is the organization responsible for managing the RF spectrum in Australia. ACMA produced a technical report that discusses different spectrum management issue in Australia and an outlook for the state in year 2019 [31]. The report focuses on generating incentives to support spectrum sharing including dynamic spectrum access and cognitive radio concepts (underlay and overlay) in 400 MHz, 800–900 MHz, and 1800 MHz bands. In New Zealand, the Radio Spectrum Management (RSM) is a business unit of the Ministry of Business, Innovation, and Employment (MBIE) and is responsible for spectrum allocation and management in the country. In 2014, it facilitated a temporary arrangement to enable access to TV white spaces without the need of a geolocation database with the only requirement that they comply with FCC rules or ETSI standard EN 301 598. Similar initiatives have been undertaken in other countries such as Canada, Singapore, China, and Korea. To summarize, the regulations with respect to spectrum sharing are not very mature in the current global networks landscape. The several initiatives discussed in this section across the world are aimed at providing some foundational knowledge that can help the regulators formulate the rules and regulations for the different spectrum sharing models.

1.4   Trials

The significance of trials in the development and maturation of wireless technologies cannot be overstated. In the area of cognitive radio and spectrum sharing, a plethora of demonstrations and trials have been conducted to assess the potential of various spectrum sharing schemes. An overview of trials and demonstrations related to the general concept of cognitive radio and spectrum sharing is presented in [32]. It summarizes the details of important trials, as well as the test beds and platforms developed during the first decade of research in this area, i.e. from 1999 to 2009. In the following subsections, we discuss key trials related to emerging spectrum sharing schemes for 5G networks, particularly LSA, LAA, and CBRS.

1.4.1   Licensed Shared Access (Authorized Shared Access)

The initial trials of ASA were demonstrated at WWRF meeting in April 2013, in Oulu, Finland, using a generic cognitive radio environment called CORE [33]. It consisted of cognitive engines for controlling different radio systems such as LTE and WARP. The motivation behind cognitive decision-making was its potential application in intelligent spectrum management and resource allocation, aimed at improving the QoS provided to the users. The CORE environment is designed to be used for spectrum sharing between mobile communication systems and other networks. A key feature of this environment is that it enables optimization of resource usage within a network, as well as across multiple networks.

The trial environment comprised of three main components, namely, central entity, a live LTE network, and a WARP-based network. The ASA-enabled eNBs are deployed in 2.3–2.4 GHz band, which is shared with the incumbent user. Trial setup involved two users connected to the eNB and receiving standard video service with 1 Mbps in the downlink and another one generating 5–10 Mbps stream in the uplink. The mobile network operator is then requested to vacate the ASA band. This request is delivered to the CE from a simulated data base. The CE responds by offloading the users to the WLAN and commands the LTE network to block the cells. When the ASA band becomes available again, the CE sends a command to unblock the cells so that the users can be moved back to the ASA band. The evacuation time, i.e. the time it takes from the request to the actual blocking, is around 25 s. Further results and measurements showed that the CORE platform is an effective tool for spectrum trials and can be used for different use cases, especially the ASA. A multitude of subsequent trials with updated features were also carried out [35, 36]. In these trials, the incumbent is the PMSE service. The staff of the incumbent broadcast company uses the spectrum to operate live cameras and related live broadcast equipment. The spectrum is shared with a TD-LTE system operating in ASA/LSA band. The trials showed that the TD-LTE belonging to a mobile network operator can successfully make use of the incumbent band and vacate it when required. More advanced subsequent trials included resource optimization apart from incumbent protection. Additional aspect of enhanced LSA resource optimization is considered in [37], in which LSA controller is implemented as a self-organizing network (SON) solution for managing the LTE network.

Other notable LSA trials in Europe include the pilot programs in Italy, France, and Spain [38]. In Italy, LSA pilot project was conducted by JRC, in collaboration with Italian Ministry for Economic Development. Trials were held in Rome from July to December 2015, where a proper architecture for enabling LSA was tested. The trials held in France were conducted by Qualcomm, Ericsson, and RED technologies [39], where the spectrum owned by ministry of defence was shared by the Ericsson RAN. It is expected that the success of this pilot will to lead to a release of high volumes of licensed spectrum. LSA trials in Spain were demonstrated at the Mobile World Congress, held in Barcelona in March 2015. These trials involved PMSE as an incumbent, sharing the spectrum with a mobile services licensee [40].

1.4.2 Licensed Assisted Access (LAA)

The field trials of LAA were completed in November 2015, in Nuremberg, Germany, by Qualcomm and Deutsche Telekom. The test equipment used in trial was designed and deployed by Qualcomm. Deutsche Telekom provided the licensed spectrum augmented with the unlicensed 5GHz band. It was reported that the trials demonstrated the ability of LAA to extend network coverage and increase capacity, while ensuring seamless mobility and fair coexistence with WiFi.

1.4.3 Citizen Broadband Radio Service with Spectrum Access System (SAS)

Recently, trials for CBRS with SAS were reported in [41], in which authors discuss a complete end-to-end architecture for realizing the three-tier spectrum management system. It includes a scalable SAS to activate dynamic exclusion zone for incumbent protection, management of primary/secondary devices, and dynamic assignment of spectrum.

Next, we discuss the potential of virtualization of spectrum for addressing these challenges in 5G networks. In particular, the focus is on network abstractions for designing an efficient spectrum sharing platform.

1.5 Virtualization-Based Spectrum Sharing Solutions

The concept of virtualization is based on the use of abstractions for an efficient sharing of available resources. It has been applied in a number of areas related to information and communication systems. Examples include virtual machines [42], virtual private networks [43], virtualized data centres [44], and virtual cloud security [45], to name a few. In particular, the potential and applications of virtualization in the context of wired networks is well established. The key idea behind network virtualization is to abstract the underlying physical resources by logical resources, thereby creating virtual networks, which can be construed as "network slices". A detailed overview focusing on virtualization in the context of next-generation Internet architecture is given in [46].

1.5.1 An Overview of Existing Work on Virtualization in Wireless Networks

Recently, similar approaches are being considered for wireless networks as well, not only due to the increased complexity and heterogeneity but also the importance of efficient resource utilization for meeting 5G requirements. Some research issues relevant to virtualization of wireless networks were introduced in [47], mainly focusing on infrastructure sharing. Moreover, it is possible to combine network virtualization with SDN through network virtualization hypervisors. A classification of network virtualization hypervisors for SDN is given in [48]. Another SDN-based spectrum sharing approach that combines sensing devices, base stations, and SDN controller is discussed in [49]. In this case, local decisions are made at base stations whereas SDN controller deals with network-wide policy-making. Virtualization for LTE networks at the level of air-interface spectrum resources is discussed in [50, 51]. The opportunities and benefits of virtualization for spectrum sharing in the context of regulatory framework for 3.5 GHz band is analysed in [52]. To this end, a number of perspectives are identified including regulation, technology, and economics.

In general, for 5G networks virtualization can enable an efficient sharing of different network resources across multiple dimensions, which include infrastructure, virtual resource slices, and spectrum [2, 4, 53]. Joint sharing along all these dimensions is often referred to as network sharing [4]. The problem of mapping physical spectrum resources to different virtual networks is considered in [54]. A virtualization-based paradigm for resource allocation is proposed in [56], for the case when user demands are uncertain. The formulated virtualized paradigm improves user satisfaction, and virtualization gains are observed to increase with the number of operators that share resources. Virtualization of RAN is discussed in [3], with an emphasis on the data processing requirements and cost analysis. In general, as network functions become software-based, their incorporation into network architectures becomes more efficient and cost-effective.

1.5.2   Overview of Existing Surveys on Cognitive Radio Networks

A number of papers survey the state of the art in spectrum sharing from different perspectives. However, we focus solely on virtualization aspects of spectrum sharing, which has not been discussed extensively. Notable existing works include [56], which discusses state of the art on green networking approaches towards cognitive radio networks. A comprehensive survey of full-duplex communication in cognitive radio networks is presented in [57]. Interference mitigation-based works include [58–60]. An in-depth survey on multimedia cognitive radio networks is give in [61].

Here, we consider only the spectrum sharing and focus on the multiple operator case, which involves the sharing of spectrum (between operators) according to some agreement or predefined scheme, e.g. exclusive access, LSA, LAA, CSA, etc. From a virtualization perspective, the ramifications of various schemes differ significantly. In particular, the constraints imposed by the inter-operator spectrum sharing may limit the scope of virtualization in the spectrum domain, thereby jeopardizing the potential gains. In order to maximize the benefits of network virtualization, all the spectra should be pooled together and considered as a single resource. It is worth noting that this requires not only full cooperation between the operators but also novel control and signalling mechanisms. On the other hand, exclusive access is the simplest to implement, but it has limited potential in terms of virtualization. In the following subsections, we discuss the potential architecture of the spectrum manager and the impact of inter-operator spectrum sharing strategies on the virtualization aspects.

1.5.3   Spectrum Management Architecture

Having in mind the nature of selected spectrum sharing strategies, as well as the current regulations and ongoing trials on simultaneous spectrum usage by various operators, it is now required to identify the key market players in the context of spectrum resource virtualization and to propose the generic spectrum management architecture. From one hand side one can say about the end-users which would like to transmit their data achieving satisfactory data rates. It technically corresponds to efficient utilization of certain frequency resources. However, the end-user would be not interested at all, on which frequency band will be used or which spectrum access technology is applied, unless it will not violate its agreed service layer agreements (e.g. cost, rate, or quality of service). Thus, the spectrum can be somehow out of its interest; hence these will be the network operators and service providers who will take care of spectrum and will select one of the spectrum sharing strategies defined in the previous sections. However, operators will need to follow the guidelines and regulations defined by the national regulation authorities (NRA), as well as will need to fulfil the constraints defined in inter-operator agreements.

From that perspective one can immediately claim that the whole processing of available spectrum resources can be virtually split in two layers of high and low level of generality. As the former one is particularly related to the spectrum management (e.g. legal constraints, influence of NRA, decisions of mobile network operators on the applied sharing strategies, etc.), the latter can be understood as the process of controlling or orchestrating (so application of high-level regulation at the lower scale). Following the computer science analogy, the connection between the controller (orchestrator) and the spectrum management plane can be understood as northbound interface, whereas the connection between the controller and the physical entities as southbound interface. On the other hand, referring the two-layer model to the microprocessors architecture, one can say about northbridge, southbridge, and central processing unit, as illustrated in Fig. 1.1.

In this figure one can note the presence of the central controller (orchestrator) who will be responsible for accurate and permanent realizations of various spectrum-related activities. It will have the knowledge of the available physical resources, and it will be able to communicate the last changes and query new requests to the engine of the spectrum manager. The spectrum manager, in order to work correctly, should possess its own processing engine, should have access to various internal and external databases (such as radio environment maps (REM), various repositories for applied policies, external repository for access network discovery and selection function, etc.) and storage (with, e.g. history, trends), and should be able to communicate with the monitoring functions (independently or via spectrum controller). Finally, as the whole spectrum management architecture can be realized in different ways (as a flat or hierarchical structure, it can have a centralized or distributed form, as shown in Fig. 1.2), the set of standardized interfaces should be defined for information exchange between various spectrum management entities existing in the entire network (e.g. belonging to other operator or dedicated third party).

As the generic spectrum management system has been defined, it is now important to answer the question: How could the spectrum resources be represented in a virtualized form? In other words, one can imagine the situation when the client (mobile user, virtual mobile network operator, etc.) would like to realize a certain type of data transmission (e.g. with some predefined reliability, data rate, and latency) and provide such a request to the controller. Then the controller analyses the whole network and decides which resources could be assigned to such a service and how it should be done. From a client perspective, it is not important if the

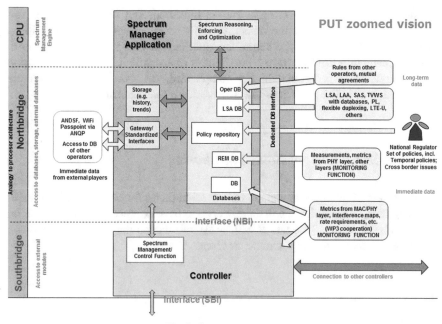

Fig. 1.1 Proposed architecture for spectrum management – analogy to computer science

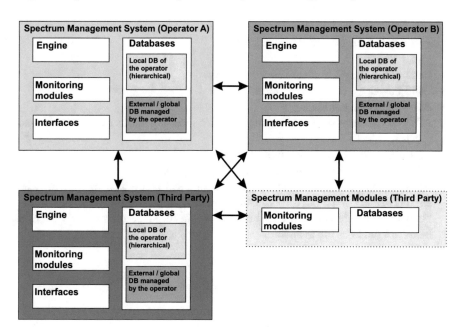

Fig. 1.2 Proposed architecture for spectrum management – high-level relations

physical resources will be located in low- or high-frequency band, if the spectrum belongs to one operator or the others, or if its service level agreements will be guaranteed. In such a case, the controller could have access to abstracted network graph, which will illustrate the true possibilities of spectrum usage. As the solution of this concept is in general highly complicated, here we concentrate on the identification of initial guidelines towards spectrum sharing abstraction.

1.5.4 Abstractions for Spectrum Sharing

All the layers of the stack model, as well as all virtualization functions, instrumentation entities, or controllers, will simply need to have access to the information on the available spectrum. The rate at which this information may change depends on the chosen spectrum sharing scheme, as discussed next.

1.5.4.1 Exempt Use

The following pieces of information would be needed by the network controller (manager, orchestrator, etc.)

Long-Term Information (Rarely Updated)

- Centre frequency
- Band ranges
- Allowed resource granularity
- Allowed transmit power and requirements for spectrum masks (related to allowed spectrum block)
- Maximal theoretic rates or capacity
- Assumed system setup, such as probability of blockade
- Allowed transmit schemes at certain geographical area (such as system type, e.g. GPRS, LTE, TETRA, modulation formats, authorization types) due to the hardware limitations (i.e. it could be a case that for some reasons only a given hardware modules are available)

Short-Term Information (Permanent Monitoring)

- Current occupancy of each resource block
- Remaining resources
- Planned handovers

1.5.4.2   License Exempt

Open access to frequency spectrum entails some specific requirements for network virtualization and abstraction. The information about the current spectrum occupancy will need to be provided continuously to the system controller (manager, orchestrator), but it will be hard or even impossible to estimate exactly the amount of remaining, vacant resources or assess the remaining system capacity. It is due to the fact that such spectrum access schemes are interference limited, and it is impossible to foresee what will be the number of users trying to access the spectrum. Of course, some daily trends could be helpful, but the real-time situation needs to be monitored permanently. In the case of exempt use, each operator will know the exact amount of resources that remain unused, also such parameters as probabilities of blockade can be considered. Here, the knowledge on the amount of users and observed interference will rely mainly on some statistical derivations. The following pieces of information would be needed by the network controller (manager, orchestrator, etc.).

Long-Term Information (Rarely Updated)

- Centre frequency
- Band ranges
- Allowed spectrum resource granularity
- Allowed transmit power and requirements for spectrum
- Masks (related to allowed spectrum block)
- Maximal theoretic rates or capacity
- Daily traffic characteristics

Short-Term Information (Permanent Monitoring)

- Current occupancy of each resource block
- Remaining resources
- Observed interference

1.5.4.3   Licensed Shared Access (LSA)

In general, the case of LSA is very similar to the exempt use one, as the information about the available resources can be estimated or even calculated in advance with very high accuracy. The key questions related to the LSA approach are twofold. The first one is associated with the type of licenses granted to the licensee by the licensor (primary spectrum holder), as various models can be used here. If, for example, the selected spectrum band will be granted for common use or so-called non-orthogonal spectrum sharing, then analogous problems appear for such a band as have been identified for license-exempt case. On the other hand, the primary license owner could further borrow the spectrum resource for exempt use; thus the observations made for that mode are valid. The second issue, which is important in the LSA case, is the time scale for which the current LSA policy is applied. In other words, from the virtualization point of view, the information on the duration of the applied LSA policy is necessarily what the controller needs to know when the rules of spectrum usage will change. If the primary spectrum holder will agree to offer its part of spectrum in the LSA mode for long-time periods, then it will reduce the information overhead delivered to the network. Otherwise, short periods for resource sharing based on LSA policy will entail frequent modification of the rules.

Besides the information types mentioned previously, the following pieces of information would need to be accessed in LSA mode.

Long-Term Information (Rarely Updated)

- LSA policy (type of renting)
- Time scale of each LSA decision

Short-Term Information (Permanent Monitoring)

- Depending on the selected LSA policy, the status of current utilization of shared resources

1.5.4.4 Spectrum Access System (SAS)

SAS architecture is more complex than LSA, and thus the abstraction of it is a challenging task. Apart from the long-term information type mentioned previously, in the SAS architecture, the following pieces of information would need to be accessed.

- SAS dynamically assigns and maintains CBRS spectrum use in real time, and there will be no fixed spectral location for PA or GAA allocation.
- CBSDs must be able to determine their geographic coordinates every 60 sec and report any changes in its position within 60 sec to the SAS.

1.5.4.5 Pluralistic Licensing

Such dynamic spectrum sharing scheme entails the need of continuous exchange of information of currently applied spectrum sharing policies. As these policies can vary in time and location, the information provided to the system controller (manager, orchestrator, etc.) has to be up-to-date and highly accurate. Besides the typical pieces of information that would be needed for network virtualization (mentioned while describing exclusive use and license-exempt cases), application of pluralistic licensing can result in the exchange of following messages:

Long-Term Information (Rarely Updated)

- Details of the applied pluralistic licensing policy (type of renting, allowed transmit powers, allowed level of interference rise, etc.)
- Time scale of each pluralistic licensing decision

Short-Term Information (Permanent Monitoring)

- Depending on the selected pluralistic licensing policy, the status of current utilization of shared resources

1.5.4.6 License Assisted Access

From the perspective of the network abstraction, the case of joint utilization of licensed and unlicensed bands can be seen as the specific mixture of the exclusive use and license-exempt approaches described above. The controller would need to know all the details related to the usage of licensed part of spectrum, and in addition to that it will require information about the possibilities of unlicensed spectrum usage.

1.5.4.7 Co-primary Shared Access

Application of CSA scheme in the virtualized network requires access to the updated information on the accurate rules that specify how the spectrum can be utilized. In particular, it has to be known in advance which operators are allowed to cooperate, what are the key agreements made by them (one may note that this information can be hard to access due to its private nature), which portion of licensed spectrum is allocated for common pool, etc. Based on that observation, the following long- and short-term data can be identified.

Long-Term Information (Rarely Updated)

- Details of the applied CSA mode (MR or LSP; amount of spectrum devoted for this purpose by each involved operator; in case of LSP the information about the agreed spectrum access mode; in both cases, MR and LSP, agreed time of renting/ using the spectrum)

- Time scale of each CSA decision
- Some details from the agreement between the operators

Short-Term Information (Permanent Monitoring)

- In both orthogonal and non-orthogonal sharing, information on the actual, temporary parameters such as interference level, number of served user (thus amount of remaining resources)

1.6   Key Challenges

In spite of the numerous benefits of inter-operator spectrum sharing for all stakeholders, its adoption in future networks faces a number of challenges. Network operators are primarily competitors, and for them adoption of spectrum sharing in their network architectures is not a straightforward decision. The promise of mutual benefits of spectrum sharing must be substantiated with sufficient proof. Moreover, the main challenges for its adoption have to be addressed. In this section, we highlight some of these main challenges and suggest potential ways to address them.

1.6.1   Service Differentiation

A main benefit of exclusive spectrum access is that the operators can provide different QoS guarantees within their networks and can differentiate from other service providers. Additionally, the operators can get a good understanding of the behaviour of their radio spectrum over time through measurements and analysis making spectrum management much easier. With inter-operator spectrum sharing, an operator serves for the benefit of a one or more competitor networks by providing more spectrum to their users. Although this works both ways, the eventual benefits of sharing may not necessarily be equal as the utilization of spectrum is highly dependent on network-specific traffic conditions. Therefore, a more inclusive and complex agreement on spectrum sharing has to be devised among the competitors. To address this challenge, the regulators have to play a main role in order to make sure that the specific interests of the operators are safeguarded through well-defined, enforceable, and fair rules and regulations.

1.6.2   Sharing of Information

Inter-operator spectrum sharing will require the network operators to expose certain network-specific information (e.g. traffic condition, resource assignment and utilization, etc.) to a management and control entity that will be responsible for equitable sharing of the spectrum among all candidates. This information can also be used for competitive advantage by network operators, and therefore operators will be reluctant to its sharing. To address this challenge, the management and control of the shared spectrum can either be provided by a third party or a regulator. This can help alleviate these concerns and allow the operators to develop a mutual trust, thereby encouraging the shared use of spectrum. However, updates on such information regarding current spectrum occupancy may be needed by system controller (manager, orchestrator). In this regard, a major challenge would be to accurate estimation of remaining resources and timely delivery of this information to the controller.

1.6.3   Need for New Network Functions

The existing mobile networks are based on exclusive spectrum use and do not have adequate functions for supporting spectrum sharing. In order to overcome such issues, a network operator would need to introduce several new control functionalities to a network-specific or a joint inter- operator spectrum manager. Essential interfaces will have to be provided to the spectrum manager for receiving the required inputs and implementing spectrum management and sharing decisions. Introducing new network services or control functions introduces new costs for network operators, and therefore the spectrum sharing benefits will have to compensate for such costs. This issue can be addressed with the new SDN and NFV paradigm which is an important and essential part of the 5G road map.

1.6.4   Long-Term Contracts

Most licenses for exclusive spectrum access are granted on a long-term/multi-year basis to different network operators. This implies that some operators will be more willing or unwilling to inter-operator spectrum sharing than others. It also complicates the issue of creating a common shared spectrum band to which different operators allocate parts of their spectrum. Although this can be a problem, the vision for 5G and the huge bandwidth, throughput and minimum latency targets, may force the operators to work towards a joint solution to some extent. Moreover, the regulators can also play a role in accommodating an agreement among the operators for creating a joint spectrum pool for inter-operator spectrum sharing.

1.6.5   Management and Control

As discussed in the previous sections, there are several ways in which inter-operator spectrum sharing can be realized. This diversity however has direct implications on the framework for spectrum management and control. Whether each operator will have its own spectrum manager having partial control over the shared spectrum without exposing internal network information or will all the operators relegate these tasks to a joint controller can be decided within the domain of specific spectrum sharing model. The unison of operators on this aspect will also be difficult to achieve. The number of trials and experimentations that have been carried out can serve to identify the most promising spectrum sharing models for the future and help in deciding the most promising spectrum management and control mechanism.

1.6.6   Responsibility Assignment

In the current landscape of independent mobile networks, the identification and placement of responsibility are simpler as the number of entities involved for enforcement of rules, quality of services, and network performance can be easily identified. Inter-operator spectrum sharing will complicate this aspect depending upon the management and control architecture adopted for this purpose. A number of issues can arise similar to license-free spectrum bands, and the operators may observe higher interference levels in their radio access domain. This issue can also be addressed once the spectrum sharing technology is matured enough to be incorporated in the network architectures.

1.7   Future Work and Conclusions

Virtualization is the key enabler for an efficient utilization of network resources such as spectrum and infrastructure. In addition, apart from cellular networks, emerging research areas related to 5G that would benefit from virtualization-based techniques include sensor networks [62] and their applications in the Internet of things [63, 64]. Studying the advantages of virtualization in such technologies constitutes an important direction for future work.

The goal of this work was to give an overview of spectrum sharing schemes that are being considered as part of the vision for 5G networks. The case of multiple operators sharing the spectrum leads to new challenges, as the underlying spectrum sharing schemes may profoundly impact the potential gains, and the granularity of network virtualization. We present the legal regulations and trial activities for these spectrum sharing schemes and look at their characteristics from the perspective of inter-operator spectrum sharing and virtualized network architectures. The rapid increase in regulatory activities and trials of spectrum sharing schemes highlights the progress currently underway in different regions and markets. Furthermore, it underscores the inherent potential of spectrum sharing for enabling the ambitious goals of 5G networks. We also discussed key challenges and potential solutions for spectrum sharing schemes. In addition, an architecture for enabling virtualization of spectrum sharing schemes is proposed, along with related abstractions, which is highly flexible and well suited for inter-operator spectrum sharing in virtualized 5G networks.

References

1. Andrews JG, Buzzi S, Choi W, Hanly SV, Lozano A, Soong ACK, Zhang JC (June 2014) What will 5g be? IEEE J Sel Areas Commun 32(6):1065–1082
2. Costa-Perez X, Swetina J, Guo T, Mahindra R, Rangarajan S (2013) Radio access network virtualization for future mobile carrier networks. IEEE Commun Mag 51(7):27–35
3. Rost P, Berberana I, Maeder A, Paul H, Suryaprakash V, Valenti M, Wbben D, Dekorsy A, Fettweis G (2015) Benefits and challenges of virtualization in 5g radio access networks. IEEE Commun Mag 53(12):75–82
4. Liang C, Yu FR (2015) Wireless network virtualization: a survey, some research issues and challenges. IEEE Commun Surv Tutorials 17(1):358–380
5. Irnich T, Kronander J, Seln Y, Li G (2013), Spectrum sharing scenarios and resulting technical requirements for 5g systems. In: Personal, indoor and mobile radio communications (PIMRC Workshops), 2013 I.E. 24th international symposium on, pp 127–132
6. Kliks A, Holland O, Basaure A, Matinmikko M (2015) Spectrum and license flexibility for 5g networks. IEEE Commun Mag 53(7):42–49
7. Kryszkiewicz P, Kliks A, Bogucka H (2016) Small-scale spectrum aggregation and sharing. IEEE J Sel Areas Commun 34(10):2630–2641
8. Mitola J, Guerci J, Reed J, Yao YD, Chen Y, Clancy TC, Dwyer J, Li H, Man H, McGwier R, Guo Y (2014) Accelerating 5g qoe via public-private spectrum sharing. IEEE Commun Mag 52 (5):77–85
9. Khan A, Kellerer W, Kozu K, Yabusaki M (2011) Network sharing in the next mobile network: Tco reduction, management flexibility, and operational independence. IEEE Commun Mag 49 (10):134–142
10. Meddour D-E, Rasheed T, Gourhant Y (2011) On the role of infrastructure sharing for mobile network operators in emerging markets. Comput Netw 55(7):1576–1591, recent Advances in Network Convergence
11. Sugathapala I, Kovacevic I, Lorenzo B, Glisic S, Fang Y (2015) Quantifying benefits in a business portfolio for multi-operator spectrum sharing. IEEE Trans Wirel Commun 14 (12):6635–6649
12. Wang N, Gao Y, Evans B (2015) Database-augmented spectrum sensing algorithm for cognitive radio. In: Communications (ICC), 2015 I.E. international conference on, IEEE, pp 7468–7473

13. Chowdhery A, Chandra R, Garnett P, Mitchell P (2012) Characterizing spectrum goodness for dynamic spectrum access. In: Communication, control, and computing (Allerton), 2012 50th annual Allerton conference on, IEEE, pp 1360–1367
14. Li Y (2015) Grass-root based spectrum map database for self-organized cognitive radio and heterogeneous networks: spectrum measurement, data visualization, and user participating model. In: Wireless Communications and Networking Conference (WCNC), 2015 IEEE, IEEE, pp 117–122
15. Denkovski D, Rakovic V, Pavloski M, Chomu K, Atanasovski V, Gavrilovska L (2012) Integration of heterogeneous spectrum sensing devices towards accurate rem construction. In: Wireless Communications and Networking Conference (WCNC), IEEE, IEEE, pp 798–802
16. Jorswieck EA, Badia L, Fahldieck T, Karipidis E, Luo J (2014) Spectrum sharing improves the network efficiency for cellular operators. IEEE Commun Mag 52(3):129–136
17. RSPG (2011) Report on collective use of spectrum (cus) and other spectrum sharing approaches. Radio Spectrum Policy Group, Tech. Rep. 11-392
18. PCAST (2012) Realizing the full potential of government-held spectrum to spur economic growth. Presidents Council of Advisors on Science and Technology Report, Tech. Rep
19. Holland O, De Nardis L, Nolan K, Medeisis A, Anker P, Minervini LF, Velez F, Matinmikko M, Sydor J (2012) Pluralistic licensing. In: Dynamic Spectrum Access Networks (DYSPAN), 2012 I.E. international symposium on, IEEE, pp 33–41
20. Chen Q, Yu G, Yin R, Maaref A, Li GY, Huang A (2015) Energy- efficient resource block allocation for licensed-assisted access. In: Personal, Indoor, and Mobile Radio Communications (PIMRC), 2015 I.E. 26th annual international symposium on, IEEE, pp 1018–1023
21. Lien S-Y, Lee J, Liang Y-C (2016) Random access or scheduling: Optimum LTE licensed-assisted access to unlicensed spectrum. IEEE Commun Lett 20(3):590–593
22. Ratasuk R, Mangalvedhe N, Ghosh A (2014) LTE in unlicensed spectrum using licensed-assisted access. In: Globecom Workshops (GC Wkshps), IEEE, 2014, pp 746–751
23. Li Y, Zheng J, Li Q (2015) Enhanced listen-before-talk scheme for frequency reuse of licensed-assisted access using LTE. In: Personal, Indoor, and Mobile Radio Communications (PIMRC), 2015 I.E. 26th annual international symposium on, IEEE, pp 1918–1923
24. Ibars C, Bhorkar A, Papathanassiou A, Zong P (2015) Channel selection for licensed assisted access in LTE based on UE measurements. In: Vehicular Technology Conference (VTC Fall), 2015 I.E. 82nd. IEEE, pp 1–5
25. Liu F, Bala E, Erkip E, Beluri MC, Yang R (2015) Small-cell traffic balancing over licensed and unlicensed bands. Veh Technol, IEEE Trans 64(12):5850–5865
26. Singh B, Hailu S, Koufos K, Dowhuszko AA, Tirkkonen O, Jntti R, Berry R (2015) Coordination protocol for inter-operator spectrum sharing in co-primary 5g small cell networks. IEEE Commun Mag 53(7):34–40
27. Agre JR, Gordon KD (2015) A summary of recent federal government activities to promote spectrum sharing. IDA Science and Technology Institute, Tech. Rep
28. ETSI (2009) Reconfigurable radio systems (RRS); functional architecture (FA) for the management and control of reconfigurable radio systems. Tech. Rep. TR 102.682 v1.1.1
29. ETSI (2010) Reconfigurable radio systems; cognitive radio concept. Tech. Rep. TR 102.802 v1.1.1
30. ETSI (2014) System architecture and high level procedures for operation of licensed shared access (LSA) in the 2300-2400 MHz band. Tech. Rep. TS 103.235, v0.0.13
31. ACMA (2015) Five-year spectrum outlook 2015-2019: the acmas spectrum demand analysis and strategic direction for the next five years. Tech. Rep
32. Pawelczak P, Nolan K, Doyle L, Oh SW, Cabric D (2011) Cognitive radio: Ten years of experimentation and development. IEEE Commun Mag 49(3):90–100
33. Matinmikko M, Palola M, Saarnisaari H, Heikkil M, Prokkola J, Kippola T, Hanninen T, Jokinen M, Yrjl S (2013) Cognitive radio trial environment: First live authorized shared access-based spectrum-sharing demonstration. IEEE Veh Technol Mag 8(3):30–37

34. Palola M, Matinmikko M, Prokkola J, Mustonen M, Heikkil M, Kippola T, Yrjl S, Hartikainen V, Tudose L, Kivinen A, Paavola J, Heiska K. (2014) Live field trial of licensed shared access (LSA) concept using LTE network in 2.3 ghz band. In: Dynamic spectrum access networks (DYSPAN), 2014 I.E. International Symposium on, pp 38–47
35. Palola M, Matinmikko M, Prokkola J, Mustonen M, Heikkil M, Kippola T, Yrjl S, Hartikainen V, Tudose L, Kivinen A, Paavola J, Heiska K, Hanninen T, Okkonen J (2014), Description of finnish licensed shared access (LSA) field trial using TD-LTE in 2.3 ghz band. In: Dynamic Spectrum Access Networks (DYSPAN), 2014 I.E. international symposium on, pp 374–375
36. Palolo M, Rautio T, Matinmikko M, Prokkola J, Mustonen M, Heikkil M, Kippola T, Yrjl S, Hartikainen V, Tudose L, Kivinen A, Paavola J, Okkonen J, Mkelinen M, Hnninen T, Kokkinen H (2014) Licensed shared access (LSA) trial demonstration using real LTE network. In: Cognitive Radio Oriented Wireless Networks and Communications (CROWNCOM), 2014 9th international conference on, pp 498–502
37. Matinmikko M, Palola M, Mustonen M, Rautio T, Heikkil M, Kippola T, Yrjl S, Hartikainen V, Tudose L, Kivinen A, Kokkinen H, Mkelinen M (2015), Field trial of licensed shared access (LSA) with enhanced LTE resource optimization and incumbent protection. In: Dynamic Spectrum Access Networks (DySPAN), 2015 I.E. international symposium on, pp 263–264
38. LSA implementation. [Online]. Available: http://www.cept.org/ecc/topics/lsa-implementation
39. Ericsson, red technologies and qualcomm inc. conduct the first licensed shared access (LSA) pilot in France. [Online]. Available: http://www.redtechnologies.fr/news/ericsson-red-technolo gies-and-qualcomm-inc-conduct-first-licensed-shared-access-lsa-
40. LSA demonstration carried out in the world congress, Barcelona. [Online]. http://www.cept. org/Documents/fm-52/27033/FM52(15)13LSA-Demonstration-carried-out-in-the-Mobile-World-Congress, – Barcelona
41. Kim CW, Ryoo J, Buddhikot MM (2015) Design and implementation of an end-to-end architecture for 3.5 ghz shared spectrum In: Dynamic Spectrum Access Networks (DySPAN), 2015 I.E. international symposium on, pp 23–34
42. Goldberg RP (1974) Survey of virtual machine research. Computer 7(9):34–45
43. Ortiz S (Nov 1997) Virtual private networks: leveraging the internet. Computer 30(11):18–20
44. Bari MF, Boutaba R, Esteves R, Granville LZ, Podlesny M, Rabbani MG, Zhang Q, Zhani MF (2013) Data center network virtualization: a survey. IEEE Commun Surv Tutorials 15 (2):909–928, Second
45. Lombardi F, Pietro RD (2011) Secure virtualization for cloud computing. J Netw Comput Appl 34(4):1113–1122, advanced Topics in Cloud Computing
46. Chowdhury NMK, Boutaba R (2010) A survey of network virtualization. Comput Netw 54 (5):862–876
47. Wang X, Krishnamurthy P, Tipper D (2013) Wireless network virtualization. In: Computing, Networking and Communications (ICNC), 2013 International conference on, pp 818–822
48. Blenk A, Basta A, Reisslein M, Kellerer W (2016) Survey on network virtualization hypervisors for software defined networking. IEEE Commun Surv Tutorials 18(1):655–685 Firstquarter
49. Akhtar AM, Wang X, Hanzo L (2016) Synergistic spectrum sharing in 5g hetnets: a harmonized sdn-enabled approach. IEEE Commun Mag 54(1):40–47
50. Zaki Y, Zhao L, Goerg C, Timm-Giel A (2010) LTE wireless virtualization and spectrum management. In: Wireless and Mobile Networking Conference (WMNC), 2010 third joint IFIP, pp 1–6
51. Wang X, Krishnamurthy P, Tipper D (2015) A collaborative spectrum sharing framework for LTE virtualization. In: 2015 I.E. conference on Collaboration and Internet Computing (CIC), pp 260–269
52. Gomez MM, Weiss MB (2016) Wireless network virtualization: opportunities for spectrum sharing in the 3.5 ghz band. In: International conference on cognitive radio oriented wireless networks, pp 232–245

53. Liang C, Yu FR, Zhang X (May 2015) Information-centric network function virtualization over 5g mobile wireless networks. IEEE Netw 29(3):68–74
54. Yang M, Li Y, Jin D, Yuan J, Su L, Zeng L (2013) Opportunistic spectrum sharing based resource allocation for wireless virtualization. In: Innovative Mobile and Internet Services in Ubiquitous Computing (IMIS), 2013 seventh international conference on, pp 51–58
55. Abdel-Rahman MJ, Cardoso KV, MacKenzie AB, DaSilva LA (2016) Dimensioning virtualized wireless access networks from a common pool of resources. In: 2016 13th IEEE annual Consumer Communications Networking Conference (CCNC), pp 1042–1047
56. Huang X, Han T, Ansari N (2015) On green-energy-powered cognitive radio networks. IEEE Commun Surv Tutorials 17(2):827–842
57. Amjad M, Akhtar F, Rehmani MH, Reisslein M, Umer T (2017) Full-duplex communication in cognitive radio networks: a survey. IEEE Commun Surv Tutorials 19(4):2158–2191
58. Kpojime HO, Safdar GA (2015) Interference mitigation in cognitive-radio-based femtocells. IEEE Commun Surv Tutorials 17(3):1511–1534
59. Ahmed F, Tirkkonen O, Dowhuszko AA, Juntti M (2014) Distributed power allocation in cognitive radio networks under network power constraint. In: 2014 9th international conference on Cognitive Radio Oriented Wireless Networks and Communications (CROWNCOM), Oulu, pp 492–497
60. Ahmed F, Tirkkonen O (2009) Local optimum based power allocation approach for spectrum sharing in unlicensed bands. In: Spyropoulos T, Hummel KA (eds) Proceedings of the 4th IFIP TC 6 International Workshop on Self-Organizing Systems (IWSOS '09). Springer, Berlin/Heidelberg, pp 238–243
61. Amjad M, Rehmani MH, Mao S (2018) Wireless multimedia cognitive radio networks: a comprehensive survey. IEEE Commun Surv Tutorials 20(2):1056–1103
62. Amjad M, Sharif M, Afzal MK, Kim SW (2016) TinyOS-new trends, comparative views, and supported sensing applications: a review. IEEE Sensors J 16(9):2865–2889
63. Rashid B, Rehmani MH (2016) Applications of wireless sensor networks for urban areas: a survey. J Netw Comput Appl 60:192–219
64. Amjad M, Afzal MK, Umer T, Kim BS (2017) QoS-aware and heterogeneously clustered routing protocol for wireless sensor networks. IEEE Access 5:10250–10262

Chapter 2
Cloud-Based Context-Aware Spectrum Availability Monitoring and Prediction Using Crowd-Sensing

Hussein Shirvani and Behrouz Shahgholi Ghahfarokhi

2.1 Introduction

Internet of Things as a new paradigm in computer science aims at connecting things to the Internet. As reference [13] predicts, the number of connected devices to the Internet increases to 50 billion by 2020. On the other hand, scarcity of radio frequency, due to its fixed allocation, is an emerging problem in the world of telecommunications. This means the remaining part of frequency bands is not sufficient for tremendous growth on millions of wireless device demands. Meanwhile, experiments show that the fixed allocated bands are mostly idle. Cognitive radio (CR) is a new paradigm which suggests reusing such frequency bands for unlicensed users, also known as secondary users (SUs) or CR nodes, at the time of licensed or primary users (PUs) inactivity. Therefore, unlicensed users must perform spectrum sensing to find the available spectrum opportunities. Each SU performs local sensing using a detection method such as energy detection, matched filter detection, or cyclostationary feature detection to detect PU signal energy level. Among the above methods, energy detection is a simple detection scheme which requires no prior knowledge about signal information and only relies on energy levels [23].

Due to fading and shadowing effects and noise uncertainty and also hidden primary user problem, stand-alone spectrum sensing may fail to detect PU activity.

H. Shirvani (✉) · B. Shahgholi Ghahfarokhi
Faculty of Computer Engineering, University of Isfahan, Isfahan, Iran
e-mail: hussein.shirvani@eng.ui.ac.ir; shahgholi@eng.ui.ac.ir

© Springer International Publishing AG, part of Springer Nature 2019
M. H. Rehmani, R. Dhaou (eds.), *Cognitive Radio, Mobile Communications and Wireless Networks*, EAI/Springer Innovations in Communication and Computing, https://doi.org/10.1007/978-3-319-91002-4_2

As the solution, SUs can help each other to form a cooperative spectrum sensing (CSS) mechanism. There are two well-known schemes to implement cooperative spectrum sensing:

- Centralized: A fusion center is responsible for gathering each SU's local decision, combine them to obtain a global decision, and publish spectrum opportunities.
- Decentralized: CR nodes share their sensing information among each other. Each CR node decides which part of spectrum to use.

By using the centralized mode, other sources could be used for acquiring sensing data, specially crowd-sensing. As sensing nodes report the spectrum status periodically, high volume of sensing data, which have some features of big spectrum data [7] and gathered by crowd of sensors, is aggregated in fusion center. Storing and processing such big spectrum data can be handled by a cloud-based service which has a centralized flexible and scalable architecture [32]. Cloud computing [3] provides three types of services, infrastructure as a service (IaaS), platform as a service (PaaS), and software as a service (SaaS) where IaaS and SaaS are used in our application.

In addition to the signal information which is acquired by spectrum sensing, some context parameters such as location and time [1] exist which may affect spectrum decision. Investigating such parameters may help spectrum sensing and decision procedures to find more accurate overview of available frequencies. Moreover, it can be useful to predict spectrum status in the future. Therefore, the present approach can help IoT devices to use the proposed service in order to find best frequency band for their instant connection using CR technology as well as predicting suitable frequency bands for future use. The preliminary version of this work has been presented in Shirvani et al. [24].

The contributions of the proposed method are as below:

- Proposing a cloud-based spectrum monitoring and prediction service which is based on processing of crowd-sensed spectrum data
- Considering the impact of contextual information such as secondary user location, time, velocity, overall weather description, and complexity of buildings around the user
- Taking the advantages of machine learning techniques to predict the future of spectrum status and suggest more accurate channel availability information based on context information, spectrum reports, and historical data

This chapter is organized as follows: The first major researches on centralized cooperative spectrum sensing, crowd-sensing-based spectrum monitoring, spectrum prediction, and cloud-based spectrum monitoring will be reviewed, respectively. Then the proposed system architecture will be presented. Finally, the chapter is concluded with discussions on open research directions.

2.2   Literature Review

In this section, we review the related works in brief. The surveyed works are categorized to recent spectrum sensing methods in CR, spectrum prediction mechanisms, and proposed solutions for cloud-based spectrum monitoring.

2.2.1   Centralized Cooperative Spectrum Sensing and Decision-Making

Centralized cooperative spectrum sensing (CCSS) has been addressed by many researchers working on cognitive radio. Shinde et al. [23] use CCSS approach to overcome the effects of fading and noise. They focused on finding the optimum number of CR users and threshold to obtain best results. SUs continuously transmit their local decision to centralized fusion center where all decisions are combined to generate global decision. Their results show that majority rule can be chosen as the optimum fusion rule to be used in fusion center. Noorshams et al. [18] denote cooperative spectrum sensing as a robust method against wireless channel impairments which degrade spectrum sensing accuracy and reliability. They compare the expectation-maximization (EM) and hidden Markov model (HMM) methods in CCSS. Their research indicates that tracking the dependencies of the activities of primary user can considerably improve the system utilization. Fu et al. [9] designed and implemented a CCSS. Their proposed approach takes the advantages of energy detection and cyclostationary feature detection methods for detecting PU energy level. Vacant spectrum opportunities around SU are uploaded to fusion center with their coordinates. Also, authors evaluated their proposed system by transmitting an H.264 encoding video from one SU to another and observed better performance using CCSS instead of individual sensing. Bi et al. [2] propose a cluster-based hierarchical fusion CSS where cooperation of nodes occurs in two levels. In the first level, each cluster head combines all collected sensing data from SUs within the same cluster and makes decision about the status of licensed spectrum at cluster-level. In the other level, fusion center gathers cluster-level decisions from all cluster heads and fuses them using a modified MAJORITY rule to make final decision about the PU status. Also, the proposed method assigns a trust degree to each cluster which varies from low to high and shows the coefficient of each cluster influence on final decision. Meanwhile, Edwin et al. [8] proposed a new approach to find optimum number of cooperating CR nodes for majority rule. Moreover, V et al. [27] take the advantages of machine learning methods, SVM, and perceptron, instead of the abovementioned rules. Their comparison shows perceptron network outperforms SVM in terms of training and classification duration. Furthermore, Gupta et al. [11] proposed an adaptive threshold for energy detection method which is based on conventional threshold and SNR of PU signal which is received at CR node. Yau et al. [30] proposed a context-aware scheme for selecting dynamic channel intelligently. The proposed scheme helps secondary users to select channels in an adaptive way to improve quality of service, throughput, and delay. It is appropriate for cognitive radio networks with mobile hosts and takes the advantage of using reinforcement learning.

2.2.2 Spectrum Monitoring with Crowd-Sensing

Chakraborty et al. [4] described the SpecSense project which aims at creating a platform to perform scalable radio-frequency spectrum monitoring with crowd-sensing mechanism and used low-power and low-cost sensors. In this project, spectrum data is gathered and processed using spectrum-aware apps. They focused on efficient answering to spectrum occupancy queries. Xiang et al. [29] proposed a method to deal with data inaccuracy and incompleteness which can occur during crowd-sensing in order to build RSS maps accurately. Ding et al. [6] described occurrence of abnormal data in crowd sensors due to random failure or malicious behaviors during spectrum sensing. In order to tackle the abovementioned issues, they analyzed the abnormal data impact on cooperative spectrum sensing overall performance as well as developing robust CSS technique which performs data cleansing to deal with abnormal data. The proposed approach greatly reduces the negative impact of abnormal data.

2.2.3 Spectrum Prediction

Tumuluru et al. [25] demonstrate the advantages of predicting channel status to improve spectrum utilization as well as preserving sensing energy. They designed two channel status predictors based on multilayer perceptron (MLP) and hidden Markov model which do not require any background knowledge about channel usage statistics. Gohider [10] in her thesis implemented an intelligent cooperative spectrum sensing which takes the advantages of contextual information including path loss, SNR, energy detector threshold, and sensing time. Her model consists of three units: spectrum sensing, context modeling, and central cognitive unit. The latter fuses all observations received from two former units using a trained neural network. Yin et al. [31] studied spectrum utilization in detail based on collected data at four different locations from 20 MHz to 3GHz spectrum band in China. They investigate channel occupancy/vacancy statistics and channel utilization within major wireless services including GSM900, GSM1800, CDMA, ISM, and TV1-4. Also, they studied spectral, temporal, and spatial correlation of the mentioned measures. Moreover, they used the information collected from spectrum correlation to develop a two-dimensional frequent pattern mining algorithm which is able to predict channel availability accurately based on historical data. Madushan Thilina et al. [16] proposed cooperative spectrum sensing mechanisms based on supervised and unsupervised learning methods. They used K-means clustering and Gaussian mixture model as unsupervised and support vector machine (SVM) and K-nearest neighbor (KNN) as supervised classifiers. They measure the classifier performance based on the duration of training, the delay of classification, and receiver operating characteristic (ROC) curves. They found supervised SVM classifier with linear kernel and unsupervised K-means clustering as promising techniques for CSS in cognitive radio networks. In another paper by Madushan Thilina et al. [15], the authors consider a cooperative spectrum sensing approach to detect primary users using pattern recognition algorithms such as SVM and weighted K-nearest neighbor (W-KNN). They exploit measured received signal strengths at SUs as the feature set for detecting PU. Meanwhile, they evaluate the classifier performance based on training time, classification time, and ROC curves. In this approach, SVM classifier preforms well in terms of high probability of detection compared to existing algorithms. Instead, W-KNN achieves lower probability of false alarm. Uyanik et al. [26] proposed spectrum prediction schemes based on correlation and linear regression analysis of previous spectrum decision results to make a prediction about channel usages in the future. Comparing to well-known decision fusion techniques such as AND, OR, and MAJORITY, the proposed correlation-based technique for spectrum prediction is more efficient. Zhao et al. [33] proposed a novel spectrum monitoring approach based on prediction. The abovementioned approach integrates high-order hidden bivariate Markov model (H^2BMM) for spectrum prediction, higher-order Markov model for user mobility prediction, and the hybrid of Monte Carlo analytic hierarchy process (M-AHP), extended version of AHP for supporting multiple users, and gray relational analysis (GRA) schemes for selecting channel. Prediction results show that cognitive radio base station is able to suggest high-quality channels to secondary users in a reasonable time. Riahi Manesh et al. [22] considered a method to improve the process of spectrum scanning. Their approach works on the basis of Bayesian model, which addresses the effect of inaccuracies may occur in spectrum sensing in terms of detection and false alarm probabilities. They found their approach works more efficient than traditional frequentist influential scheme. Concluding the above review, it is obvious that previous prediction methods do not consider context parameters adequately.

2.2.4   Cloud-Based Spectrum Monitoring

Rawat et al. [20] proposed a dynamic spectrum access framework driven by global positioning system for cyber-physical transportation systems with the help of cloud. Each vehicle is equipped with a GPS as well as two transceivers. One transceiver is responsible for querying the spectrum database in order to find available channels, and the other is used for communications. In this approach, each vehicle finds the best route along its destination using its GPS and idle channels in the selected route by querying spectrum database. Meanwhile, along its path, cognitive radio user queries the database periodically to retrieve update list of available channels which results in minimum interference with primary user. CR user will switch to another idle channel when its current channel is not going to be idle anymore. Rawat et al. [21] has also proposed an approach based on game theory for allocating resources in a cloud-assisted cognitive radio network. It performs dynamic spectrum access based on secondary users and available frequency band geolocation. Also, active secondary users use the abovementioned approach to adjust their transmit power adaptively based on network conditions. Li et al. [14] investigated a spectrum monitoring approach based on cloud computing. Software-defined radios (SDRs), distributed geographically, are connected to cloud and receive their configurations. In this approach, every sensor monitors its electromagnetic environment and transmits spectrum monitoring data to cloud. The cloud analyzes sensor network behavior based on its feedback and tries to make better decision for the future. Wu et al. [28] characterizes the new features of spectrum sharing in the future wireless networks including heterogeneity in sharing frequency bands, different patterns of sharing, high level of intelligence in devices that are used for the purpose of designing efficient spectrum sharing methods, and increasing density of infrastructure of sharing networks. The authors aimed at introducing a new cloud-based architecture for managing the Internet of spectrum devices (IoSD) including spectrum monitoring devices and vast number of spectrum utilization devices. It also plans to enabling spectrum sharing in efficient manner for future wireless networks. Rawat et al. [19], in another paper, study a near real-time spectrum sensing approach giving the ability of spectrum access opportunistically in a wideband structure. The proposed scheme uses National Instruments (NI) USRP devices and interacts with a cloud for processing sensed data and storing idle channel geolocation. Table 2.1 reviews the advantages and disadvantages of surveyed researches at a glance.

Considering the researches that have been reviewed in this section, we conclude that a cloud-based spectrum monitoring service, which considers adequate contextual information in spectrum sensing and prediction taking the advantages of crowd-sensing technology, is necessary. This chapter aims at introducing a new architecture covering aforementioned idea.

2.3   Proposed Method

2.3.1   System Model

Figure 2.1 shows system model related to proposed approach in a real-world situation which consists of several base transceiver stations (BTSs), PUs, SUs (e.g., IoT devices), crowd-sensing users (some PUs that participate in spectrum sensing), several obstacles such as buildings as in a real city, and a cloud server which takes the responsibility of processing received sensing data from crowd-sensing users/SUs as well as responding those SUs that request for available frequency band(s).

Table 2.1 Reviewed researches

Research	Goal	Decision method	Evaluation method	Regarded contextual information	Using crowd-sensing
[2, 23]	Spectrum decision	*AND, OR, MAJORITY*	Simulation	–	–
[18]	Spectrum decision	Expectation-maximum/ *HMM*	Simulation	–	–
[9]	Spectrum decision	–	Implementation	–	–
[30]	Spectrum decision/ context awareness	Reinforcement learning	Simulation	Transmission range, packet error rate, PU utilization level	–
[4]	Spectrum decision	–	Implementation	–	√
[29]	Creating RSS map	–	Implementation	–	√
[6]	Spectrum decision	Likelihood ratio/equal gain combining	Simulation	–	√
[25]	Spectrum decision/ spectrum prediction	*ANN/HMM*	Simulation	–	–
[10]	Spectrum decision/ spectrum prediction/context awareness	*ANN*	Simulation	Path loss, sensing time, SNR, energy detection threshold	–
[31]	Spectrum decision/ spectrum prediction	Frequent pattern mining	Implementation	–	–
[16]	Spectrum decision/ spectrum prediction	*SVM/KNN/* Gaussian mixture model	Simulation	–	–
[15]	Spectrum decision/ spectrum prediction	*SVM/W-KNN*	Simulation	–	–
[26]	Spectrum decision/ spectrum prediction	Correlation	Simulation	–	–
[33]	Spectrum decision/ spectrum prediction	Advanced H^2 *BMM*	Simulation	–	–
[22]	Spectrum decision/ spectrum prediction	Frequentist inference/ Bayesian inference	Implementation	–	–
[20]	Spectrum decision/ context awareness/ cloud-based	Online database	Simulation	Location	–

(continued)

Table 2.1 (continued)

Research	Goal	Decision method	Evaluation method	Regarded contextual information	Using crowd-sensing
[21]	Spectrum decision/ context awareness/ cloud-based	Cosine similarity	Simulation	Location	–
[14]	Spectrum decision/ cloud-based	Suggest machine learning	Simulation	–	–
[28]	Spectrum decision/ spectrum predic-tion/context aware-ness/cloud-based	Suggest online learning/statis-tical learning	–	Spectrum state, channel state, location, energy	√
[19]	Spectrum decision/ context awareness/ cloud-based	Online database	Simulation	Location	–

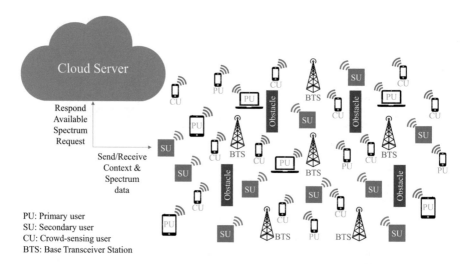

Fig. 2.1 System model

2.3.2 Proposed Architecture

The goal of this chapter is introducing a novel cloud-based spectrum monitoring architecture which collects contextual information as well as spectrum data from the SUs and crowd-sensing contributors. We are taking the advantages of using contextual information to achieve more accurate knowledge about channel conditions. Moreover, the proposed architecture aims at utilizing historical data to have the chance of predicting future behavior of PUs. Figure 2.2 shows the system

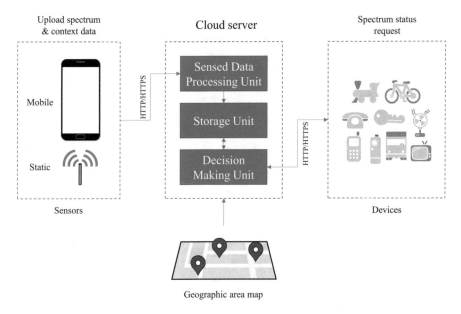

Fig. 2.2 Proposed system architecture

architecture which consists of stationary and mobile nodes (as SUs or crowd-sensing contributors) that perform spectrum sensing. Each node sends its sensing reports through REST API (HTTP/HTTPS protocol) to the cloud server for further processing. The cloud service consists of three main components, i.e., processing, storage, and decision-making units. Other secondary users or monitoring users that wish to find available spectrum opportunities (now and the future) should send their request for the decision-making service using HTTP/HTTPS protocol. The proposed architecture will be discussed in the following subsections in detail.

2.3.3 Spectrum and Contextual Sensors

In order to take the advantage of crowd-sensing, affordable devices which are equipped with appropriate sensors such as smartphones are necessary. In the proposed architecture, a developed Android application senses the RSS (dBm) from registered BTS/access point (AP) and its neighbors, as well as its global position, date, time, weather conditions, and device information. Weather condition can be captured using an online weather information service such as OpenWeatherMap [5] according to user location and time. The abovementioned information will be sent to the cloud server periodically. It is worth mentioning that the density of buildings around the sensor can be determined from the area map which is uploaded to cloud server before. Table 2.2 shows spectrum and contextual information used in our proposed architecture.

Table 2.2 Spectrum and contextual information exploited in proposed architecture

	Variable name	Variable notation	Measurement unit	Domain
Context information	Latitude and longitude	Latitutde and longitude	UTM and geo	–
	Date	D	–	–
	Time	T	UTC/GMT +4:30 hours	HH/mm/ss
	Weather description	W	–	{Rain, snow, clear}
	Humidity	H	Percent	{Low, med, high}
	Density of buildings	C	–	$0 < C < 1$
	Distance from BTS	D_{BTS}	Meters	–
Spectrum data	Network tech.	–	–	{2G, 3G, 4G, wireless}
	Network type	–	–	{GPRS, EDGE, HSPA, HSPAP, LTE, Wi-Fi}
	Signal strength	–	dBm	−120 to −50

2.3.4   Data Processing and Storage Units

This unit eliminates invalid records from input stream and stores valid records for further process. Invalid records are those which have out of range spectrum data or contextual information. In spectrum data, signal strength can be out of defined range due to unpredictable issues in smartphone connectivity to registered BTS/AP. On the other side, since the proposed architecture retrieves some of the contextual data based on user location, GPS error may lead to invalid data. In addition to eliminating invalid data, it is better to integrate sensed data in order to reduce the volume of sensed data. The spectrum and contextual data do not change frequently, and subsequent samples could be averaged if the values are close to each other. Figure 2.3 shows the process of integrating sensed data.

In the next step, density of buildings around the spectrum sensing node must be calculated. Therefore, sensor location is mapped on a map using well-known geographical software packages ESRI ArcGIS as can be seen in Fig. 2.4. For each coordinate of user, a circular polygon with 100-meter radius will be clipped. Dividing the area of buildings within the polygon by the whole area of polygon indicates the density. Moreover, in order to sense node's velocity, the distance between previous and current consecutive coordinates is obtained using the haversine formula [12] and will be divided by time interval. The refined data will be delivered to the decision-making unit.

Latitude	Longitude	Date	Time	Cell ID	RSS
32.60029621	51.66702811	Mon Sep 04 2017	14:23:05	41008	-83
32.6003826	51.6670136	Mon Sep 04 2017	14:23:13	41008	-83
32.60050565	51.66700485	Mon Sep 04 2017	14:23:22	41008	-83
32.60050565	51.66700485	Mon Sep 04 2017	14:23:29	30323	-71
32.60050565	51.66700485	Mon Sep 04 2017	14:23:31	30323	-71
32.60067184	51.6669349	Mon Sep 04 2017	14:23:34	30323	-83
32.60070072	51.66691937	Mon Sep 04 2017	14:23:36	30323	-83

Latitude	Longitude	Date	Time	Cell ID	RSS
32.60039482	51.66701552	Mon Sep 04 2017	14:23:13	41008	-83
32.60059597	51.66696599	Mon Sep 04 2017	14:23:32	30323	-77

Fig. 2.3 Sensed data integration procedure

2.3.5 Decision-Making Unit

The purpose of a decision-making unit is reaching proper decision on channel condition as well as predicting the future of channel usage. For this purpose, machine learning techniques have been taken into account. A learner has been considered to indicate the current state of a channel regarding the context parameters which is known as classifier. The input vector of this learner consists of latitude, longitude, date (day of week), time (hour of day), weather description (clear/rain/snow), humidity (above/below 80%), surrounding buildings' density, velocity, channel, and signal strength as depicted in Fig. 2.5. Moreover, for predicting the future of the spectrum bands, another learner, which is known as predictor, has been regarded where its inputs are latitude, longitude, and the channel of interest as shown in Fig. 2.6. In order to train the channel state indicator and the channel state predictor, some of the gathered dataset has been digitized using appropriate membership functions to be readable by learning techniques. Figure 2.7 shows possible inputs of membership functions and the related digital values. For training the learners, a dataset of channels that is assigned to the registered BTSs has been attained from the cellular operator for a period of time. This dataset is the set of channels which are exploited by each cell ID (related to a BTS) and are changed over the time based on cellular network operator's policies. Figure 2.8 shows an example of the channel lists that are occupied in each location (their RSS is much greater than −120 dB).

For creating a classifier and predictor, there are various supervised learning methods discussed in [17] such as SVM, artificial neural networks (ANN), KNN, etc. which can be used in proposed architecture.

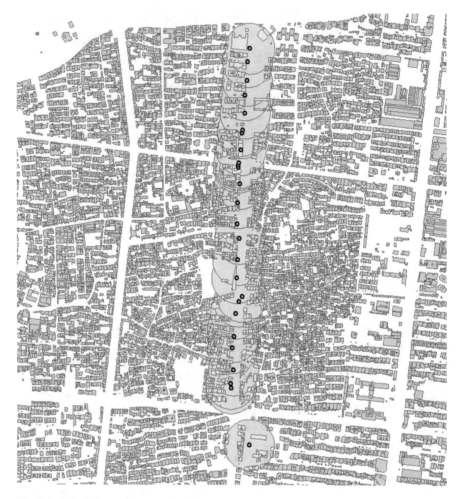

Fig. 2.4 Calculating environment density using ArcGIS

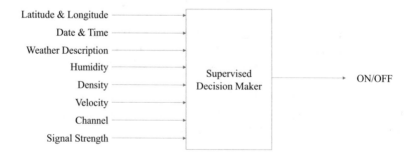

Fig. 2.5 Proposed classifier

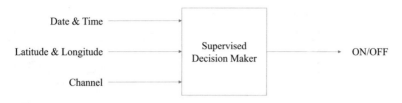

Fig. 2.6 Proposed predictor

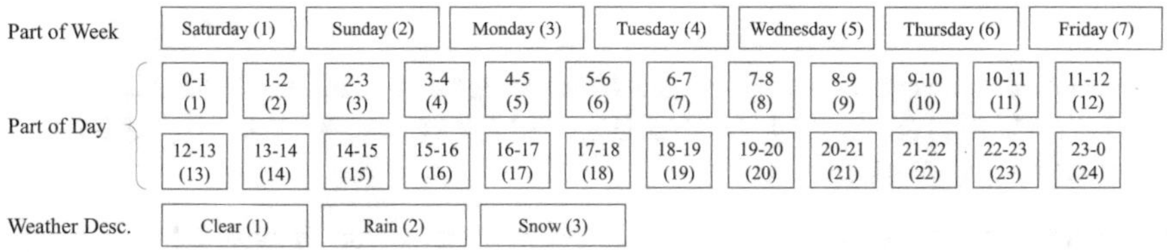

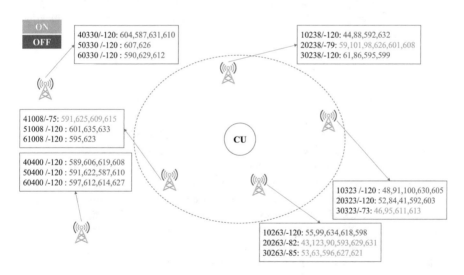

Fig. 2.7 Possible inputs of membership functions and their related number value

Fig. 2.8 Neighbors (including registered cell) of secondary user and its channels' status

After training the learner/predictor, each secondary user can query the cloud server using a GET request as below with the help of designated parameters which can be observed in Table 2.3:

Table 2.3 Designated parameters for queries

	Parameter name	IN_PARAM	VALUE (sample)
1.	Latitude	lat	35.7148995
2.	Longitude	long	51.4163083
3.	Time	t	11
4.	Date	d	2
5.	Humidity	h	Low
6.	Channel number	c_n	409
7.	Weather	w	Clear
8.	Received signal strength	rss	−67

http://www.opensamp.com/osamp_data/osamp_retrieve_data.php?PARAM_
   1=VALUE_1&PARAM_2=VALUE_2&IN_PARAM_3=VALUE_3&IN_
   PARAM_4=VALUE_4&..."

After this request, based on selected parameters, status of channel(s) or set of available channels' number will be returned to user. Proposed predictor helps users to escape from network latency when access time to the cloud is considerable.

It is notable that the implementation of proposed architecture consists of two phases: (1) data gathering and (2) spectrum decision and prediction. Currently the first phase has been completed, and a preliminary set of spectrum and contextual data are available. Further evaluation will be considered as the future work.

2.4   Conclusion

Scarcity of radio-frequency bands is a challenging problem in wireless communications that is due to fixed allocation of some spectrum bands to existing wireless services. Existing researches show these frequency bands are idle most of the time. Therefore, there is a chance that unlicensed users (SUs) reuse these bands at the time their PUs are unavailable. Cognitive radio as a new concept in wireless communication era discusses about the abovementioned opportunity. In order to find idle frequency bands, SUs must perform spectrum sensing. Due to fading, shadowing, and noise uncertainty effects, PU activity may not be detected by stand-alone sensing. For better results, more SUs can cooperate with each other. In this chapter, a context-aware cloud-based spectrum monitoring system has been reported which aims at introducing a new centralized architecture that takes the advantages of crowd-sensed data of static and mobile users and uses it for both spectrum decision and spectrum prediction. This data consists of spectrum data as well as contextual information such as location, time, and complexity of buildings around user which may improve decision and prediction accuracy. Supervised learning methods such as SVM, ANN, and KNN can be used to make a decision about current status of a channel as well as predicting its condition in the future. Each user can send its request to cloud using GET method based on its demand. Moreover, the proposed predictor can be supportive for a user to find the channel status when accessing the cloud may amount considerable time due to network latency. The data-gathering phase of proposed architecture has been implemented, while the complete implementation is considered for the future.

2.5 Future Research Directions

Our suggestion for future researches is working on suitable learning methods for streamline processing of spectrum and contextual data aiming at spectrum decision and prediction. Investigating other existing parameters that affect spectrum decision quality is also in our future work. Moreover, implementing crowd-sensing applications for various platforms in addition to present Android version is a suggestion. Furthermore, comparing the performance of the proposed approach to a decentralized method used in cognitive radio is of our future work.

Acknowledgments We would thank like to Mobile Telecommunication Company of Iran – Isfahan Branch – for their support that assisted data collection for this research.

References

1. Akhtar F, Rehmani M, Reisslein M (2016) White space: definitional perspectives and their role in exploiting spectrum opportunities. Telecommun Policy 40:319–331. https://doi.org/10.1016/j.telpol.2016.01.003
2. Bi Y, Jing X, Sun S, Huang H (2016) Hierarchical fusion-based cooperative spectrum sensing scheme in cognitive radio networks. In: 16th international symposium on Communications and Information Technologies (ISCIT), Qingdao, IEEE, pp 579–583
3. Botta A, de Donato W, Persico V, Pescapé A (2016) Integration of cloud computing and internet of things: a survey. Futur Gener Comput Syst 56:684–700. https://doi.org/10.1016/j.future.2015.09.021
4. Chakraborty A, Rahman M, Gupta H, Das S (2017) SpecSense: crowdsensing for efficient querying of spectrum occupancy. In: IEEE conference on computer communications (INFOCOM 2017), Atlanta, IEEE, pp 1–9
5. Current weather and forecast – OpenWeatherMap. In: Openweathermap.org. https://openweathermap.org/. Accessed 28 Nov 2016
6. Ding G, Wang J, Wu Q et al (2014) Robust spectrum sensing with crowd sensors. IEEE Trans Commun 62:3129–3143. https://doi.org/10.1109/tcomm.2014.2346775
7. Ding G, Wu Q, Wang J, Yao Y-D (2014) Big spectrum data: the new resource for cognitive wireless networking. http://arxiv.org/pdf/1404.6508.pdf
8. Edwin K, Walingo T (2016) Optimal fusion techniques for cooperative spectrum sensing in cognitive radio networks. In: International Conference on Advances in Computing and Communication Engineering (ICACCE), Durban, IEEE, pp 146–152

9. Fu Y, Li Z, Liu D, Liu Q (2014) Implementation of centralized cooperative spectrum sensing based on USRP. In: International conference on Logistics, Engineering, Management and Computer Science (LEMCS), Atlantis Press

10. Gohider N (2016) Context augmented spectrum sensing in cognitive radio networks. Master Thesis, University of Waterloo

11. Gupta M, Verma G, Dubey R (2016) Cooperative spectrum sensing for cognitive radio based on adaptive threshold. In: Second international conference on Computational Intelligence & Communication Technology (CICT), Ghaziabad, IEEE, pp 444–448

12. Haversine formula. In: En.wikipedia.org. https://en.wikipedia.org/wiki/Haversine_formula. Accessed 12 Aug 2017

13. Hotel News Resource (2017) IoT devices installed base worldwide 2015–2025|Statistic. In: Statista. https://www.statista.com/statistics/471264/iot-number-of-connected-devices-world wide/. Accessed 10 Apr 2017

14. Li R, Li J (2014) A novel clouds based Spectrum monitoring approach for future monitoring network. In: 2nd international conference on systems and informatics (ICSAI), Shanghai, IEEE, pp 520–524

15. Madushan Thilina K, Choi KW, Saquib N, Hossain E (2012) Pattern classification techniques for cooperative spectrum sensing in cognitive radio networks: SVM and W-KNN approaches. In: IEEE Global Communications Conference (GLOBECOM), Anaheim, IEEE, pp 1260–1265

16. Madushan Thilina K, Choi KW, Saquib N, Hossain E (2013) Machine learning techniques for cooperative spectrum sensing in cognitive radio networks. IEEE J Sel Areas Commun 31:2209–2221. https://doi.org/10.1109/jsac.2013.131120

17. Mitchell T (2013) Machine learning. McGraw-Hill, New York [u.a.]

18. Noorshams N, Malboubi M, Bahai A (2010) Centralized and decentralized cooperative Spectrum sensing in cognitive radio networks: a novel approach. In: IEEE 11th international workshop on Signal Processing Advances in Wireless Communications (SPAWC), Marrakech, IEEE, pp 1–5

19. Rawat D, Bajracharya C, Grant S (2017) nROAR: near real-time opportunistic spectrum access and management in cloud-based database-driven cognitive radio networks. IEEE Trans Netw Serv Manag 14:745–755. https://doi.org/10.1109/tnsm.2017.2730201

20. Rawat D, Reddy S, Sharma N et al (2015) Cloud-assisted GPS-driven dynamic spectrum access in cognitive radio vehicular networks for transportation cyber physical systems. In: IEEE Wireless Communications and Networking Conference (WCNC), New Orleans, IEEE, pp 1942–1947

21. Rawat D, Shetty S, Raza K (2014) Game theoretic dynamic spectrum access in cloud-based cognitive radio networks. In: IEEE International Conference on Cloud Engineering (IC2E), Boston, IEEE, pp 586–591

22. Riahi Manesh M, Subramaniam S, Reyes H, Kaabouch N (2017) Real-time spectrum occupancy monitoring using a probabilistic model. Comput Netw 124:87–96. https://doi.org/10.1016/j.comnet.2017.06.003

23. Shinde S, Jadhav A (2016) Centralized cooperative Spectrum sensing with energy detection in cognitive radio and optimization. IEEE international conference on Recent Trends in Electronics, Information & Communication Technology (RTEICT), Bangalore, IEEE, pp 1002–1006

24. Shirvani H, Shahgholi Ghahfarokhi B (2017) A cloud-based context-aware spectrum monitoring platform. In: 1st international conference on internet of things; applications and infrastructure, Isfahan

25. Tumuluru V, Wang P, Niyato D (2010) Channel status prediction for cognitive radio networks. Wirel Commun Mob Comput 12:862–874. https://doi.org/10.1002/wcm.1017

26. Uyanik G, Canberk B, Oktug S (2012) Predictive spectrum decision mechanisms in cognitive radio networks. In: IEEE Globecom Workshops (GC Wkshps), Anaheim, IEEE, pp 943–947

27. Balaji V, Nagendra T, Hota C, Raghurama G (2016) Cooperative Spectrum sensing in cognitive radio: an archetypal clustering approach. In: International conference on Wireless Communications, Signal Processing and Networking (WiSPNET), Chennai, IEEE, pp 1137–1143

28. Wu Q, Ding G, Du Z et al (2016) A cloud-based architecture for the internet of spectrum devices over future wireless networks. IEEE Access 4:2854–2862. https://doi.org/10.1109/access.2016. 2576286
29. Xiang C, Yang P, Tian C et al (2016) CARM: crowd-sensing accurate outdoor RSS maps with error-prone smartphone measurements. IEEE Trans Mob Comput 15:2669–2681. https://doi. org/10.1109/tmc.2015.2508814
30. Yau K, Komisarczuk P, Teal P (2009) A context-aware and intelligent dynamic channel selection scheme for cognitive radio networks. In: 4th international conference on Cognitive Radio Oriented Wireless Networks and Communications (CROWNCOM '09), Hannover, IEEE, pp 1–6
31. Yin S, Chen D, Zhang Q et al (2012) Mining spectrum usage data: a large-scale spectrum measurement study. IEEE Trans Mob Comput 11:1033–1046. https://doi.org/10.1109/tmc. 2011.128
32. Zhang Q, Cheng L, Boutaba R (2010) Cloud computing: state-of-the-art and research challenges. J Internet Ser Appl 1:7–18. https://doi.org/10.1007/s13174-010-0007-6
33. Zhao Y, Hong Z, Luo Y et al (2017) Prediction-based spectrum management in cognitive radio networks. IEEE Syst J:1–12. https://doi.org/10.1109/jsyst.2017.2741448

Chapter 3
Cooperative Spectrum Handovers in Cognitive Radio Networks

Anandakumar. H and Umamaheswari. K

3.1 Introduction

The present advanced wireless communication system is a replacement of the conventional cellular systems. The increased mobile usage services have created a large requirement for higher data rates; nonetheless the whole spectrum is assigned only to available users. Furthermore the frequency spectrum is assigned to licensed users by limiting a particular band of spectrum to specific users, thus separating utilization by other users. Therefore vacant frequency bands are made available. This frequency range is further expanded to channels with particular encoding and modulation schemes that do not allow interference between its users. The above regulation works only for limited frequencies, and they are presently a regeneration of telecommunication standards.

Wireless technology has constantly been emerging and its vigor has increased because more users are using it and spectrum allocation is more by the users for the reason that of spectral crowding. The limited bandwidth marks overcrowding inescapable. Mitola [23] proposed the cognitive radio (CR) concept in line with the following reasons:

The CR is an innovation with an enhancing field of research, where any device can spontaneously sense the environmental conditions, and the interrelation parameters [18] could be accepted applicably. The foremost components of CR networks

Anandakumar. H (✉)
Department of Computer Science and Engineering, Sri Eshwar College of Engineering,
Coimbatore, Tamil Nadu, India

Umamaheswari. K
Department of Information Technology, PSG College of Technology, Coimbatore,
Tamil Nadu, India
e-mail: uma@ity.psgtech.ac.in

© Springer International Publishing AG, part of Springer Nature 2019   47
M. H. Rehmani, R. Dhaou (eds.), *Cognitive Radio, Mobile Communications and
Wireless Networks*, EAI/Springer Innovations in Communication and Computing,
https://doi.org/10.1007/978-3-319-91002-4_3

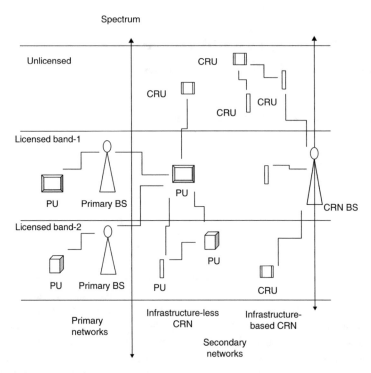

Fig. 3.1 CRN architecture

are primary networks and cognitive networks [17]. The primary network consists of the primary user (PU), and the cognitive network comprises the secondary user (SU) [20]. Principal users of the network are identified as the licensed users, and the users at the next level are recognized as the unlicensed SU [21].

The CR technology paves way for SUs to use the unutilized PUs frequency space, and the structure of CR technology relies on software-defined radio (SDR). CR networks are divided into two types, centralized and distributed. The centralized network is referred as infrastructure-oriented network, and the distributed network is known as infrastructure-less network where the secondary base station user manages the CR users.

CR network architecture is represented in Fig. 3.1, which is classified into structure-oriented CR networks and structure-less CR networks. The CR user has a base station (BS) in infrastructure-based CR network, which is a central network unit in cellular networks; here, the CR user is well ordered by medium access controller (MAC) unit.

The CR users communicate in an ad hoc style in the infrastructure-less CR network, with each other on both spectral bands, namely, licensed and unlicensed frequencies. The Federal Communications Commission (FCC) states that the spectrum operation is either underutilized or overcrowded in frequency bands. The cognitive radio networks put forward the dynamic spectrum access method which

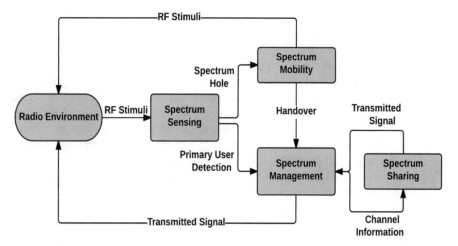

Fig. 3.2 Cognitive radio cycle

is fully reconfigurable. In this technique, the network finds out the environment by design and satisfies the user's demands by means of which the communication parameters can be altered.

CR network encounters many research trials which include network initialization factors, concealed with unavoidable problems, and spectrum provision issues. The available spectrum can be further reprocessed by CR technology toward effective utilization. The major factor that limits spectrum reuse is interference, which occurs due to noise during the transmission of other radio signals [26]. The functions of CR are spectrum detection, spectrum management, spectrum mobility, and spectrum distribution.

In CR, (1) spectrum detection is a key function where unused spectrum is detected and is used speculatively; (2) spectrum channel is preferred based on the necessities of user communication followed by detecting the spectrum holes during spectrum management in the CR; (3) in spectrum mobility, when the qualified user is not in range, then the spectrum with lower priority effortlessly moves to the next accessible vacant channel; and (4) with CR spectrum distribution facility, every CR will get an opportunity to use the spectrum. Figure 3.2 detects the CR cycle which is the edifice of the whole process.

3.2 Literature Survey

Mishra et al. [22] stated that energy detection is a simple and valuable sensing method, which was commonly used by CR; however, its sensed performance may be degraded by the lower conventional signal-to-noise ratio (SNR). It had been revealed that a network of supportive CRs with different vanishing states from the target would have a better chance of detecting the PU if they adopted cooperative spectrum sensing. Each CR sensed PU autonomously and then advanced its observed energy statistic to a coordinator that would combine all the received statistics from CRs by decision or data fusion with the presence of the PU.

Zhao and Swami [33] explored the spectrum handover procedure and have suggested four metrics to show both short-term and long-term spectrum handover performance; they are (1) link maintenance probability, (2) the number of spectrum handover, (3) switching delay, and (4) non-completion probability. The probability mass function (pmf) and an average number of spectrum handover were built in addition by exploring opportunistic and negotiated spectrum access strategies. Spectrum handover techniques are proposed to evaluate the primary and secondary user's co-existence and support for representation and optimization of CR network.

Akyildiz et al. [2] intrinsic properties and current research tests of spectrum organization in CR networks were surveyed. In particular, they examined novel spectrum administration functionalities such as spectrum sensing, spectrum resolution, spectrum sharing, and spectrum mobility. Many investigators are presently involved in developing the communication technologies and etiquettes required for CR networks. However, to perform spectrum reactive communication, more research is necessary along with the spectrum location.

Quan et al. [27] premeditated the result of a large-scale measurement and found that the wireless spectrum is nonstationary, the background probabilistic model is varying continuously, and the error of prediction is unavoidable. In order to progress the performance of dynamic spectrum access by leveraging spectrum usage, the prediction error also has been contemplated.

Yucek and Arslan [31] suggested a double-threshold cooperative detection algorithm to build on weighing and improved the detection probability of CR. Both the decision fusion and data fusion are applied in the proposed scheme in which the energy statistics observed by CRs. The cooperative detected performance of the CRs in the erroneous area is encouraged by the weighed data fusion, although the cooperative detected performance of the CRs. In this work, both the periodical single band and wideband supportive findings were examined, and the probabilities of the spectrum functioned by these two configurations were individually investigated.

Anandakumar and Umamaheswari [4] worked over an information-sharing model for dispersed interference detection systems. In this model, each individual ID in the network uses cumulative sum (CUSUM) along with its local data and relates it with a local inception to check if an irregularity has been detected. One of the benefits of using the CUSUM algorithm was that it can consume less computing resources, making it very suitable for the characteristics of resource-constrained.

Suganya and Anandakumar [28] research focused on classical block-based recognition schemes such as energy detection, feature detection, or coordinated filtering. Surrounded by these schemes, the cognitive users always gather a sequence of interpretations within a fixed detecting time window and then compute their test data for decision. Most of them put pressure on maximizing the prospect of detection while upholding a satisfactory level of false alarm rate. The detection problem was introduced by successive change-point detection (also called quickest detection) with a suitable framework for energetic sensing in CR. Quickest finding is a branch area of successive-type detection, and some inclusive studies can be found. The theoretical imprint was to detect changes in spreading of comments as quickly as possible. Spectrum under min-max formulation, which was planned as well-known page is increasing CUSUM algorithm and has been proved optimal in the sense of diminishing the mean delay of revealing while maintaining a certain level of false alarm rate.

Pacheco et al. [24, 25] had associated packet-switched handover (PSHO) to reduce handover interruption because the spectrum sensing was not essential in the handover procedure. In addition, it was easier to have a consensus transmission on their target channel. However, when the spectrum handover process was in progress, the PSHO had determined that the amount of unavailable channel was increased. Thus, one challenge for the PSHO was to verify the target channels to reduce packet loss probability (overall or individual) and reduce bandwidth fragment ratio.

Anandakumar and Umamaheswari [5] stated that without exact detection method trials, the influence of missed detection and false alarm recommends an improved threshold-based spectrum access policy that reaches close to optimal performance. In addition they develop and assess a planned scheme that allows multiple SUs to jointly protect the PU while adjusting to behavioral changes in PU usage design.

Anandakumar and Umamaheswari [6] had worked on user cooperation for CR systems that have been studied, and mutual finding was employed among all the collaborative users. The process of assembling the whole data received at one place may be very complex under practical communication. Moreover, any user can assist along with others only if there are other users in its surrounding area monitoring under the same frequency as itself; in this case collaboration between the CR users cannot be defined in general. A more feasible system like the hard-decision strategy was considered, where the individual SUs make self-determined decisions about the presence of the PU in the frequency range that they are observe and connect their evaluation to a fusion center. The final decision is made by fusion center by combining the all cooperating radios decisions. In observe the fusion center is a centralized organizer that administers the channel obligation requirements for the secondary users.

3.3 Handover Procedure for CRN

The process of dynamic handover is appropriate for all secondary users in CRNs. Here when the primary user arrives, the secondary user need to vacate that occupied channel, and this process forced the termination of secondary user. This approach prevails over by a method known as fraction guard channel assignment; this results in increasing throughput of unlicensed users but cannot adjust the value adequately.

The effective and dynamic channel handover process is given in Fig. 3.3 for a CRN. The CHP (channel handover process) in cognitive radio network is a time-consuming process. To identify the optimal exchange, an effective handover strategy is exposed between the CHP duration and user activities. The evaluation to initiate

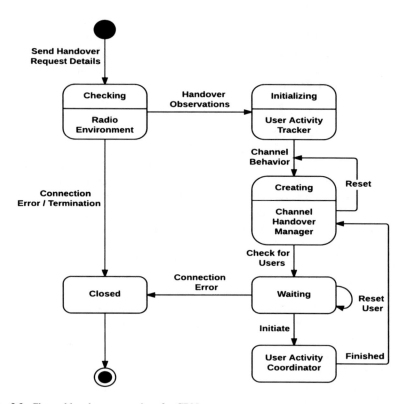

Fig. 3.3 Channel handover procedure for CRN

the channel handover procedure or terminate the activity of ongoing secondary user's nodes is done by tracking set of confined information. According to self-motivated circumstance of the channels, this structure varies, thereby increasing the throughput resultant in the decrease of handover occurrence.

3.3.1 Spectrum Management

The cognitive radio techniques like dynamic spectrums provide mobile users with high bandwidth by having access to a heterogeneous wireless environment. Cross-layer design approach is mostly recommended for infrastructure networks where the need of central network entities and impromptu networks depend on distributed coordination. There are certain novel and unique spectrum management functions such as spectrum sharing, decision, sensing, and mobility.

The integration of functions in many layers of the protocol stack remains a greatest challenge in CR networks, which will allow the secondary users to communicate reliably over dynamic spectrum environments [12]. Spectrum holes are identified effectively by detecting the received primary user data which are available within a range. In real time, it is certainly a complex process wherein the channel between a primary and transceiver needs to be measured. Hence, focus is shifted to primary transmitter detection by describing many sensing methods, which are classified as follows:

3.3.1.1 Cooperative Detection

Cognitive radios in the network will initiate to send sensed data to centralized location called fusion center, in order to increase each CRs sensing capabilities. The status about the channel is synched and sent to the central fusion center by all CRs [9]. The channel keeps monitoring all CRs for the free spectrum incessantly, while the transmitter is transmitting for the licensed user.

The core station combines with the cognitive radio networks by aligning signals that are received from the different radio networks. All the CRs in the system will send the information to the fusion center when the free spectrum becomes available and decisions regarding the availability of spectrum will be made. The false alarm probability (Q_f) and missed detection probability (Q_m) can be done by considering cooperative sensing with the same threshold (λ) level at each CR. In cooperative detection, different radio networks within a cognitive system will complete cognitive radio sensing.

3.3.1.2 Noncooperative Detection

On the other hand, CRs sense the spectrums independently and send the data in the noncooperative detection, while they do not acquire any data of other CRs in the locality. The position of each CR is different, and the channel itself is imperfect where all CRs have varying signal-to-noise ratios as well as threshold levels [11] leading to an ambiguous situation at the fusion center based on correctness of the situation.

This results in sensing difficulty about the primary receiver's status. The primary user's transmission is detected by sensing the signals from primary transmitter, and this category of spectrum sensing is known as primary transmitter detection. Here the spectrum sensing ensures that when a cognitive radio configures based on the signal, it senses the data, statistics, and information with which it was preloaded.

3.3.1.3 Spectrum Assignment

The spectrum that is not utilized from the primary user to the secondary user in cognitive radio is consigned, and the interference can be inadequate by the spectrum management between primary user and secondary user units; thereby numerous difficulties arise while allocating the spectrum to the secondary users.

3.3.2   Spectrum Utilization

The spectrum allocation and management issue is due to disproportionate and unequal spectrum utilization [7]. The network entry, initialization, and the hidden incumbent problem cannot be overcome by prevailing protocols. The cognitive radio mobility is an issue that is still unexplored and can be overcome by "LEO satellite" which is a CR-assisted architecture.

3.3.3   Sharing of Spectrum

The process of spectrum sharing is done between primary and the secondary user where spectrum allocation is carried out based on two main goals: (1) maximizing the system bandwidth reward and (2) increasing secondary user's access fairness using band-limited AWGN method. Band allocation technique optimizes the overall optimization problem performance of the system [13], and a fair trade-off between secondary users access in the existing method is provided.

3.4   The Proposed Cooperative Spectrum Handover

The cognitive radio cooperative spectrum sensing occurs when cognitive radios as a network share their sensed gain information. A better representation about spectrum usage over the area where the CRs are situated is obtained. Generally, abrupt changes at a greater speed than expected in any phenomenon like amplitude, mean, variance, frequency, etc. are known as cooperative change detection. Cognitive radio spectrum sensing becomes challenging technology to implement during handover due to investigation and instance varying in multipath fading effects.

3.4.1   Threshold Optimization Based on Cooperative CUSUM

The channel band increases from the PU to SU become inefficient when the SNR values become minimal, and hence the task of primary user detection based on the single secondary user observation becomes very complex. The cooperative spectrum sensing can be applied in order to overcome this problem, and multiple secondary users are allowed to cooperate by levering the spectrum variance inherent in the radio environment. Figure 3.4 depicts the scenario for handover management that is followed for estimation of efficient spectrum analysis. The following methodology is required in any CRN: sensing frequency band, spectrum management, sharing, and

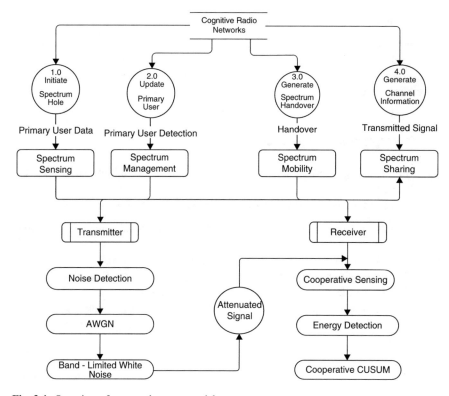

Fig. 3.4 Overview of proposed system model

mobility. Actually a certain amount of noise during transmission of radio signals is vital for estimation of efficiency.

Additive white Gaussian noise (AWGN) signal is added for analyzing the signal efficiency, and the detection technique uses the band-limited white noise [8]. Cooperative CUSUM, a spectrum sensing method, is followed to analyze the SNR level as the attenuated noise signal is passed to the receiver side called cooperative CUSUM algorithm.

3.5 Cooperative Spectrum Sensing During Handover

3.5.1 Spectrum Sensing Approach

The purpose and challenge [10] of the secondary user are to decide the existence of PUs in a spectrum and then to instantly leave that frequency band. The resulting primary radio will emerge in avoiding any interference to licensed users.

There are two spectrum sensing approaches:

- *Direct frequency-domain approach* – Direct signal-based spectrum allocation is carried.
- *Indirect time-domain approach* – Spectrum allocation performed using autocorrelation of the signal.

3.5.1.1 Algorithm for Spectrum Sensing

Spectrum sensing and allocation performed in the initial steps of CR implementation.

The steps involved in spectrum sensing shown in Fig. 3.5 are as follows:

- *Initialization* of message frequency, carrier frequency signals, and sampling frequency.
- *Amplitude modulation* of existence band relevant to user data.
- *Addition* of entire modulated frequency signals to generate a transmitting signal.
- *Estimation* of the received signals power spectral density.
- *Detection* of available vacant spectrum holes for allocation to new users.
- *Quantity* of attenuated noise about *15 Percentage* of Attenuation is introduced.

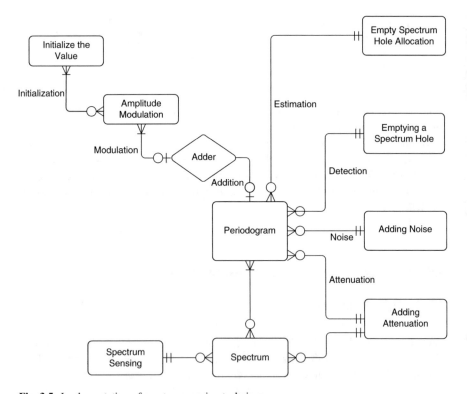

Fig. 3.5 Implementation of spectrum sensing techniques

3.5.2   *Energy Detection over AWGN Channels*

The comparative energy detection [13] of noise in any AWGN channel is described below:

- Transmits the initial signals from all primary users.
- The signal range is calculated by adding all received signals at the receiver.
- Meanwhile, the power spectrum density of signals is estimated by using period gram function.
- Finally, the average power in a signal is calculated by taking the integral of the PSD over a certain frequency band.
- After comparison, the detector perceives and designates that a primary user is present as the power of received signal is greater than the threshold signal power.
- Range of spectrum detection is calculated using MarcumQ function.

3.5.3   *Spectrum Detection*

The hypothesis test considers the formulation of spectrum sensing problem. The secondary user band-limited signal, which is being sensed, is indicated as $X(t)$ [15]. The presence or absence of users is detected by cognitive radio (CR) network using any of the spectrum sensing techniques [29].

Hypothesis (H) is indecate as the channel gain and distinct as the hypotheses of not receiving any signal from the licensed primary user in the target frequency band, and $N(t)$ is denoted as additive noise.

The two basic hypotheses H_0 and H_1 are considered in Eqs. (3.1) and (3.2).

$$H_0 : \text{Power of primary useris absent} \tag{3.1}$$

$$H_1 : \text{Power of primary user during 't' is known} \tag{3.2}$$

CR detects the primary user signal at time "t" and bandwidth "w," and H_1 is observed during spectrum "S" and compared with the threshold "λ" as in Eqs. (3.3) and (3.4).

$$H_0 : X(t) = n(t) \tag{3.3}$$

$$H_1 : X(t) = h(t) + n(t) \tag{3.4}$$

where:

$t = 0, 1, 2, 3, 4 \ldots N$
$X(t) = \text{SUs received signal}$
$H(t) = \text{PUs signal}$
$N(t) = \text{noise variance of} \rightarrow \text{AWGN with PSD } \mathbf{N_0}$

3.5.4 Probabilities of Detection

The performance of spectrum sensing is measured using the following parameters:

- Detection probability (Pd) – indicates that the licensed user is available and should be at its maximum factor, and it helps to secure the PU from interference.
- False alarm probability (Pfa) – indicates the presence of primary user, but actually, there is no primary user in reality. In order to increase the spectrum utilization factor, this should be at its minimum.

Spectrum sensing is considered the highly preferred method for energy detection due its low computational and implementation intricacies [30]. It detects the licensed user signal based using FFT that proselyte a signal from time variance to frequency variance domain [32]. The energy known as power spectral density (PSD) is calculated for any signal.

The detection parameters are revealed as a series of fast Fourier transform (FFT) segment derived in Eq. (3.5):

$$D(x) = \frac{1}{N} \sum_{t=1}^{N} \Theta[X(t), H(t)] \underset{<}{\overset{>}{=}} \lambda \qquad (3.5)$$

Where,

$D(x)$ = detection of probability
N = number of samples
λ = detection threshold value

The various probabilities based on the 'λ' can be defined as probability of detection measured using Eq. (3.6),

$$P_d = P_r \left(D(x) > \frac{\lambda}{H_1} \right) \qquad (3.6)$$

The probability of false alarm can be calculated using Eq. (3.7),

$$P_{fa} = P_r \left(D(x) > \frac{\lambda}{H_0} \right) \qquad (3.7)$$

The probability of missed detection can be calculated using Eq. (3.8),

$$P_m = 1 - P_d = P_r \left(1 - D(x) > \frac{\lambda}{H_1} \right) \qquad (3.8)$$

3.5.5 Signal-to-Noise Ratio Selection

The amount of signal power and respective amount of variation in noise conflicts is considered as the signal-to-noise (SNR) band. The proposed algorithm works on the basis of this SNR [15] channel estimation.

This ratio is based on Gaussian distribution for the AWGN [18] and is generalized in Eq. (3.9) as follows:

$$\text{SNR} = \frac{\rho}{\sigma_n^2} \tag{3.9}$$

Here "σ" is denoted as standard power on autocorrelation.

3.5.6 Selection of Threshold

The threshold for the primary user can be selected based on the low SNR, by any one of the following probabilities as in Eqs. (3.10, 3.11, 3.12, and 3.13)

$$P_f = Q\left(\frac{\lambda - \mu_0}{\sigma_0}\right) \tag{3.10}$$

$$P_d = Q\left(\frac{\lambda - \mu_1}{\sigma_1}\right) \tag{3.11}$$

where Q = complementary error function.

$$Q(x) = \frac{1}{2\lambda} \int_{-\infty}^{\infty} \text{Sxx}(w)dw \tag{3.12}$$

where $F-1(\text{Sxx})$ = autocorrelation function = $\text{Rxx}\ (w)$

$$\lambda = \sigma_i Q(P_f) + \mu_{i+1} \tag{3.13}$$

Thus the obtained threshold value needs enrichment in order to reduce the SNR value by interprets various probabilities based on environmental conditions and number of users available.

3.5.7 Cooperative CUSUM Algorithm

The CUSUM (cumulative sum) algorithm is considered as a sequential analysis method used for monitoring changes in spectrum detection. The sequential detection techniques perform better than the novel rule based on a single observation [1].

These techniques utilize the past observations gainfully along with the current values, and recently these algorithms are being used in a decentralized setup. It can efficiently be implemented online and are constant in nature and require nominal computations at each trace. Energy at the secondary channel transmission is minimized, and also primary channel interference is reduced when it has already started transmission.

The CUSUM performance is measured through false alarm in case the quality of handover is achieved. But when the handover quality is poor, necessary action is taken to measure delay. Finally CUSUM uses all the past data from all the secondary nodes from the fusion node to optimally make a decision.

An efficient CUSUM algoritham, which can discriminate between two hypotheses using lowest number of samples (N), is major challenges where false alarm and missed detection probability constraints prevail of Amjad et al. [3] prevail. The discovery algorithm is based on minimal threshold of any spectrum, and for assessing these changes cumulative sum (CUSUM) algorithm is a unique standardized technique [3]. A cumulative sum of positive and negative change estimation in spectrum frequency is required compared with threshold (λ) for implementation. When there is a gain in the reveal cumulative sum, it resumes from zero.

The proposed algorithm is based on SNR channel assessment as the variance of noise changes the SNR channel. Hence the detected threshold can observe changes in projected SNR that varies based on network uniqueness. Licensed user interference is reduced by increasing spectrum usage of the secondary user. CR effectively utilized the power and bandwidth, efficient decentralized protocols are required since any CRN consists of distributed secondary nodes geographically located.

Algorithm

1. Threshold value $\lambda > 0$ is initialized.
2. SNR for current received signal $X(t)$ is estimated.
3. The total number of samples (N) that is required for achieving desired probabilities Pd and Pfa is computed.
4. Threshold value of receiver based on estimated SNR and number of samples must be set depending on.
5. Test statistics results are analyzed.
6. Test statistics of computed and proposed threshold value is compared to decide the presence or absence of the licensed user.

3.6 Evaluations and Discussion

This chapter presents the basic concepts of spectrum sensing and handover which is one of the fundamental prerequisites for the successful deployment of cognitive radio networks. It also reviewed the most common spectrum sensing techniques with which the secondary users are able to monitor the activities of the primary users.

Spectrum sensing and handover techniques are followed in the research build upon how much information about the primary signal is present to the secondary users. The spectrum sensing technique can generally be classified as energy-based sensing, cyclostationary feature-based sensing, matched filter-based sensing, and other sensing techniques. Energy detection is the most commonly followed technique for spectrum sensing since it has low computational and implementation intricacies, and here previous perceptive of the primary user signal is not needed.

In order to address the limitations of the spectrum sensing techniques by secondary user, cooperative spectrum sensing and its main elements have been discussed. Thus the proposed cooperative sensing is an effective technique to improve sensing accuracy by exploring multiuser diversity at the expense of cooperation overhead in terms of increased sensing time, energy consumption, or reduced opportunistic throughput during handovers.

3.7 Summary

This chapter had analyzed the energy detection and spectrum sensing ability using (1) fast Fourier transform (FFT) within the specified frequency band by introducing specific amount of AWGN and then (2) predicting SNR value on various threshold levels. The proposed algorithm called cooperative CUSUM is based on improvising the overall channel throughput during handover by conclusive the power spectral density (PSD) of the channel. During handovers this approach is used to identify the vacant spectrum holes or gaps that can be utilized by the new incoming users (SU). On any cognitive network, the proposed algorithm outperforms all traditional CUSUM method of noise prediction and removal for supporting efficient handover. The cognitive radio networks are capable of working dynamically and at the same time changing the frequency band from one to another, and that has also been proved.

References

1. Akhtar F, Rehmani MH, Reisslein M (2016) White space: definitional perspectives and their role in exploiting spectrum opportunities. Telecommun Policy 40(4):319–331
2. Akyildiz IF, Lee W-Y, Vuran MC, Mohanty S (2008) A survey on spectrum management in cognitive radio networks. IEEE Commun Mag 46(4):40–48. https://doi.org/10.1109/mcom. 2008.4481339
3. Amjad M, Akhtar F, Rehmani MH, Reisslein M, Umer T (2017) Full-duplex communication in cognitive radio networks: a survey. IEEE Commun Surv Tutorials 19(4):2158–2191
4. Anandakumar H, Umamaheswari K (2014) Energy efficient network selection using 802.16G based GSM technology. J Comput Sci 10(5):745–754. https://doi.org/10.3844/jcssp.2014.745. 754

5. Anandakumar H, Umamaheswari K (2017a) An efficient optimized handover in cognitive radio networks using cooperative spectrum sensing. Intell Automat Soft Comput 1–8. https://doi.org/10.1080/10798587.2017.1364931

6. Anandakumar H, Umamaheswari K (2017b) Supervised machine learning techniques in cognitive radio networks during cooperative spectrum handovers. Clust Comput. https://doi.org/10.1007/s10586-017-0798-3

7. Anandakumar H, Umamaheswari K (2017c) A bio-inspired swarm intelligence technique for social aware cognitive radio handovers. Comput Electr Eng. https://doi.org/10.1016/j.compeleceng.2017.09.016

8. Arulmurugan R, Sabarmathi KR, Anandakumar H (2017) Classification of sentence level sentiment analysis using cloud machine learning techniques. Clust Comput. https://doi.org/10.1007/s10586-017-1200-1

9. Bayhan S, Gur G, Alagoz F (2007) Satellite assisted spectrum agility concept. MILCOM 2007 – IEEE military communications conference. https://doi.org/10.1109/milcom.2007.4454876

10. Cabric D, Mishra SM, Brodersen, RW (2004) Implementation issues in spectrum sensing for cognitive radios. In: Conference record of the Thirty-Eighth Asilomar Conference on Signals, Systems and Computers, 2004. https://doi.org/10.1109/acssc.2004.1399240

11. Cabric D, Tkachenko A, Brodersen RW (2006) Spectrum sensing measurements of pilot, energy, and collaborative detection, military communications conference, 2006. MILCOM 2006. IEEE, 1–7, 23–25. https://doi.org/10.1109/MILCOM.2006.301994

12. Crow BP, Widjaja I, Kim JG, Sakai PT (1997) IEEE 802.11 wireless local area networks. IEEE Commun Mag 35(9):116–126. https://doi.org/10.1109/35.620533

13. Diego PP, Pla V, Martinez-Bauset J (2009) Optimal admission control in cognitive radio networks. Department of Communication, Universidad Politecnica de Valencia (UPV)

14. Gao Z, Zhu H, Li S, Du S, Li X (2012) Security and privacy of collaborative spectrum sensing in cognitive radio networks. IEEE Wirel Commun 19(6):106–112. https://doi.org/10.1109/mwc.2012.6393525

15. Gozupek D, Bayhan S, Alagoz F (2008) A novel handover protocol to prevent hidden node problem in satellite assisted cognitive radio networks. In: 2008 3rd international symposium on wireless pervasive computing. https://doi.org/10.1109/iswpc.2008.4556298

16. Hassan MR, Karmakar GC, Kamruzzaman J, Srinivasan B (2017) Exclusive use spectrum access trading models in cognitive radio networks: a survey. IEEE Commun Surv Tutorials 19(4):2192–2231

17. Haykin S (2005) Cognitive radio: brain-empowered wireless communications. IEEE J Sel Areas Commun 23(2):201–220. https://doi.org/10.1109/jsac.2004.839380

18. Jeongkeun L, Sung-Ju L, Wonho K, Daehyung J, Taekyoung K, Yanghee C (2009) Understanding interference and carrier sensing in wireless mesh networks. IEEE Commun Mag 47(7):102–109. https://doi.org/10.1109/mcom.2009.5183479

19. Khan AA, Rehmani MH, Rachedi A (2017) Cognitive-radio-based internet of things: applications, architectures, spectrum related functionalities, and future research directions. IEEE Wirel Commun 24(3):17–25

20. Lu D, Huang X, Zhang W, Fan J (2012) Interference-aware spectrum handover for cognitive radio networks. Wirel Commun Mob Comput 14(11):1099–1112. https://doi.org/10.1002/wcm.2273

21. McHenry M, Livsics E, Nguyen T, Majumdar N (2007) XG dynamic spectrum sharing field test results. In: 2007 2nd IEEE international symposium on new frontiers in dynamic spectrum access networks. https://doi.org/10.1109/dyspan.2007.90

22. Mishra S, Sahai A, & Brodersen R (2006) Cooperative sensing among cognitive radios. In: 2006 I.E. international conference on communications. https://doi.org/10.1109/icc.2006.254957

23. Mitola J (1999) Cognitive radio for flexible mobile multimedia communications. In: IEEE international workshop on Mobile Multimedia Communications (MoMuC'99) (Cat. No.99EX384). https://doi.org/10.1109/momuc.1999.819467

24. Pacheco-Paramo D, Pla V, Martinez-Bauset J (2009a) "Optimal admission control in cognitive radio networks", Department of communication, Universidad Politecnica de Valencia (UPV)
25. Pacheco-Paramo D, Pla V, Martinez-Bauset J (2009b) Optimal admission control in cognitive radio networks. In: 2009 4th international conference on cognitive radio oriented wireless networks and communications. https://doi.org/10.1109/crowncom.2009.5189133
26. Pei QQ, Li Z, Ma LC (2013) A trust value-based spectrum allocation algorithm in CWSNs. Int J Distrib Sens Netw 9(5):261264. https://doi.org/10.1155/2013/261264
27. Quan Z, Shellhammer SJ, Zhang W & Sayed AH (2009) Spectrum sensing by cognitive radios at very low SNR. In: GLOBECOM 2009 – 2009 I.E. global telecommunications conference. https://doi.org/10.1109/glocom.2009.5426262
28. Suganya M, Anandakumar H (2013) Handover based spectrum allocation in cognitive radio networks. In: 2013 international conference on Green Computing, Communication and Conservation of Energy (ICGCE). https://doi.org/10.1109/icgce.2013.6823431
29. Tragos EZ, Zeadally S, Fragkiadakis AG, Siris VA (2013) Spectrum assignment in cognitive radio networks: a comprehensive survey. IEEE Commun Surv Tutorials 15(3):1108–1135. https://doi.org/10.1109/surv.2012.121112.00047
30. Yonghong Z, Ying-chang L (2009) Eigenvalue-based spectrum sensing algorithms for cognitive radio. IEEE Trans Commun 57(6):1784–1793. https://doi.org/10.1109/tcomm.2009.06.070402
31. Yucek T, Arslan H (2009) A survey of spectrum sensing algorithms for cognitive radio applications. IEEE Commun Surv Tutorials 11(1):116–130. https://doi.org/10.1109/surv.2009.090109
32. Zhang B, Hu K, Zhu Y (2010) Spectrum allocation in cognitive radio networks using swarm intelligence. In: 2010 second international conference on communication software and networks. https://doi.org/10.1109/iccsn.2010.23
33. Zhao Q, Swami A (2007) A survey of dynamic spectrum access: signal processing and networking perspectives. In: 2007 I.E. international conference on acoustics, speech and signal processing – ICASSP'07. https://doi.org/10.1109/icassp.2007.367328

Chapter 4
Network Coding-Based Broadcasting Schemes for Cognitive Radio Networks

Khaqan Zaheer, Mubashir Husain Rehmani, and Mohamed Othman

4.1 Introduction

Rapid development in the field of wireless communication and widespread use of mobile devices to access the Internet from anywhere and anytime result into the scarcity of the available licensed spectrum for communication owned by the mobile operators. However, according to a survey of the Federal Communications Commission (FCC), only 15–85% of the spectrum is being utilized which compels the researcher to develop technologies for opportunistic spectrum utilization [59]. Moreover, the spectrum management is the most important task in wireless network to provide good quality of service (QoS) to the users. To solve these problems and for efficient spectrum utilization, the emerging field of cognitive radio networks (CRNs) is coined by Mitola which can not only solve the fixed spectrum assignment policy problem but also utilized the unused spectrum of licensed users effectively and efficiently [71]. According to FCC's policy, CR nodes or secondary users (SUs) can use the licensed spectrum of primary radio users (PU) when it is not utilized by PUs. Moreover, the PUs have highest priority for utilizing the licensed spectrum, and CR nodes can only access this available idle spectrum when it is vacant, and their transmission do not cause the harmful interference to PR nodes [71]. SUs can utilize the available idle spectrum opportunistically for the transmission of messages across the network with very low complexity and cost. Finally, the FCC approval has solved the problem of fixed spectrum assignment policy and has enabled the SUs to

K. Zaheer (✉) · M. Othman
Department of Communication Technology and Network, Universiti Putra, Malaysia 43400,
Selangor, Malaysia

M. H. Rehmani
Telecommunications Software and Systems Group, Waterford Institute of Technology,
Waterford, Ireland

© Springer International Publishing AG, part of Springer Nature 2019
M. H. Rehmani, R. Dhaou (eds.), *Cognitive Radio, Mobile Communications and Wireless Networks*, EAI/Springer Innovations in Communication and Computing,
https://doi.org/10.1007/978-3-319-91002-4_4

access the unused spectrum opportunistically without interfering the PU transmission. These temporary idle bands of spectrum are also known as 'spectrum holes' or 'white spaces' which are owned by mobile operators [3]. In this chapter we will use spectrum holes and white spaces interchangeably as they refer to the same concept. Moreover, we will also use the SU and CR nodes interchangeably as they also refer the same concept.

4.2 Cognitive Radio Network (CRN)

The key motivation behind CRN is that the limited spectrum resources are effectively and efficiently utilized by SUs. Moreover, the problem of fixed assignment policy can be solved by allowing the SUs to access the available white spaces of spectrum without interfering the PUs. This section provides comprehensive insights into CRN in the following subsections.

4.2.1 Definitions and Basic Concepts

In this current era of information technology, the most important field of cognitive radio network (CRN) has emerged as an exciting solution to utilize the unused licensed spectrum opportunistically that enables coexistence of PUs and SUs using dynamic and hierarchical spectrum sharing. In CRN, there are two types of users: one is the primary user (PU) who has a licensed spectrum, while the other users are secondary users (SUs) or CR users who do not purchase the licensed spectrum, but they can utilize the licensed spectrum as per FCC approval with the condition that the spectrum is not utilized by PUs. Moreover, SUs must also ensure that during the utilization of licensed spectrum, if PR users return to that band, then SUs have to vacate the channel and switch to another channel without interfering the PUs' communication. This process is called spectrum handoff [71]. The cognitive radio technology is an intelligent technology in which CR nodes actively sense the environment and collect all the information regarding the available spectrum bands, power requirement for channel selection, service discovery, etc., regularly to tune themselves according to the configuration of the network. This technique is known as dynamic spectrum access (DSA) [71]. These functionalities enable the SU nodes to access the idle channels dynamically and intelligently and switch to another channel when PU node starts its transmission on that channel, causing little or no interference to PR nodes. Therefore, CRN technology enables the dynamic utilization of spectrum by CR or SU users which increase the spectrum utilization efficiency and reduced the spectrum scarcity problem [60]. In this chapter, the mobile device refers to CR node, and we use them interchangeably as they refer to the same concept in CRN.

4.2.2   Architecture

The previous literature [4, 60] has clearly shown that CRN is deployed as infrastructure-based, ad hoc or hybrid network for efficient utilization of spectrum. These three different architectures are deployed as per scenario or users' needs to fulfil the requirements of both licensed and unlicensed users as shown in Fig. 4.1.

Infrastructure-Based CRN Architecture In this architecture, all the CR nodes communicate with each other using the central entity which is known as base station or access point as shown in Fig. 4.1a. All the tasks related to routing and communication are performed by the base station or access point. For instance, the wireless access point or router serves as a base station in wireless local area network. All the communication and routing tasks are controlled by this access point. Similarly, the base station in cellular network acts as an access point that enables the communication between the mobile devices. This base station implements different communication protocols to enable the communication between different mobile devices. Therefore, all the decisions related to spectrum sensing, allocation and management are performed by the access point based upon the acquired spectrum information from all the CR nodes to enable SUs to utilize the unused spectrum without interfering the PU for better utilization of spectrum resources.

Ad hoc Architecture In this architecture, the CR nodes establish a direct point-to-point communication link to exchange the information between two CR nodes as shown in Fig. 4.1b. Moreover, this architecture does not rely on central relay station for communication between them. The mobile nodes usually used the existing wireless ad hoc communication protocols on unlicensed bands, e.g. Wi-Fi, Zigbee, etc., or used white spaces opportunistically to communicate with each other. In this architecture, all the tasks related to spectrum sensing, spectrum allocation and spectrum management are managed by the CR nodes. Basically, each CR node has all the cognitive functionalities and performs the decisions based on its local observations.

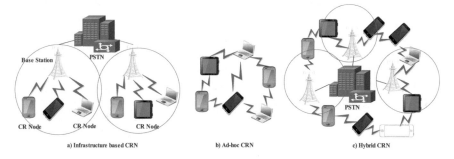

Fig. 4.1 Basic architecture of cognitive radio network

Hybrid Architecture The hybrid architecture is the combination of both infrastructure-based and ad hoc architectures. In this architecture, CR nodes communicate with each other by either using the access point or other CR nodes as multi-hop relay station as shown in Fig. 4.1c. Basically, the base station or access point works like a wireless router which enables the communication between two CR nodes, or these CR nodes can communicate with other CR nodes using intermediate CR nodes as multi-hop relay nodes.

4.2.3   Fundamental Working Rules

The concept of CRN was coined by Mitola who wanted to add brain to software-defined radio (SDR) by using the artificial intelligence. He had proposed the CRN with the vision that it will be aware of both user and radio environment, capable of using different communication technologies and advance features can be added to it [58]. Haykin [27] further enhanced his basic idea for using the spectrum opportunistically. Their work enables the CR node to sense the spectrum holes, identify the channel state information and perform dynamic spectrum management [27]. Their concept enables all the CR nodes to perform major tasks of spectrum sensing, spectrum management, spectrum sharing and spectrum mobility as shown in Fig. 4.2.

Initially, the CR nodes sense the spectrum and identify the unused spectrum or white spaces in the first phase [60, 84]. The second phase involves the management and selection of the best available channel among numerous channels [60, 86]. In the third phase, the CR nodes communicate with each other about the best available channel for communication. Finally, in spectrum mobility phase, the CR vacates the channel on the arrival of PU and maintains its seamless connectivity with other CR nodes by switching on to the next available channel.

All these tasks formulate the cognition cycle through interaction with the RF environment. The cognition cycle is composed of different CR nodes which can perform three major functions i.e. they can intelligently sense and analyse the RF environment, make decision or perform spectrum management and adjust their parameters for allocation of spectrum resources accordingly [60, 65] as shown in Fig. 4.3. Among these functionalities intelligent sensing of CR node includes spectrum sensing, spectrum sharing, service discovery, network discovery and location identification. The decision-making includes the management task of CR node including mobility management, trust management, radio resource management and issues related to security. Finally, the CR nodes are also capable of adjusting their parameters including dynamic frequency selection, adaptive modulation, transmit power control, security and frequency adjustment.

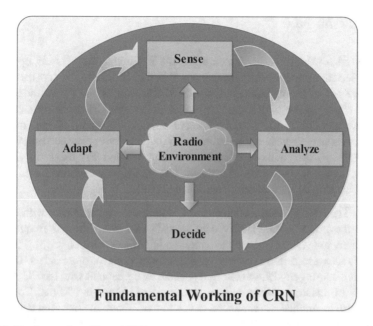

Fig. 4.2 Fundamental working of CRN

Fig. 4.3 Fundamental
functionalities of CRN

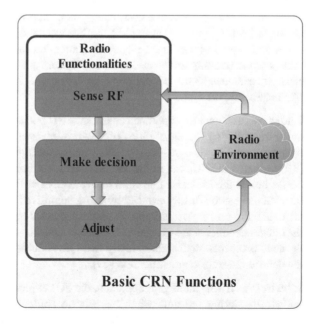

4.2.4 Techniques

The basic idea behind the implementation of CRN is that the SUs can opportunistically select the unused white spaces for effective utilization of spectrum. White spaces are the available frequency bands in licensed or unlicensed spectrum or different signal space which can be exploited by the CR users by sensing the spectrum continuously for the effective and efficient utilization of spectrum. Basically, white spaces represent the vacant radio frequencies of the spectrum which are utilized by CR user without effecting the transmission PR users. In CRN, the spectrum occupancy of white spaces by CR nodes is categorized into overlay, underlay, interweave and hybrid techniques [3, 29].

Overlay The overlay model enables the CR nodes to communicate with other CR nodes in the presence of PUs by continuously observing the PU frequency and appropriately tune their frequency to avoid the interference with PUs [3]. This model is further categorized into cooperative and non-cooperative approaches. In case of cooperative overlay model, the CR nodes make an agreement with the PU nodes for the usage of spectrum white spaces for an agreed amount of time. The coding schemes can be applied to protect the data and to differentiate between data streams of different mobile nodes. While on the other hand, the SUs sense and analyse the spectrum independently in non-cooperative model and then determine the available white spaces which can be utilized by CR nodes without causing interference to PUs. The throughput of SUs is increased significantly when they operate in full duplex mode, but it will increase the cost, energy consumption and interference between antennas [29]. CR users can exploit the licensed and unlicensed bands for radio communication, but they must pay special attention in terms of role and responsibilities while utilizing unlicensed band as all users are considered PUs, as they can cause interference to the critical communication which is not tolerable in systems like health support system.

Underlay The underlay approach enables the SU nodes to send the data to other SU nodes by utilizing the capabilities of low-power devices having limited range on licensed band in the presence of PUs within acceptable PU interference limit [3]. Basically, the SUs can utilize the licensed spectrum by keeping its transmission power below the PU noise threshold to avoid any harmful interference to PUs. The interference can be further reduced by using cognitive femtocell that can adaptively allocate radio resources and improve spatial reuse for cellular coverage. The need of the hour is to conduct the extensive research in this area of study for the development of new protocols that enable low-power transmission and utilize the existing unsilenced band for communication [29].

Interweave In this intelligent approach, the SUs explore the spectrum holes opportunistically during the time when there is no communication conducted by PUs [3]. During transmission, when the PU becomes active, then the CR user will stop its transmission or switch to another frequency band to continue its communication. The DSA techniques can be deployed by the SUs to maintain the quality of transmission during channel-switching process. In this technique, the basic parameters of time and frequency are used to determine the occupancy of frequency band by PU to categorize these bands into idle, fully occupied or underutilized states. After analysing each band status or state, the SUs can utilize them if they are in the idle state until the arrival of PUs. Time frequency resource conversion highlights the allocation of spectrum resources according to behaviour or context of user. This means that if the users get the information about the application which has the highest user usage and focus, then the spectrum utilization is managed properly by applying time frequency resource conversion (TFRC) [3]. This method will manage resource allocation and freeing up of many virtual spectrum holes for enhancing the spectrum utilization and management [3]. Finally, this approach enables the SU base station to continuously sense the spectrum and maintains the record of sensing in a list of viable frequencies according to different application requirements.

Hybrid Underlay-Relay In underlay-relay hybrid paradigm, the SU nodes can transmit simultaneously along with PUs without transmission power limitation when they facilitate the PUs for relaying their data to other PU nodes in the network [3]. This technique is very useful in CRN as it can increase the throughput of the network by relying the PU data to other PUs in the network when there is no direct link available for communication. In this way the issue of weaker or no direct link between PU is solved, and more idle frequencies or spectrum holes are also created in the network for future utilization.

Hybrid Overlay/Underlay In this approach, the SU nodes can use both the nonutilized and underutilized spectrum of CRN-based femtocell networks. Spectrum mapping for underlay networks is achieved by using the radio environmental map (REM) which creates a detailed map of urban and rural areas by using multidomain environmental data from geolocation databases, past experiences, spectral regulations and relevant policies [3].

4.3   Broadcasting in CRN

Broadcasting is the process of transmitting control and local measurements among nodes of the network that facilitates them to synchronize for the performance of basic tasks like neighbour discovery, channel discovery, route discovery, etc. Basically, the wireless transmission is broadcast by nature because the signal sent by the source to the destination node could also be heard by any receiver placed in close proximity to the path [70]. Moreover, the transmitted signal undergoes reflection, refraction or diffraction, and different copies of this transmitted signal reach the receiver in different paths that can be heard by different nodes in their path. Single-hop broadcasting is used to send the control information to one-hop neighbour of the sender node or centralized system, while multi-hop broadcasting transmits the control information across the network with the help of relay nodes that forward the information to each node of the network. In CRN, the distributed nature of ad hoc network requires multi-hop broadcasting as the sender node is unable to transmit the information to all nodes of the network due to no direct communication link or no centralized controller associated with the nodes. Therefore, the sender node transmits the broadcast information to its neighbour nodes who work as relay nodes to forward the information to their neighbours and so on to transmit it throughout the network.

Broadcasting is the most viable solution for exchanging the control information and synchronizing the neighbours throughout the network. Broadcasting is started when a new node enters the network or changes the channel for communication. In CRNs, the SUs select the best channel opportunistically from their pool of channels. The selection of best channel is accomplished based on its available bandwidth and PR occupancy for communication because every SU does not have the availability of the same set of frequency for communication. When a SU tries to switch to a new channel based on the switch pattern, the false detection or miss-detection about PU activity will lead to collision of control information with PU that will result in loss of control information at the end of the allocation period [71]. Thus, an optimal broadcasting scheme is required which can calculate the probability that the SU will retrieve the information from all users without any collision [70]. All the SUs require the broadcasting of control information among other SUs to perform all the important networking tasks like neighbour discovery, channel assignment, etc. The challenges in broadcasting include broadcast storm problem, collision among packets, redundant broadcasts by CR nodes and congestion in the network. Moreover, diversity and dynamicity of channels in CRN make it more challenging for effective and efficient data dissemination in the network.

4.3.1 Broadcasting Key Characteristics in CRN

The key characteristics of broadcasting strategy in CRN include:

Efficient Data Transmission This aspect is related to the effective broadcast of message between the SU nodes. This attribute is the most important factor as the broadcasting will enable the SU nodes to retrieve the information from all users without collision. This feature not only decreases the probability of collision but also increases the throughput of network as well.

Effective Spectrum Utilization This aspect is more related to channel selection for broadcasting the control information to the SU nodes for effective and efficient data transmission. This parameter also considers that dedicated common control channel which is not utilized by CR node for broadcasting as it will result into throughput degradation and spectrum utilization inefficiency. This attribute guarantees that different broadcasting strategies for channel selection in CRN are used in an intelligent way.

PU Constraint Modelling This aspect ensures that SU does not interfere with the PU to avoid collision. Moreover, it is also ensured that CR nodes select the best channel opportunistically based on its available bandwidth and PR user occupancy for communication. Finally, the efficient broadcasting strategy must be implemented which guarantees that when CR node switches to new channel based on the switch pattern, the false detection or miss-detection about PU activity must be mitigated at the end of the allocation period.

Minimum Time Consumption This characteristic is related to the total time taken by the packet to reach its destination in minimum time. This aspect is related to the selection of best channel that will guarantee the minimum time for message propagation.

Reduced Overhead This aspect is related to the opportunistic utilization of idle spectrum to broadcast data to CR nodes and minimize transmission overhead. This attribute guarantees that the spectrum holes are exploited properly to minimize the network overhead.

Reduced Channel Switching This characteristic is related to channel switching during the message transmission which results in more delay as the network performance is decreased. The main reason for low throughput is the frequent channel switching of CR nodes that results in more delay and network congestion.

PR User Occupancy This aspect is related to the selection of the best channel in which the PR occupancy and their returning probability on same channel is very low. This attribute is also related to utilization and reselection of the channel by PR user. Therefore, the best channel selection strategy should be used in which the PR occupancy and its returning probability must be low so that the CR nodes can easily utilize that channel for long period of time. This will result in less channel switching and the throughput maximization of network.

4.4 Network Coding in CRN

4.4.1 Definitions and Basic Concepts

Efficient propagation of data is accomplished by network coding (NC) technique in which the data is encoded by the sender node and the receiver node decodes that received data to maximize the throughput of network, reduce latency and create a robust network [22]. NC is a technique in which the data from different nodes arrives at intermediate node and is encoded and forwarded to the next neighbouring node. NC improves the throughput of the network as it combines or encodes the data packets coming from different sender nodes and forwards these encoded packets to different receiver nodes by utilizing only a single transmission. The major benefit of

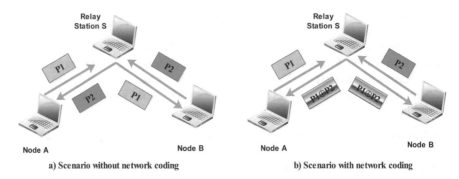

Relay
Station S

a) Scenario without network coding   b) Scenario with network coding

Fig. 4.4 Simple scenario of network coding example

using network coding for sending data is that it will send more information on less bandwidth, which will result in less transmission. Moreover, there is no need for retransmission of packets which will not only save the bandwidth but also reduce the delay on the network. The usage of NC not only prevents the packet loss but also avoids the link failure by utilizing the network resources efficiently. Finally, this evolutionary technique also provides security and optimal routing facility by encoding the packets which will improve the overall network performance [2, 28].

The basic concept of NC is explained with the help of a practical example shown in Fig. 4.4. In this figure it is clearly seen that node A and node B try to exchange the data packet through relay node station S. In the first case of data exchange without network coding, node A transmits a data packet P1 to node B through the relay node S. Similarly, node B transmits the data packet P2 to node A through relay node S. The transfer of packets between node A and node B requires four transmissions as shown in Fig. 4.4a. In second case of data exchange with network coding, node A and node B transmit data packets P1 and P2, respectively, to the relay node S. Then this relay node combines or encodes these two data packets into a single packet, i.e. P1 $\oplus$ P2, by using XOR function and then broadcasts XORed single data packet to both node A and node B. Both nodes A and B received the encoded packet and decode their respective packet by simply doing P $\oplus$ P2 and P $\oplus$ P1, respectively, as shown in Fig. 4.4b. The NC approach will require only three transmissions for data exchange and result into high throughput of the network.

In CRN, SUs can intelligently sense the available frequency bands and then make the decision to adapt network coding for efficient and effective data transmission. Basically, NC is done at packet level in which the data is compressed by the nodes using the network coding techniques like linear network coding, subspace coding, etc., at physical, network, MAC, transport or application layer [2]. Moreover, NC also enables the multiple CR nodes to utilize the spectrum simultaneously for more effective spectrum utilization. NC not only provides the security in CRN but also protects the CR nodes against different types of jamming and denial of service attacks. Moreover, NC also facilitates the interference mitigation by effectively allocating the network resources and neighbour discovery in CRN. Finally, NC also provides efficient spectrum management functions in CRN and increases the reliability of the network. The multipath NC uses network relay nodes to combine the messages using XOR between two packets and broadcast a single packet to destination. This will result in reduction of transmission and delay. Moreover, the source node can also include the size, time and transmission rate of each packet by using NC which encodes the packets and transmit them to their destination at required transmission rate by meeting the deadline for decoding at destination. The network coding techniques are classified into random NC, vector NC and linear NC [22] which are explained in detail in the following sections.

4.4.2   Network Coding Key Characteristics

The usage of NC in CRN has numerous characteristics, but some of them are discussed below.

Efficient Data Transmission   When NC is used in CRN, the CR node will encode the data which will result in reduction of data transmission time. This aspect also ensures the efficient and effective data transmission. Moreover, in [109], the authors have shown that NC can increase the spectrum availability for SU nodes, and conflict between SUs and PUs is also decreased.

Throughput Maximization   The usage of NC in CRN will result in throughput maximization of both PUs and SUs which result in more efficient message delivery. In [24], the author showed that the usage of NC with relay techniques will result in throughput maximization. In PUNCH, the authors have applied NC to encode data packets which will result in maximization of network throughput and reduce the retransmission of packets that will create more spectrum opportunities. Finally, the throughput of relay network is also improved by using NC techniques with the joint power and rate-adaptive QAM in CRN [39].

Overhead Reduction   This aspect is related to multicast transmission of data over available spectrum holes. In [67], the author showed that the NC can effectively and efficiently reduce the transmission overhead by utilizing the white spaces opportunistically to multicast the data using the scheduling algorithms. This framework employs NC that guarantees the error control and data recovery by multiple SUs in CRN.

PU Interference Reduction   When NC is employed in CRN, it will result in low interference to PUs. Moreover, this scheme also results in transmission reduction and usage of time scheduling algorithms can further reduce PU interference [98]. Basically, this coded message contains all the information regarding the SU and PU channel occupancy which enables the other SUs to switch to new available channels that result in interference avoidance with PU. In [18], the authors have proposed an adaptive NC scheme for cognitive relay network that utilizes NC to improve network performance and increase the throughput of the network.

Security Enhancement The usage of NC in CRN will increase the security of both PU and SU transmission by encoding the packets together. Moreover, it also prevents various jamming and primary user emulation attacks in CRN (Xie et al. 2013; [8]).

Efficient Neighbour Discovery This attribute guarantees the efficient neighbour discovery process using NC in CRN. In JEENA [8], the author proposed a neighbour discovery algorithm which is based on random linear combination of packets in CRN. Moreover, this proposed scheme is completely distributed and protects the CR nodes from jamming attacks.

Efficient Spectrum Utilization This is one of the key characteristics of NC as it results in efficient utilization of spectrum resources in CRN. In [52], the author proposed a scheme which is the combination of NC, multicasting and orthogonal frequency division multiplexing to improve the spectrum efficiency of future 5G cellular network. They have proposed the cooperation scheme in which SUs can cooperate with two PUs simultaneously for better spectrum utilization.

Efficient Spectrum Sensing The usage of NC in CRN will increase the spectrum sensing opportunities for CR nodes. In [106], the author has proposed the solution for efficient spectrum sensing and access. This approach enables the SUs to detect the idle PUs efficiently which will result into throughput maximization. Moreover, the transmission utilization efficiency is further increased when NC is applied on PU channels which results in increased SU channel availability and their ability to detect the idle spectrum.

Efficient Selection of Common Control Channel This aspect is the most important aspect in wireless communication as all the control information, e.g. channel availability status, spectrum management, etc., about communication is distributed via a common control channel. The usage of NC in CRN enables SUs to efficiently select the common control channel for communication to increase the throughput of the network. In [9], the authors have proposed virtual control channel scheme which utilizes NC for better effective coordination between PUs and SU. This scheme also improves the dissemination performance. Moreover, the usage of NC along with a pseudorandom channel pattern technique enables the creation of virtual control channel which provides the tolerance against link failure and packet losses. Finally, this robust technique do not require any reserved spectrum for control information exchange.

4.5   Broadcasting in CRN

This section highlights the most important task of broadcasting in CRN. We present comprehensive details about the broadcasting protocols for exchanging the control information in CRN. Moreover, the broadcasting techniques for broadcasting in CRN are also discussed in detail. Finally, the issues and challenges related to broadcasting in CRN are presented in the following section.

4.5.1   Broadcasting Protocols

Broadcasting is the most important task in wireless networks for exchanging the control information between the neighbouring nodes, but it is highly challenging in CRN due to spectrum diversity and heterogeneity. The previous literatures showed that the research community has used numerous protocols for broadcasting which are listed in Table 4.1. In this table, we have provided the comprehensive details of these protocols and compared them for better understanding of readers. For example, in [37, 48, 57], the authors have presented the random broadcasting scheme in which SUs can randomly select the channel from channel availability pool. Similarly, the receiver also selects the channel randomly for reception of data. In this technique, the success rate is very low and delay is very high because both the sender and receiver select channels randomly. In [66, 79]), the authors have proposed a fully distributed multi-hop broadcasting scheme in which the set of channels are selected from the pool based on their availability to send the data on these channels. They have showed that their proposed scheme increased the success rate and reduced the delay when the set of best channels are selected for broadcasting the packets. The complete broadcasting scheme is presented in [103] for single-hop receivers. This centralized technique provides power control and optimal scheduling in CRN. In [16], the authors have presented the group-based scheme for broadcasting in the multi-hop CRN. The results have clearly shown that this scheme works efficiently for cluster of nodes and provides optimal sharing of spectrum among SUs through coordination and learning. In [7, 34, 51, 80], the authors have proposed the metric-based schemes which broadcast the packets based on virtual matrix. These schemes can provide reliability, low latency, optimal resource sharing, robust data transmission, less delay and redundancy in CRN. The vital characteristics of all these broadcasting protocols are that the CR nodes can work in conjunction with the frequency spectrum of nonoverlapping channels of current cellular network including LTE, UMTS and TV. Moreover, the SU can also utilize the unsilenced frequency bands, e.g. Zigbee, WLAN, etc., for effective and efficient communication. In any available spectrum, CR nodes can tune their frequency and modify the transmission power as per FCC regulation. In short, they have different transmission characteristics as per the requirement of the spectrum and can utilize a wide range of spectrum resources for broadcasting the control information across the network.

The previous literature clearly showed that CR nodes can utilize the spectrum efficiently and effectively which is not utilized by PU. Moreover, CR nodes have to select the best channel based on its availability, PU activity and spectrum quality. Basically, PR nodes traffic pattern and channel availability vary from standard to standard, and CR node needs to sense and analyse these parameters for the selection of best channel among numerous channels. Moreover, the selection of the best channel is the most important factor as it decreases the channel switching and increases the SUs which results in throughput maximization and delay minimization in CRN. Therefore, it is suggested in many previous literatures to carefully design the protocol for broadcasting in which channel selection is the key criteria for

Table 4.1 Broadcasting protocols for CRN

Ref no	Name of protocol	Scheme used for broadcast	Optimization parameters	Topology/ hop
32.	BRACER	Broadcasting over a set of channels	Average broadcast delay, successful broadcast ratio	Distributed/ multi-hop
56.	Distributed broadcast scheduling and collision avoidance scheme	Broadcasting over a set of channels	Average broadcast delay, successful broadcast ratio	Distributed/ multi-hop
46.	Selective broadcasting	Broadcasting over a set of channels	Broadcast delay and redundancy	Distributed/ multi-hop
57.	Primary receiver-aware opportunistic broadcasting	Broadcasting over a set of channels	Successful broadcast ratio and average broadcast delay	Distributed/ multi-hop
59.	Broadcast scheduling with latency and redundancy analysis	Broadcasting over a set of channels	Broadcast latency and redundancy	Distributed/ multi-hop
36.	Heuristic algorithm based on polynomial time	Metric-based broadcasting	Fair resource utilization and delay minimization	Distributed/ single hop
47.	Counter-based broadcasting	Metric-based broadcasting	Message transmission speed, total receivers and number of transmissions	Distributed/ multi-hop
37.	Tree-based broadcasting protocol	Metric-based broadcasting	Link quality, reliability, redundancy and latency	Distributed/ multi-hop
40.	MBS-UDG	Metric-based broadcasting	Redundancy and latency	Distributed/ multi-hop
55.	Protocol design and performance issue	Complete broadcasting	Average broadcast delay and redundancy	Centralized/ single hop
34.	Distributed TDMA	Complete broadcasting	Power control and optimal SU scheduling	Centralized/ single hop
49.	Distributed coordination protocol for CCC	Group-based broadcasting	Average broadcast delay, latency	Distributed/ multi-hop
51.	Joint routing and channel selection protocol	Group-based broadcasting	Average broadcast delay, redundancy and latency	Distributed/ multi-hop
52.	Swarm intelligence-based routing scheme	Group-based broadcasting	Broadcast delay and redundancy	Distributed/ multi-hop
31.	Channel selection using random broadcasting	Random broadcasting	Delay and packet delivery ratio	Distributed/ single-hop
29.	Gel'fand-Pinsker coding	Random broadcasting	Delay and packet delivery ratio	Distributed/ multi-hop
30.	Tree-based local healing protocol	Random broadcasting	Latency and nodes connectivity	Distributed

effective and efficient communication. To summarize it, we can conclude that channel selection is the most important parameter for successfully delivery of broadcast packet in CRN.

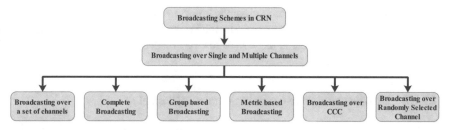

Fig. 4.5 Broadcasting schemes for CRN

4.5.2   Broadcast Schemes in CRN

In CRN, the broadcasting schemes are basically classified into two major categories of single-channel broadcast and multiple-channel broadcast schemes according to the number of channels which are used for broadcasting the packets. There are six basic schemes for broadcasting over single and multiple channels that include broadcasting over random channel, broadcasting over dedicated common control channel (CCC), metric-based broadcasting, group-based broadcasting, complete broadcast and broadcasting over set of channels as shown in Fig. 4.5. In the next section, we provide a detailed discussion about each scheme.

4.5.2.1   Broadcasting over Randomly Selected Channel

In this scheme, SU sender node can randomly select the channels from the available pool of channels for transmission of packets. Similarly, the SU receiver node also randomly selects the channel from the available channel pool. When the number of channel is increased in this scheme, then the packet delivery ratio will decrease, and delay will increase [71]. This is because when the number of channel pool is increased, then the probability that both sender and receiver node select the same channel will be decreased which will degrade the performance of network. While on the other hand, this scheme is effective for fewer channels as the probability that both sender and receiver select the same channel will be increased [57]. Moreover, the PU activity affects fewer channels more severely as compared to large channel pool. In [48], the authors have presented the scheme for channel selection in single-hop scenario which will result in low computational complexity, but there is a trade between the number of channel and packet delivery ratio. The packet delivery ratio decreased when a number of channels are increased while it is increased as the number of channels in pool decreases. In [37], the authors have utilized the concept of NC for broadcasting the packets randomly on distinct relay channels in multi-hop CRN. They have explored the limits of broadcast channels in CR relay nodes and calculate the achievable rate regions in CRN. In [57], the authors have presented the scheme for recovering the broken link of the broadcasting tree which is caused due to hidden terminal problem. The previous literatures have clearly shown that the selection of the best channel is the key factor for broadcasting.

4.5.2.2   Broadcasting over Common Control Channel

In this scheme, CR nodes select the common control channel for broadcasting the control information among them. The usage of common control channels between SUs facilitates numerous operations including transmitter-receiver handshake, neighbour discovery, channel access, change in network topology and updates about routing information [53]. The biggest problem in CRN is that due to PR activity, the CCC is not unique nor always available to CR nodes. In CRN, out-of-band and in-band control channel approaches are proposed to design control channel. In out-of-band approach, dedicated spectrum is used for information exchange, while there is no dedicated channel for control information exchange in in-band approach. The biggest problem of PU activity, channel diversity and heterogeneity does not enable the CR nodes to reserve the dedicated control channel for communication and hence the out-of-band control channel approach is the feasible solution. Therefore, in-band control channel is preferred in the distributed CRN. The in-band control channel design in the distributed CRN faces many challenging issues including the PU activity, control channel saturation and control channel security that leads the SU to switch the control channel available in the cluster for further communication.

In sequence-based CCC schemes, the CCC is allocated according to a random or predetermined channel-hopping sequence to diversify the CCC for minimizing the impact of PU activity [53]. In permutation-based sequence scheme [10], the channel-hopping sequence is designed according to the availability of the common channel between two SUs. In this scheme the PU activity and CR user coordination will result into interference and delay for two CR users to rendezvous in a channel. The main drawback of this scheme is longer CCC establishment time and no coverage beyond rendezvous pair of nodes. Finally, the channel-hopping sequence is designed according to quorum system in quorum-based schemes to improve the overlapping of multiple sequences and minimize the time to rendezvous between two SUs [11]. The challenges in all these sequence-based schemes include sequence design for PU activity and CCC coverage that will create the overhead of interference and delay in communication. Abdel-Rahman et al. propose quorum-based frequency-hopping (FH) algorithms that mitigate DoS attacks for unicast and multicast on the control channel of an asynchronous ad hoc network. Furthermore, the multicast algorithms maintain the multicast group consistency. However, this algorithm does not consider PU activity which will result into interference and switching overhead in communication [1]. Tessema et al. [83] proposed a protocol that enables the SUs to design an asynchronous channel-hopping sequence which can establish the link between SUs without the support of global synchronization clock. This intelligent scheme enables the successful transmission in predefined time interval [83]. Kondareddy and Agrawal [41] designed a dynamic control channel-hopping scheme in which the SUs can randomly meet on different bands and agree on a common control channel or rendezvous channel to exchange their data. The selection of this random channel is based on the best common channel between them to exchange the data through the transmission.

4.5.2.3 Metric-Based Broadcasting

In this scheme, CR node computes link quality matrix based on local information of tree for efficient broadcasting in CRN. Basically, the broadcast tree was built on the basis of the channel weights and then the global broadcast tree built by using message-passing technique. Finally, the functions related to link quality and interference measurements are defined in the metric for the better utilization of spectrum resources for broadcasting. This matrix can store the channel information and calculates the interference in the network. The information stored in the matrix helps CR nodes to select the best channel for broadcasting. Moreover, the channel selection criteria for broadcasting using the matrix parameters also depend on the scheme used for communication, e.g. the CR nodes who wish to transmit the control information must consider the spectrum usage or link quality matrix for efficient transmission of message. In [90], the authors have proposed heuristic algorithm to reduce the delay in transmission and fairly utilize the spectrum resource. This polynomial-time greedy algorithm is also capable of reducing the computational complexity and delay in broadcasting by using the node ID ratio as a metric. In [34], the authors have proposed an exciting broadcasting scheme which utilizes both unicast and broadcast algorithms to create the broadcasting tree in unit disk graph model. This matrix-based mixed broadcasting and scheduling scheme efficiently calculates the spectrum opportunities and minimizes the latency and redundancy. In Kondareddy and Agrawal [41], the authors have proposed a counter-based algorithm for selecting the random channels for broadcasting. In this straightforward scheme, the SUs will initiate the counter in the packet which contains the information about the packet broadcasted by each node on the network. On the reception of packet, every SU checks the countervalue stored inside the received packet. If the value is non-zero, then SU decrements the counter of packet before rebroadcasting and will eventually discard it once its value becomes zero. In [80], the authors have proposed the tree-based broadcasting scheme from the local information of nodes for efficient communication. Moreover, they have also defined the link quality matrix for calculating the interference during the transmission. This matrix-based efficient scheme is used to calculate the link quality, reliability, redundancy and latency of the network. In [51], the authors have proposed a scheme to solve the optimal power allocation problem for maximizing the ergodic sum capacity of SU broadcast channel. This scheme enables those SUs to broadcast the control information which has the highest power capacity in CRN.

4.5.2.4   Group-Based Broadcasting

The group-based broadcasting schemes are the most suitable distributed schemes for broadcasting in CRN as they have grouped the CR nodes into small clusters in which the cluster head is responsible for efficient spectrum resource allocation and management. These cluster heads periodically analyse the spectrum resource and assign the set of best channels to the cluster member based on spectrum sensing. Moreover, they are also capable of restricting the CR nodes to communicate with the neighbouring nodes as per predefined policy or schemes. The main objective of clustering includes the establishment of common control channel, enhancement of cluster stability, energy efficiency, enhancement on cooperative tasks and minimizing number of clusters which is discussed in detail by Yau et al. [100]. In this paper, the authors have presented a detailed survey on different algorithms used for clustering. Moreover, the paper also highlights the key functionalities of clustering including clustering metrics and intra-cluster distance. Zhao et al. proposed the protocol which enables the distributed coordination among SUs by using the local channel information from neighbouring SU. This channel availability information is regularly updated by each SU on the basis of PU activity and coordinated channel information update is regularly exchanged between them for effective and efficient transmission (Zhao et al. [107]). The authors have proposed the method of creating numerous small clusters in the situation when the PR activity is high. Lazos et al. [45] have proposed an efficient scheme based on spectrum opportunity for calculating the best common control channel for cluster-based CRN. More recently Saleem et al. [73] proposed an exciting scheme that enables the cluster-based joint routing and channel selection scheme for CRN. Moreover, this scheme not only facilitates the SUs to form clusters in the network but also calculates the optimal path for SUs to broadcast the message to its destination node in clustered CRN. Finally, the deployment of reinforcement learning algorithms also maximizes the network performance that includes the minimization of SU-PU interference and increased packet delivery ratio. Huang et al. [32] propose swarm intelligence-based routing scheme for cluster formation which is capable of calculating and selecting the best common channels for intercluster communications. In [82], the authors have proposed the efficient routing scheme based on weighted clustering metric that calculates the node degree and mobility levels. Moreover, they have also calculated the set of the available channels efficiently for each node to increase the connectivity and reduce signal-to-interference ratio. This scheme intelligently utilized the spectrum resources which results in higher packet delivery and lower delay and routing overhead. Salameh et al. proposed a distributed group-based spectrum-aware coordination scheme which enables the neighbouring CR users to create optimal virtual clusters by utilizing the spectrum opportunities asynchronously. Moreover, this intelligent scheme also enables the neighbouring nodes to coordinate their communications in the cluster by using locally available CCCs [72]. Liu et al. [52] proposed a cluster-based architecture based on bipartite graph that allocates different channels for control information broadcasting at various clusters in the network. Moreover, this intelligent technique also guaranteed the desirable number of common channels for agile channel switching when primary radio (PR) activity is detected, without the need for frequent re-clustering.

4.5.2.5   Complete Broadcasting

In complete broadcasting scheme, SU nodes broadcast the packets on all available channels of CRN for the reception packets to all nodes in the network. Basically, due to potentially large channel set, the SUs usually listen to different channels and are unable to receive the message until both sender and receiver are tuned to the same channel. Therefore, SU utilizes this scheme to transmit the packet on every channel to ensure that every neighbour node will receive the packet. However, when a SU broadcasts the packets on all the available channels, then it will occupy all the spectrum resource which results into network congestion, contention among SU and packet collision problem. These problems become more severe when the number of available channels increased because the SU node must transmit on every channel randomly. Moreover, this inefficient technique also results in large delay, increased redundancy and higher energy consumption by the SUs for the reception of packets due to potentially large channel set. Finally, the simultaneous broadcast on all channels will result into minimization of broadcast problem as the channels are fully occupied by the SU sender node. In [40], the authors have implemented the complete broadcast scheme for calculating the delay, congestion, redundancy and energy consumption by SU in CRN. This study also shows the comparison of complete broadcasting with selective broadcast scheme which clearly indicates that complete broadcasting is the most inefficient scheme which results in throughput minimization and delay maximization of CRN. In [103], the authors have utilized complete broadcasting to enable the multiple accesses by CR nodes for calculating the ergodic sum capacity of CR-MAC. Basically, complete broadcasting is performed to calculate the ergodic sum capacity of CR-MAC for backhaul network. Moreover, the proposed scheme also enables the CR nodes for optimal resource allocation and utilization by using the interference-power control and scheduling algorithm which protects PU from interference.

4.5.2.6   Broadcast over Set of Channels

In this scheme the SU selects a downsized set of channels for broadcasting that will cover all the neighbouring node on all selected channels. This is the most effective scheme for broadcasting in CRN as it caused less message overhead, less channel contention and congestion in the network. Moreover, the previous literature also showed that the usage of this technique results into less channel switching, delay tolerant and less energy consumption due to the downsized channel set. Intuitively, this approach seems quite simple and reasonable, but the selection of downsized channel set from big pool of channel should be performed efficiently as it plays a major role in broadcasting. For example, if the SU forms a downsized channel set which contains many frequency agile channels, then it will result into more channel switching and retransmission of packets on channels. Moreover, this bad selection of frequency agile channel set also causes the harmful interference to PU nodes which will degrade the performance of the network. In [79], the authors have proposed broadcast protocol in multi-hop cognitive radio ad hoc networks with collision avoidance (BRACER) that intelligently downsize the channel set for multi-pairs of neighbouring nodes. In this paper, they have considered the practical scenarios in which CR users are not aware of the spectrum and network parameter, global network topology, user spectrum availability and time synchronization information for broadcasting the message in CRN. This protocol minimizes the average delay and avoids collision by intelligently downsizing the original available channel set. Moreover, they have also proposed the solution to design the sequence for broadcasting and utilize the scheduling schemes for effective and efficient transmission between SUs [79]. In [78], they have also proposed solution for broadcasting under blind information in multi-hop CRN that consider the practical scenario in which the SUs are unaware of network topology, the spectrum availability information and time synchronization information. Al-Mathehaji et al. [6] propose a distributed broadcast protocol for ad hoc CRN that eliminates the interference and collision risk to the PU communications. Moreover, this algorithm also facilitates the protection of PU receivers which are not detected during spectrum sensing and provides a high successful broadcast ratio. Finally, they have also designed a technique where no global network topology is known or no common control channel is assumed to exist, and they provide receivers' protection by alleviating the interference impact of the cognitive radio node transmissions on the primary protected zone which is a distinctive feature in CR networks. In (Yau et al. 2014, [7]), the global network topology and the available channel information of all SUs are assumed to be known for the creation of downsized channel set. The proposed solution in these studies enables the SU to discover their neighbour node accurately for creation of downsized channel set which will minimize the redundancy as well as minimize the network overhead in CRN. In [31], a common signalling channel for the whole network is employed which is also not practical. In [35], the authors have proposed an algorithm for broadcast scheduling which can minimize the latency and redundancy of CRN.

4.5.3 Issues and Challenges of Broadcasting in CRN

In CRN, the broadcasting has become quite complex due to dynamic and agile radio environment. The major issue and challenges related to broadcasting in CRN include:

4.5.3.1   Channel Diversity and Heterogeneity

In CRN, the biggest challenge for broadcasting is the management of heterogeneous and diverse spectrum to exchange the control information among the neighbouring nodes. Moreover, the PR activity makes the channel availability more complex and uncertain for broadcasting the messages in CRN as explained in previous sections. The channel availability for CR nodes is less in case of high PR occupancy. Moreover, PR nodes have varying channel occupancy at different channels and location. Therefore, it is quite challenging to predict the channel availability due to channel diversity, heterogeneity and PR activity. The need of hour is to design the joint channel sensing and broadcasting schemes which calculates the PR activity efficiently for better utilization of spectrum in CRN.

4.5.3.2   Agile Nature of Channel Availability

The second biggest issue for broadcasting CRN is the agile nature of channel availability. It is very challenging for the CR nodes to determine the availability due to large channel set which is spread over a wide frequency range as mentioned in previous sections. In order to broadcast the control information to neighbouring nodes, SU selects the best channel among the set of available channels. The selection of best channel for broadcasting is the most challenging task as it is required that the PR occupancy should be calculated in advance before selecting that particular channel. Otherwise, the bad channel selection may lead to interreference with PR nodes, and SUs have to retransmit the packet again to its neighbour which will result in more delay and less packet delivery ratio. Moreover, this channel switching also caused the channel congestion, broadcast collision and heavy traffic generation on network. The need of the hour is to design the efficient channel selection schemes for CRN which can efficiently cater the agile nature of channel availability in the presence of PR activity.

4.5.3.3   Common Control Channel Selection Challenges

The broadcasting of control information on single or dedicated control channel is not the feasible technique due to channel diversity, heterogeneity and PU activity. The biggest problem in CRN is that due to PR activity, the CCC is not unique and not always available to CR nodes. As discussed in previous Sect. 4.3.2, the control channel should be selected dynamically by CR node for effective and efficient dissemination of control information in CRN. The need of the hour is to design a dynamic channel selection scheme which will select the best virtual common control channel between CR nodes.

4.5.3.4 Challenges Related to Neighbour Discovery

The challenges related to neighbour discovery arise whenever a new node joins the network. The neighbour discovery is the process of identifying the new neighbours by exchanging the control information via broadcasting in CRN. This process enables the neighbouring nodes or relay nodes to forward the packets using common channel to their respective destination as they already get all the information related to their two-hop neighbours via control information exchange. However, this process of discovering the common channel between the neighbouring nodes is quite complex and requires a lot of time to scan each channel for its availability which results into enormous delay. Moreover, the availability of these common channels between neighbouring CR nodes varies due to PR activity. Therefore, it is of utmost importance to design the intelligent adaptive schemes which can solve all these problems for broadcasting in CRN.

4.5.3.5 Challenges Related to Neighbour Channel Selection

The challenges related to neighbour channel selection for broadcasting are quite complex in nature. The previous literatures have showed that the selective broadcasting scheme usually selects the channels which have higher neighbour ratio or degree. The problem arises when PR nodes occupy that selected channel which has the highest degree. Due to this agile nature of channels, all the neighbours have to select the new channel for broadcasting among them. The previous literature clearly showed that the best channel for broadcasting must be selected not only on the basis of its number of neighbours but also on the basis of traffic pattern and channel occupancy. The need of hour is to design the efficient algorithms that can analyse the traffic pattern, channel occupancy and degree of CR node collectively for effective and efficient control information broadcasting in CRN.

4.5.3.6 Challenges Related to Collison Avoidance

The broadcast collision issue is quite complex in CRN because CR nodes used multiple channels for broadcasting and the timing for receiving the broadcast message by neighbouring nodes may also become identical which results in the collision of packets. Moreover, the neighbouring nodes also compete to occupy the available channel for transmission of packets which causes the contention among the nodes. Therefore, it is required that the neighbouring nodes are synchronized with each other on a common channel along with timing information for broadcasting the packet successfully in CRN. As discussed in previous sections, the issue of broadcast collision can be mitigated by using efficient scheduling schemes to achieve successful broadcast ratio and short broadcast delay. The need of the hour is to design the protocols that can schedule the broadcasting and lessen the contention among neighbours for achieving high successful broadcast ratio and short delay in CRN.

4.5.3.7   Challenges Related to Route Selection

The route selection is also a challenge in CRN as it requires the intelligent decision-making by CR nodes to select the best channel along the path of communication. Basically, the current schemes are still lacking of cooperative intelligent decision-making between the path and channel selection in CRN. The selection of nearest neighbour in CRN may not result into optimal routing decision because it may result into more hop counts during transmission. Moreover, it is also possible that PR activity also effects the route that will eventually result into more channel switching and more delay in transmission. Therefore, it is essential to correlate the route with channel selection for effective and efficient broadcasting in CRN. The need of the hour is to design the intelligent decision-making scheme which is capable of correlating the route and spectrum selection to reduce the broadcasting delay in CRN.

4.5.3.8   Challenges Caused by Rapid Channel Switching

In CRN the available spectrum for communication is wide which provides a large number of available channels for each CR node which are diverse and heterogeneous. This heterogeneity and diversity provides numerous characteristics including channel switching, delay, PR interference, etc. The previous section of complete broadcasting clearly showed that broadcasting on all available channels results in frequent random switching on every channel which will increase the delay and minimize the throughput of network. Moreover, complete broadcast also requires higher channel switching between data channels and CCC for CR nodes which are equipped with single radio. This frequent channel switching also degrades the channel quality. Moreover, PR activity also increases this probability of channel switching in CRN. The need of the hour is to design group-based intelligent protocols that can facilitate the selection of fewer best channels for CR nodes and also implement the scheduling algorithms which can intelligently synchronize the CR node in case of PR activity.

4.6   Network Coding in CRN

This section demonstrates the coding procedure in detail with the help of practical example. Moreover, the classification of NC techniques is discussed in detail. Finally, the practical implementations of NC schemes at different layers are presented in the following section.

4.6.1 Illustration of NC Using Simple Example

The intelligent scheme of NC is very effective to increase the broadcasting efficiency by encoding or combining as many as possible packets to minimizing the redundant transmission in CRN. For example, the source node S picks the first packet P1 from its output buffer and checks the coding opportunity of P1 with other packets in its buffer to encode them together. Normally, the probability of encoding the packets is small which is usually bounded by node degree. Figure 4.6 shows a wireless network of four nodes in which the source node S wants to send packets P1 to P4 to all the nodes in the network. In this wireless network, node T is in the range of source node S, and hence it will require four transmissions to send the packets to node T. Node T will act as a forwarder and require four transmissions to forward packets P1 to P4 to node U. Similarly, four transmissions are required to forward these four packets to node V. Hence, source node S will require 12 transmissions to deliver packets P1 to P4 to all the nodes in the network.

Now consider a simple XOR-based network coding approach. In Fig. 4.6, the source node S will look at its output queue that contains four packets for node T so it will send the native packets to node T as there is no coding opportunity available. Now, node T will act as forwarder and tries to encode the packets in its output queue, but due to the lack of packets for decoding at nodes U and V, it will transmit native packet P1 to node U at first transmission. Similarly, node T will transmit the packet P2 to node V during second transmission, and it also maintains the copy of packets in its virtual queues for further utilization and coding. The transmission of packet and reception reports to all nodes enables forwarder node T to maintain the neighbour table which contains information about its neighbours and packets they have. Hence, node T will broadcast the encoded packet P1 $\oplus$ P2 in the next transmission to both nodes U and V, and they will recover their packets P2 and P1 by simply doing P $\oplus$ P2 and P $\oplus$ P1 respectively. During the next two transmissions, node U will receive native packet P3, and node V will receive packet P4 as there is no coding

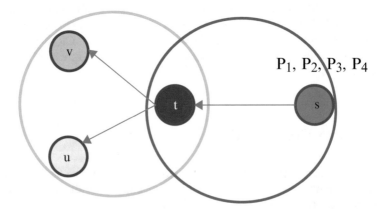

Fig. 4.6 Broadcasting using network coding in CRN with transmission ranges

opportunity at node T. Finally, in the last transmission, the forwarder node T will broadcast the encoded packet P3 $\oplus$ P4 to both nodes U and V, and they will recover their packets P4 and P3, respectively. Hence, source node S will require ten transmissions to send four packets to all nodes using XOR-based network coding algorithm. Therefore, we can conclude that using NC for broadcasting can save two transmissions which results in coding gain of 1.33.

Discussion The whole process of coding can be explained as the nodes exchange their neighbour information through overhearing and reception reports so that each node knows the network topology within its two-hop neighbourhood. In our example, node T is a forwarder which determines if all its neighbours have already received the packets based on its neighbour table, and then it does not rebroadcast the packets. Otherwise, node T will dequeue the packets from the head of the output queue and see if there exists a network coding opportunity, and then it will generate one or more encoded packets and schedule the transmission. If not, then it will simply forward the packet or the node will buffer the packet for a random amount of time and process it later to create more coding opportunities. Moreover, it will send the reception reports and add acknowledgement to the header of packets and send to the wireless node U or V.

On the receiver side, the node first extracts the acknowledgement and updates neighbour information based on reception reports. If the received packet is not encoded, then it is added to packet pool otherwise it will be decoded. For decoding, when node U receives an encoded packet which includes P native packets, the receiver checks all the native packets in its buffer to decode the encoded packet by retrieving the corresponding native packet from its packet pool. Finally, the node U XORed the P1 packets with the received encoded packet to retrieve the missing packets P2 and P4 by simply doing P $\oplus$ P2 and P $\oplus$ P4. The same rule can be followed to decode packets P1 and P3 at receiving node V by doing P $\oplus$ P1 and P $\oplus$ P3. Finally, the node will check whether it's the destination node or it has to forward the packet further to its destination and is enqueued in output queue accordingly.

4.6.2 Classification of NC Schemes for CRN

The NC coding schemes are classified into the following major categories.

4.6.2.1 Intersession NC Scheme

In this NC scheme, the relay node combines or encodes the packets from two or multiple CR source nodes having different flows in CRN [68]. The encoding of packets not only decreases the number of transmissions but also increases the throughput and decreases the interference between the links in CRN. This can be

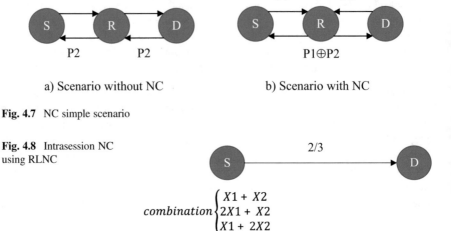

a) Scenario without NC b) Scenario with NC

Fig. 4.7 NC simple scenario

Fig. 4.8 Intrasession NC using RLNC

explained with the help of example in Fig. 4.7, where nodes S and D want to exchange their packets P1 and P2, respectively. In first case, the two CR nodes have to perform four transmissions for exchanging the packets, two transmissions for sending the packets to relay node R and two transmissions for relaying the packets to their respective destination as shown in Fig.4.7a. While on the other hand, if we utilize the capabilities of XOR-based NC scheme then only two transmissions are required to send the packets to relay node by nodes S and D. On reception of packets, the relay node encodes the packet by XORing the packets P1 $\oplus$ P2 as shown in Fig. 4.8b [64, 68]. The receiver node will decode their respective packets as discussed in previous sections. This technique will reduce the transmissions to three by using NC technique for different flows of data.

4.6.2.2 Intrasession NC Scheme

In this NC scheme, the relay node linearly encodes the packets from the same source. This encoding scheme exploits the diversity of CRN network and results into reliability and efficient utilization of network resources [64]. Moreover, the K coded packets combined by relay nodes do not need to store the information of successfully received packets out of transmitted coded packets by the destination node. For example, in Fig. 4.8, the source node S wants to deliver packets X1 and X2 to node D. In this scenario, the reliability of the link between S and D will be $\frac{2}{3}$. While on the other hand, when S wants to send three coded packets, $X_1 + X_2$, $2X_1 + X_2$ and $X_1 + X_2$, then the destination node will receive two of the three packets. Therefore, the destination will receive both packets which will result in more reliability and less transmission of packets [64, 68]. However, in case of no network coding, the source node needs to transmit each packet twice and use the feedback mechanism as well.

4.6.2.3 Linear NC Scheme

In linear NC scheme, the relay node receives the source packets X_i of size S bits from different sender nodes and encodes them by multiplying them linearly with a coefficient selected from the finite field whose size is typically equivalent to Gallio's field GF(2^s) [8]. In this scheme the transmitted packets P are the linear combination of source packets on which the addition and multiplication operation are applied over Gallio's field GF(2^s). Mathematically it can be represented by the equation

$$P = \sum_{i=1}^{n} v_i.X_i$$

In this equation v_i represents the global encoding vector which is required by the receiver nodes to decode the messages. In our practical scenario, the GF = {0,1}, i.e. the packet size is X_i. Therefore, the relay node R combines both packets P_1 and P_2 as a linear combination, i.e. P = P_1 + P_2. Moreover, the encoding is performed linearly by using the summation of packets for every packet by using bitwise XOR operation. This scheme even encodes the already encoded packets by choosing a set of coefficients $\gamma_i, \ldots . \gamma_j$ and calculating the linear combination:

$$P_i = \sum_{i=1}^{n} \gamma_i.Y_i$$

where Y_i represents the received encoded packet whose global encoding vector is also known to the relay node. Now, the new global encoding vector is given by:

$$v_i = \sum_{i=1}^{n} \gamma_i.v_{i,j}$$

On the receiver side, the packets are recovered by using this global encoding vector by performing the Gaussian elimination on the received encoded packet matrix [8].

4.6.2.4 Random Linear NC Scheme

In this scheme, the relay node encodes the source packets by multiplying them linearly with a random coefficient selected over Gallio's field GF(2^s) [10, 64]. Basically, this scheme is quite similar to linear NC except the difference that the encoded packet is the random linear combination of packets by the source node. Initially, suppose that P packets are stored in memory of source node and when it receives the packets by source node, then it performs RLNC on received packets [10].

The encoded packets are the random linear combination of the packets from sender node memory. Let us suppose that the relay node has K packets $P_1, \ldots P_k$ in its memory, then the encoded packet is random linear combination which is:

$$P_0 = \sum\nolimits_{k=1}^{K} \gamma_l . P_l$$

where γ_l is selected according to uniform distribution over the elements of Gallio's field $GF(2^s)$. Similarly, the global encoding vector is given by:

$$v_i = \sum\nolimits_{i=1}^{n} \gamma_i . v_{i,j}$$

This technique is most appropriate in CRN as the probability of linearly dependent packets is related to Gallio's field which is usually very low. This efficient scheme reduces the packet loss and increases the spectrum availability for the CR nodes.

4.6.2.5 Physical Layer NC

In this scheme, the CR nodes utilize the characteristics of electromagnetic waves to encode the received packets from different nodes into linear combination of a single encoded packet [91]. Basically, this scheme is applied at physical layer of wireless medium in which the relay node encodes the bits of data by performing bitwise XOR operation on packets. This encoding scheme not only helps to remove the noise on the transmitted signal but also boosts up the signal at physical layer to improve the performance and increase the spectrum utilization in CRN. The decoding of the packet is performed by subtracting the received encoded packet with the packets in its own packet buffer as discussed in previous section.

4.6.2.6 Analogue NC Scheme

In this scheme, the relay node utilizes the characteristics of signals to encode the packets received by different source nodes into network-coded XORed signal [49]. This scheme works with the signals instead of bits, and network coding is performed in the form of phases. In initial phase, the source nodes transmit the packets to the relay node, and relay node tries to decode the XORed superimposed signals. Then, in next phase, the relay node encodes the signals by XORing them together and transmits on the wireless medium again. The decoding procedure is the same as in the previous scheme in which the encode packet is XORed with the already received packet to recover the original packet from the encode signal. This exiting technique utilizes the electromagnetic properties of the signal at physical layer by encoding the signal to reduce the interference and noise level of the signal.

4.6.2.7 Rateless NC Scheme

In this scheme, the relay node utilizes the properties of rateless encoder which can generate unlimited number of encoded packets from the received packets P packets of source node [74]. Basically, this scheme is the class of error control codes which can encode unlimited number of packets with low computational complexity for effective utilization of spectrum resources. On CR receiver node, the decode is performed on the received encoded packet $P.\gamma$, where γ represents the NC overhead which is slightly greater than one. This cooperative scheme enables the receivers to receive enough packets that can be utilized for decoding of received packet regardless of which packet it has received. The exiting technique can reduce the delay and improve the throughput of CRN. Moreover, the cooperative rateless NC scheme also provides all the functionalities of automatic repeat-request technique which facilitates the retransmission of lost packets.

4.6.2.8 Fountain Code NC Scheme

In this scheme, the relay node encodes the set of source symbols into unlimited symbols in a way that they can be decoded from any subset of encoded symbols [43]. This coding scheme is efficient for encoding large block of packets but does not perform well with short block lengths because of coding overhead associated with them. The received P packets by the relay node encode these packets together into unlimited packets in such a way that there are $P(1 + \varepsilon)$ packets required to be received successfully for decoding. This scheme ensures the redundancy by combining the source symbols linearly which require fewer XOR operations per generated symbol for both encoding and decoding. This innovative technique increases the spectrum utilization and reduces the PU interference.

4.6.2.9 Asymmetric NC Scheme

In this scheme, the relay node encodes the most appropriate received packets from source CR nodes using NC technique that can fulfil the encoding requirement through splicing combination of received packets in CRN [108]. Basically, this scheme encodes combination of different splices of packets into a single packet that can fulfil the NC requirement to increase spectrum efficiency and reliability. Moreover, the usage of NC technique also ensures that PU's activity does not create interference to SU's. Finally, the asymmetric data received by relay nodes is encoded using linear combination technique so that it can maximize the throughput and minimize the bit error rate.

4.6.2.10 Adaptive Dynamic NC Scheme

In this scheme, the adaptive approach is used in which the NC is performed on the basis of system current state. Moreover, the packets are not decoded at relay node; rather only the destination nodes can decode them usually at the end of transmission [50]. Basically, the packets are encoded linearly based on channel variation adaptation method in which data packet or block size is modified according to the channel conditions and availability. The packets are encoded based on linear combination of received packets at intermediate nodes and utilize the Gaussian elimination method for decoding the packets. This scheme enables the CR nodes to intelligently sense the channel as well as packet size information and then dynamically utilizes the adaptive NC technique to encode the packets according to the system current state and channel requirements for effective utilization of resources. This intelligent scheme also enables the SUs to recover the source packets, perform load balancing, increase throughput and decrease the delay of network.

4.6.2.11 Differential Mesh Information Coding Scheme

In this scheme, the differential information coding is utilized for the selection of basic mesh among numerous meshes in the network. Moreover, it also facilitates in grouping the homogeneous meshes in the network. Basically, in wireless network, the geographical regions are divided into small meshes, and then the network information is broadcasted on each of these small meshes for the improvement of network information accuracy. The differential mesh information coding scheme is based on frequency/occupancy graph and mesh selection strategy which are capable of reducing the duplicate network information in homogeneous meshes and improvement in NC efficiency by quantizing the differential information among meshes (Zhang et al. 2012). This intelligent grouping and coding technique enables the wireless nodes in mesh network to broadcast the encoded packets in small meshes for the improvement of accuracy and efficiency of heterogeneous network information delivery [105].

4.6.2.12 Multiple Description NC Scheme

In this scheme, the stream of data packet is divided into s substreams ($s \geq 2$) which are known as descriptions by using NC scheme by the relay nodes. These substreams are then routed over disjoint paths to reach all the destinations [99]. Any description can be utilized by the receiver nodes to decode the data stream information. This innovative scheme can be used for multimedia application in CRN to improve the spectrum efficiency and avoid the collision of media streams. Basically, this coding scheme divides the original media stream into small streams by utilizing NC functionalities and utilizes any description for decoding purpose which will result in less network congestion, error resilience and higher packet delivery ratio [14]. Moreover, this scheme also facilitates to reduce the packet loss occurred by PU traffic by efficiently utilizing the spectrum resources.

4.6.3 Cross-Layer Design of NC Schemes

The biggest challenge in CRN includes PU activity, channel heterogeneity, diversity and its broadcast nature as discussed in previous sections. In CRN, the CR nodes' PU activity results in interference between the links and produces unwanted multiple copies of the same packet. However, if we utilize the intelligent NC technique, then the relay node will efficiently and effectively utilize the available spectrum. The NC coding techniques presented in the previous section can be applied at different layers of CRN to improve the spectrum efficiency, throughput spectrum capacity and robustness of the network. The NC coding techniques applied at different layers include:

4.6.3.1 Physical Layer NC (PHY-NC)

The physical layer is responsible for transmitting and receiving the packets into electromagnetic (EM) waves. Basically, this layer is responsible for handling the signal and works on bit level to remove the noise and amplify the signal physically. When NC is applied on physical layer, the signals or bits of packet data are added up physically. Basically, the relay node implements NC at physical layer to encode the received packets by using bitwise XOR operation as discussed in previous section in detail. Similarly, the decoding procedure is performed by receiver nodes by XORing the received packets with already available packets in its buffer [38, 47, 102].

In [91], the authors have successfully applied NC scheme at physical layer which enables the CR nodes to detect the threats and attacks by malicious nodes. Basically, they have implemented the concept of node positioning and signal strength of EM waves in wireless network to detect the presence of malicious attack. At this layer, the manipulation of signal characteristics is first analysed, and then fine-tuning and amplification of signal are performed. In [92], the authors have presented physical layer-based cooperative model which increases the network throughput and also reduces the transmission time. In [85], the authors have proposed the scheme to successfully transmit the signals to destination node in a single time slot which utilizes physical layer characteristics of relay node. They have utilized the full duplex mode which enables the SUs to fully utilize the spectrum opportunities of relay node which results into robust transmission of packets across the network. In [87], the authors have proposed PHY-NC scheme that has enabled the robust packet recovery and multiple access simultaneously in CRN. In [21], the authors have presented the solution which is capable of encoding and decoding PU signals effectively by applying NC at physical layer for secure transmission of packets in CRN. In [26], the authors have applied NC at physical layer in two phases. The first phase involves simultaneous transmission of packets by source node to relay node, while second phase involves the encoding of packets by XORing them together. Moreover, they have also proposed PHY-NC protocol which enables the CR nodes to share the spectrum resources efficiently with multiple PUs.

4.6.3.2 MAC Layer NC

The MAC layer is responsible for managing the spectrum resources, timing and coordination between mobile nodes. The implementation of NC at this layer enables the CR nodes to encode the packets opportunistically for effectively and efficiently managing the spectrum resources [104]. NC equipped this layer with intelligence as it enables the relay node to opportunistically encode (i.e. XOR) the packets which can be easily decoded by its neighbouring nodes based on their buffer estimation. The previous literature ([44], Zhao et al. 2007, [36]) has extensively proposed numerous MAC layer protocols to enable effective and efficient communication in CRN.

In [9], the authors have proposed MAC layer NC protocol which enables the CR nodes to effectively and efficiently disseminate the control information across the CRN. This MAC protocol enables the CR nodes to dynamically access the spectrum on multiple channels. Moreover, it also enables the SUs to detect PU presence and allows them to coordinate with each other for effective communication and better spectrum utilization. In [81], the authors have proposed power management strategy at data link layer which effectively increases the performance of secondary network.

4.6.3.3 Network Layer

The network layer is responsible for managing the logical addressing, selection of communication route and connectivity between sender and receiver nodes. Basically, this layer deals with data packets, their routing, topology management and function related to data management at packet level. The deployment of NC schemes enables this layer to encode the packets at relay node. When NC is applied at this layer, it enables the relay nodes to route the packets efficiently to increase the throughput of CRN.

In [9], the authors have proposed a virtual channel for exchanging the control information by utilizing the NC scheme for efficient and effective packet dissemination. The proposed intelligent scheme based on NC enables the SUs to utilize the spectrum resources effectively and efficiently. Moreover, this scheme also enables the SUs to coordinate with each other for efficient resource allocation. In [88], the authors have proposed a NC-based routing protocol which reduces the delay and maximizes the capacity of the spectrum utilization. In [75], the authors have presented a comparative analysis of three routing schemes which include NC-based routing, shortest path routing and multipath routing schemes. The results have clearly showed that NC outperforms from other routing schemes in terms of performance even without calculating the shortest path between the sender and receiver nodes. This intelligent NC scheme enables the PUs and SUs to coordinate with each other even without any prior knowledge of PU nodes.

4.6.3.4 Transport Layer NC

The transport layer is responsible for data dissemination, data segmentation, reliable delivery, error control and data management function in wireless network. The implementation of NC at transport layer guarantees the packet recovery and resolves the problem of transmission error. Basically, NC enables the receivers to calculate the packets they received after transmissions. In case, if some packets are lost during the transmission then the receiver apply intelligent NC scheme to re-encode the packets for their successfull recovery at receiver node. Moreover, the implementation of NC at transport layer enables the multiple flow of data which results in maximization of spectrum utilization.

In [5], the authors have proposed the RLNC-based global selective acknowledgement algorithm for delay tolerant networks (DTN) which can improve the performance of reliable transport. The method of mixing the packets and acknowledgement information together using NC can solve the problem of randomness and finite capacity limitation of DTN. Moreover, the usage of RLNC results in the improvement in fairness among flows and minimizes the delay in the network. In [111], the authors have proposed NC-based DGSA algorithm which works along with TCP protocol to improve its performance in CRN. The results showed that the deployment of NC-based DGSA algorithm in conjunction with TCP protocol will reduce the retransmission and delay in CRN. Moreover, these proposed algorithms also enhance the PU behaviour, spectrum sensing and reshaping for multi-hop CRN. Finally, the implementation of NC on proposed algorithms also maximizes the bandwidth utilization, increased throughput and reduces the delay in CRN. In [69], the authors have calculated the performance degradation of TCP in CRN and proposed an intelligent joint NC-based TCP algorithm which is called as TCP joint generation network coding (JGNC) for multichannel multi-radio multi-hop CRN. In this algorithm, the CRN environment is first sensed and analysed, and then adjustment of packets according to the environment is performed to achieve better decoding probability. Moreover, they have also modified the original TCP algorithms and apply NC on it so that it can fit CRN environment to achieve better throughput, spectrum efficiency and minimum delay.

4.7   Network Coding-Based Broadcasting Techniques in CRN

This section is dedicated to discuss about appropriate NC scheme for successful broadcasting in CRN. We provide detailed survey of intersession and intrasession NC schemes for broadcasting applications.

4.7.1   Intersession NC

4.7.1.1   CODEB

In wireless network, the broadcasting is performed by using the deterministic approaches (the sender node preselects the neighbouring nodes for rebroadcasting the packets) or probabilistic approaches (each relay node rebroadcasts the packet with given probability). In CODEB, the deterministic rebroadcast approach is used for the reduction of transmission in the network [46]. In this scheme, the authors have proposed two NC algorithms, i.e. XOR-based NC algorithm and Reed-Solomon NC algorithm that relay on the local two-hop topology information and utilize opportunistic listening to reduce the number of transmissions. Basically, this proposed approach consisted of three techniques, i.e. opportunistic listening, forwarder selection and pruning and opportunistic coding technique to reduce the number of transmissions. In first phase, the opportunistic listening is responsible for creating a two-hop neighbour graph by periodically listening to all two-hop neighbours in their communication range. Moreover, they also keep record of previous hops along with next two-hop neighbours in neighbour reception table. In this scheme, when there is no coding opportunity for the packets, then it will be stored in buffer or interface queue for a limited time until it finds some coding opportunities with the newly arrived packets. In second phase, they have implemented PDP algorithms [54] which select the forwarder nodes from a subset of forwarder nodes. Basically, the forwarder selection process is based on the number of transmissions required by the nodes to deliver a packet to all its covered nodes, divided by the number of covered nodes by each forwarder. This algorithm also attached the forwarder set in the packet header, and only the node which included in the forwarder list will forward the packets. In the last phase, opportunistic coding is implemented in which each node analyses the information in its neighbour table through opportunistic listening to calculate how it can opportunistically encode the packets in its buffer. If it finds the related packets to successfully encode the packets, then it will send the packets or otherwise wait for coding opportunity. In this research they have used two algorithms for coding packets:

XOR-Based NC It is a simple XOR-based algorithm that encodes the packets in the buffer using XOR in a way that it will enable the maximum number of nodes to decode the packet. Basically, this algorithm retrieves P packets from buffer and sequentially searches for the packets which can be combined with P to enable all neighbours of node U to decode the packet as discussed in previous section of NC. If the operation is successful, then these packets are added to set S. Otherwise, the algorithm is unable to decode the packet, and we have to use the neighbour table to confirm that all the neighbours have already received at least |T|−1 of the packets in set T to successfully decode that coded packet.

Reed-Solomon-Based NC It is an optimal coding scheme which uses Reed-Solomon code as the coefficients for combining native packets linearly. The algorithm constructs coded packet set Q= Θ P. Basically, this technique includes the information about the set of native packet IDs in the encoded packet. Moreover, it also adds the index number utilized codes in every coded packet. This intelligent algorithm is very fast and fault tolerant as it can rapidly decode all the remaining packets when the receiver node gets at least P-K distinct original packets. Finally, the literature showed that agile nature of wireless link can be easily managed by sending more than K-encoded packers and implementing negative acknowledgement (NACK)-based scheme for keeping the record of lost packets in CRN.

4.7.1.2   CROR

In CROR [110], the authors have proposed intelligent opportunistic routing protocol which broadcasts the packets by utilizing the feature of NC for effective spectrum utilization in multi-hop CRN. The major contributions of this paper include proposed routing metric, candidate selection and prioritization, proposed XOR-based coding techniques and timer adjustment. Firstly, they have proposed a new routing matrix called successful packet delivery ratio matrix which is used to calculate the number of packets required for successful packet delivery to destination SU nodes. Basically, this matrix contains information related to packet lost and available time for selecting the forwarder node which is useful in calculating the required packet for efficient transmission. This matrix along with the expected transmission count matrix enables the SUs to successfully complete the transmission successfully for achieving better performance in CRN. Moreover, every node contains a forwarding node set which contains the information related to the available channel set and required information for decoding the packets successfully. These forwarding set nodes are neighbours of each other and have common channel available between them for effective transmission of encoded packet. They have utilized the XOR-based NC method in which they also consider the channel availability and average coding opportunities for efficient packet dissemination of packets across the CRN. Moreover, they also consider that the destination SUs have enough packets to decode the encode packets successfully. Finally, they have implemented the timer in which the received packet at every node is set with the timer and its value is inversely proportional to the product of coding matrix and coding opportunities (number of packets that can be coded successfully), while it is directly proportional to the order of each node in forwarding set. This timer is the most important part of the protocol as it reduces the redundant transmission and increases the overall throughput of the network. In this paper they have calculated effect of PU activity on throughput and packet collision. The results clearly showed that their proposed scheme not only increases the throughput of CRN by efficiently utilizing the bandwidth but also reduces the PU-SU packet collision probability as compared to previous schemes.

4.7.1.3 Directional Antennas

The efficient broadcasting using NC and directional antennas is also an active area of research in wireless network for efficient and effective transmission. In [94], the authors have proposed an efficient NC-based broadcasting scheme which utilizes the capabilities of directional antennas to reduce the number of transmissions by each relay node. Basically, in this technique, the NC is used to reduce the number of transmissions by relay nodes, and directional antennas are used to reduce the energy consumption by selecting the specific sector from a subset of sectors for broadcasting. In this intelligent scheme, the authors have divided the transmission area into small sectors, and nodes utilize the capabilities of directional antennas to broadcast the coded or uncoded packet into a specific sector which can decrease the energy consumption of relay node. They have also proposed the static and dynamic relay node selection algorithms and implemented the NC for both selection approaches. Moreover, they have created a backbone network in which they have defined a connected dominating set that contains the nodes which are directly connected with each other or the node is connected with the member of this set [95]. They also ensure that the coding must be performed by the relay node on the packets which are sent on the same sector of the network. The simulation results have clearly shown that this approach not only saves the node energy consumption but also mitigates the interference and improves the node mobility in the network. This proposed scheme can be successfully applied to CRN with some modifications as the proposed method assumed the links to be perfect which is not feasible in CRN due to diversity, heterogeneity and PU activity.

4.7.1.4 Deadline Aware

In deadline-aware technique, the packets need to be transmitted by the relay node before the expiration of deadline. In [19], the authors have proposed the NC-based deadline-aware scheduling algorithm that minimizes the number of packets which miss their deadlines for reducing delay and increasing the packet delivery ratio of the network. They have utilized the intelligent graph model to describe the relationship between the packets in the client buffer. Basically, they have implemented encoding algorithm based on maximal weight clique graphs in which the node vertex has assigned a weight as decreasing function of packet deadline, which facilitates the

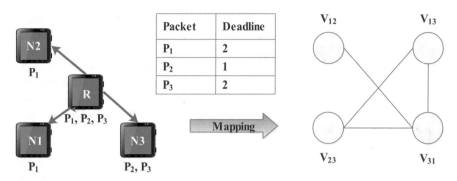

Fig. 4.9 Delay aware NC

sender to encode the packet and broadcast it on the network. This approach helps to find that clique which contains the vertex having the smallest deadline for transmitting the packet which is urgent to send in the network. The simulation results clearly showed that the utilization of maximum clique can reduce the missed deadlines and increased packet delivery in network. This scheme can be explained with the help of practical scenario as shown in Fig. 4.9. In this scenario, node N1 contains packet P_1, node N2 contains packet P_1, and node N3 contains packets P_1 and P_3. We apply this graph technique to map the nodes to their corresponding vertex and create the link between them as per their packet requirements as shown in Fig. 4.9. As both nodes N1 and N2 do not have packets P_2 and P_3, the vertex corresponding to them includes v_{13} and v_{23}, respectively. Moreover, there is a direct link between these two vertices. On the other hand, node N3 does not have packet P_1; therefore vertex v_{31} represents that it needs this packet. Moreover, node N3 contains packet P_2, and node N2 contains packet P_1 in its queue; the vertices v_{12} and v_{31} are also connected with each other. Finally, the weights are assigned to each vertex which are proportional to deadlines of packets, after performing the matching function. In [19], the authors have utilized the greedy algorithm to calculate the maximal clique in a weighted graph. The packets are encoded by the relay node using linear combination of packets which belongs to the vertices in the clique. This scheme is useful for single-hop scenario and is unable to calculate the delay for global deadline. However, the work in [101] calculates the effects of weight functions for better transmission of packets in the network.

In [15], the authors have proposed NC-based broadcasting technique which encodes the packets based on RLNC technique to efficiently minimize the decoding delay at receiver node. Basically, they have implemented the RLNC scheme for both systematic and nonsystematic scenarios and derive an equation for calculating the minimum number of packets required to minimize the decoding delay at receiver node to meet the deadline constraints. The simulation results clearly showed that the field size has a nominal impact on decoding delay and minimum packets required for transmission. Moreover, nonsystematic NC has achieved higher gain as compared to systematic NC. The proposed work also helps to generate optimal field size and minimum packets efficiently to minimize the energy consumption and minimum resource utilization. The proposed scheme also handles the decoding delay to minimize the energy broadcast by transmitting node. For instance, when a transmitting node broadcasts the fixed number of packets during a time period, then the receiver could switch to energy collection mode until the packets are recovered. The expressions derived in this paper facilitate to calculate the deadline, and if the feedback channels are utilized, then the transmission could be paused before the deadline is reached. Basically, the expressions derived in this paper efficiently calculate minimum number of packets required for transmission in terms of field size, deadline constraint and erasure probability.

In [63], the authors have proposed the scheme for multichannel radio which can enable the subset of nodes to broadcast the deadline-aware packets successfully to the receiver nodes. In this scheme, the authors have utilized the deterministic PDP approach along with XOR-based NC technique to minimize the deadline constraints and efficiently transmit the encoded packets within their deadline. In their proposed scheme, the relay node will immediately encode and transmit the packet if it has shorter deadline which will result in less coding opportunities. To solve this problem, the authors in [63] have proposed different schemes to calculate the average waiting time of packet at relay node to overcome the deadline constraint. They have also calculated the marginal or extra time that a node can get for successful transmission which is time left to deadline minus time required for packet to hop to the destination node by assuming that each transmission requires only one time slot. They have calculated the average waiting time based on velocity distribution of each relay node which is equal to marginal time divided by number of hops left to arrive at destination node. Moreover, they also showed that there are more coding opportunities when there are multiple flows of packets on the relay node. Finally, the last scheme proposed by them is based on random distribution in which the relay node can randomly select the waiting time of each node from predefined time ranges.

In [63], the authors have proposed a spanning tree-based algorithm that ensures the successful delivery of packets within their deadlines. Basically, they have selected the subset of nodes as source node and construct a broadcasting tree for broadcasting the encoded packets using RLNC technique. Moreover, they have utilized the features of heuristic algorithms to calculate the waiting time at relay node for improving the NC efficiency without missing the deadlines of each packet. These heuristic algorithms calculate the waiting time of packets at relay node according to velocity, random selection and proportional distribution as discussed in [63]. This scheme reduces the transmission by using spanning tree and heuristic algorithms that enable the relay nodes to efficiently calculate the waiting time for achieving more coding opportunities.

4.7.2 Intrasession NC

4.7.2.1 Single Hop

The broadcasting schemes using network coding are implemented in CRN to improve the spectrum utilization. In [112], the authors have proposed intelligent social-based scheme that utilizes RLNC-based coding scheme for broadcasting the packets in CRN. Basically, they have implemented RLNC coding scheme on data that belongs to group of people or community having common interest to encode the lost packets for efficient and effective transmission of data packets. This scheme consists of three phases that enable the SU to intelligently select the available channels, utilize the RLNC coding scheme and perform transmission analysis. Initially, the SUs select the best available channel and transmit the packets to all neighbouring nodes by utilizing this best channel. Basically, the sender node collects all the information related to channel set availability and then selects the best channel based on the transmission time and total time of channel availability for a specific node. The sender node also maintains the acknowledgement matrix which keeps the record of packet sent and lost during transmission. In retransmission phase, the packets are first encoded and sent to all the receiver nodes according to the coding vector of matrix. Each sender node updates its matrix upon the reception of ACK feedback and keeps on transmitting the packets until all packets are recovered at receiver node. The results clearly show that the usage of RLNC scheme for encoding the packets significantly decreases the transmission and also efficiently utilizes the spectrum resources.

In [62], the authors have proposed broadcasting scheme based on NC and machine learning for error-prone transmission in wireless network. Basically, the sender nodes transmit the packets to the receivers and store the feedback from the receivers by ACK/NACK messages/signals in transmission phase. The sender node stores this historical feedback data set with its features and labels. This scheme then utilizes the NC along with machine learning technique on received set of feedback signal by the transmitter to train the classifier. The data encoding is performed by the sender based on this classifier which predicts the states of each transmitted data packet at different receivers by using the feedback signals NACK which are collected previously. Basically, the classifier creates a map for packet states for all data packets which are transmitted in transmission phase, and then based on this information, it will encode the packets which are lost during transmission phase. These encoded packets are then broadcasted in retransmission phase by using NC and machine learning techniques for maximizing the throughput of the network. The simulation results clearly showed that the proposed classifier has efficiently classified 90% of the states of transmitted data at receiver which will result in a more effective utilization of spectrum resources.

Fig. 4.10 Single-hop
scenario

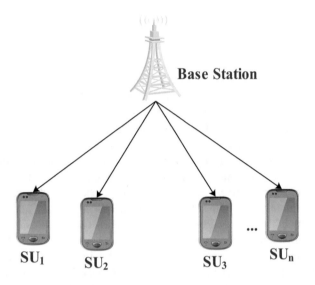

In [61], the authors have proposed acknowledgement-based broadcasting schemes which utilized the NC scheme to reduce the number of retransmission in case of packet loss during transmission phase as shown in Fig. 4.10. They have proposed two schemes based on NACK method to inform the relay node that the packet is lost during the transmission and resend that packet again to achieve one-hop broadcasting reliability. In first scheme, they have used NC for broadcasting the packet at relay node, and if the packet is lost during transmission, then it will send NACK to the relay node for retransmission of packet. The relay node maintains the record in the table about the lost packets at reach receiver node and does not resend the packet immediately. The relay node waits for predefined interval of time and tries to find the maximum coding opportunities for coding the data. The relay nodes then retransmit the packet until all the destination nodes will receive their lost packets. On the other hand, the second scheme utilized the feedback functionalities of the network after each retransmission, and source node can dynamically change the coded packet according to the network conditions. Hence, this intelligent scheme can change the source packets in the encoded packets after each retransmission to improve the performance of the network.

In [55], the authors have used the same method of NACK for retransmission of lost packets but the only difference is that in this scheme the transmission of packets is performed by the base station in two phases. In first phase, the base station broadcasts K packets to the nodes in its coverage area and receives the NACK of the lost packets from the mobile nodes. In second phase, the base station uses XOR-based NC technique for retransmitting the lost packets to the destination nodes. Moreover, they have also proposed the algorithm to find out the maximum number of lost packets for a destination and then add all the lost packets to its coding set. This algorithm also provides the functionality to sort out the lost packets in increasing order so that the coding constraint in the coding set can be fulfilled. This approach has used the coding set which has minimum number of lost packets and tries to encode and retransmit these packets. After successful retransmission, the sender node updates its coding set by adding the remaining lost packets to it. In case sender node does not have any coding set remaining, then a new encoding set will be created to send the remaining lost packets. Finally, the base station encodes all the packets from the coding set and broadcasts it to the destination node.

In [77], the authors have proposed an efficient broadcasting technique for transmitting single file to numerous receivers by using RLNC technique. Basically, they have divided a single file into small pieces and then apply NC and scheduling techniques on nonoverlapping subsets of file to reduce the number of transmissions in the network with unreliable links. Moreover, they have also presented the scheduling policy for least received chunk of data in which least receiver chooses the useful batch with minimum ID at each time slot for efficient file transfer completion time. Basically, they send a single file into small chunks from the base station by using NC and calculate the file completion transfer time from feedback of last receiver node. This scheme also handles the delay constraints by managing the minimum coding window size for better coding opportunities. The results clearly shows that this scheme reduces transmission, minimizes delay and needs less acknowledgement from the users.

In [20], the authors have proposed the greedy heuristic algorithm to solve the packet lost problem by mapping it to graph colouring problem. Basically, they have added a vertex to the graph for each lost packet with the condition that two neighbouring nodes do not have the same vertex colour and add the link between the vertices if the packet is lost by the same destination node. In this scheme, the vertices having the same colour cannot be coded together because of the coding constraint. Moreover, the neighbouring nodes do not have the same colour; otherwise it will result into coding constraint, and the graph colour is dependent on the number of transmissions which is NP complete. In [89], the authors have implemented the greedy algorithm to solve the mapping problem. Basically, the proposed algorithm first sorts the vertices according to their degree and then performs colouring of the nodes. The nodes which are not connected to each other have assigned the same colour, and the connected have assigned the distinct colours until every node is coloured according to this rule.

In [17], the authors have presented the scheme to broadcast the packets of data using NC and graph theory in on-demand broadcast environments. Basically, they have used the adaptive NC along with clique graph to intelligently capture the relationship among request and then combine the cached packet data as per user request for implementing flexible coding technique. Therefore, the optimal adaptive coding technique based on graph utilizes the spectrum resources efficiently. Moreover, they have also presented the two optimal scheduling algorithms which are known as ADC1 and ADC 2. In ADC1 algorithm, the client request with highest priority is identified, and then adaptive coding scheme is applied on identified maximum clique of priority client in a single broadcast. While on other hand, NC and scheduling are performed simultaneously in ADC2 to reduce the computational complexity of the algorithms.

In [42], the authors have proposed interlayer coding scheme that can enhance the performance of the network. Basically, this scheme utilized the inter-coding technique which enables the relay nodes to access more useful layers from more combinations of packet reception. Moreover, they have utilized triangular schemes [23, 25] along with heuristic algorithms to improve the gain. The results showed that the usage of heuristic algorithms with triangular schemes significantly reduces the coding strategies and increases the gain of the network.

4.7.2.2 Relay Aided

In this scheme the relay node facilitates to retransmit the lost packets during transmission. This scheme is also implemented in CRN to enable both SUs and PUs to transmit the packets and infect SU's works as relay node to facilitate the PU transmission for effective and efficient utilization of spectrum resources. In [93], the authors have proposed an intelligent NC scheme based on fountain codes to efficiently broadcast the packets of PU or base station to the receiver PU nodes as shown in Fig. 4.11. In this proposed scheme, the SU acts as a relay station and encodes the PU packets to efficiently utilize the spectrum resources. Basically, in this scheme the base station utilizes the functionalities of the fountain code to transmit the packets to both SU relay nodes and PU nodes simultaneously. The base station stops its transmission when the SU relay node received enough packets to decode the original packet efficiently. Finally, the SU will transmit the encoded packets based on dirty packet coding technique until all the PU receiver nodes recover the original packets successfully. The SU's relay nodes are selected for remaining transmission as they have more reliable link as compared to base station links. The simulation results

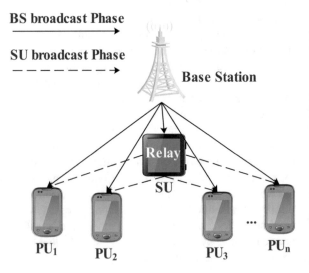

Fig. 4.11 Relay-aided broadcast using NC

have shown that this intelligent protocol efficiently utilizes the spectrum resources, increases the throughput and saves energy expenditure of BS in CRN.

In [97], the authors have proposed an intelligent broadcasting scheme based on adaptive dynamic network coding for effective utilization of spectrum resources in CRN. Basically, in this scheme, the PUs transmit the packets to common base station with the help of SU which acts as relay station for forwarding the packets to the base station. The SU or relay station utilizes the functionalities of ADNC and adaptive turbo trellis modulation scheme (ATTCM) for recovering the original packet from the encoded packet. Basically, the NC encodes the packets, and ATTCM facilitates not only NC but also enables the SU to adjust the modulation and code rate of channel coding according to near instantaneous channel conditions. Moreover, the feedback mechanism through ACK also facilitates the SU to retransmit the lost packets until all the packets are recovered by the base station. The simulation results have clearly shown that this scheme efficiently utilizes the spectrum resources, reduces the transmission and increases the throughput of the CRN. Moreover, the reduced transmission also creates more opportunities for SUs to transmit their packets. Additionally, the activation of adaptive encoder based on feedback flag related to success and failure of transmission also facilitates the SUs to extend its transmission in CRN for effective and efficient utilization of network resources. Finally, the simulation results also showed that the proposed scheme released almost 40% of PU bandwidth as compared to non-cooperative schemes.

In [56], the authors have presented the scheme in which the relay node facilitates the nodes to retransmit the lost packets during transmission. Basically, they have extended the work presented in [93] in which there is a direct single-hop link between the relay node and the destination nodes for transmission of packets with the exception that link quality between base station and relay node is better than lossy link between base station and receiver nodes. This proposed scheme is implemented in three phases. The base station transmits packets to both destination nodes and relay station in the first phase. Upon the reception of packets, the destination nodes and relay node send the ACK packet to the base station to inform it about packet lost. The encoded packets are retransmitted by base station to increase the efficiency of the second phase. Upon reception of packets, the user and relay node provide the acknowledgement about the lost packets again. Finally, in the third phase, the relay node only retransmits the coded packets which are lost in previous transmissions. The relay node is chosen in this phase because it has better links and connectivity as compared to direct link between base station and destination nodes.

4.7.2.3 Multi-hop

In [30], the authors have proposed adaptive NC (AdapCode) scheme that utilizes the functionalists of linear network coding for reducing transmission of packets and reliable packet transmission in multi-hop wireless network. This scheme exploits more coding opportunities for a large number of neighbours. Basically, they have shown that when there are more neighbours having available channels, then their scheme can adaptively increase the segment size that will exploit more coding opportunities for packet transmission. This feature allows more packets to be coded together which will result in less transmission of packets, and the delay is also decreased in the network. Therefore, AdapCode enables the node to change their segment size dynamically according to neighbours and topology of network to exploit more coding opportunities which results in fewer transmissions and effective utilization of network resources.

In [76], the authors have proposed the method to efficiently discover the neighbour node by using beacons for optimal creation of segments that will utilize the linear coding technique for efficient transmission of packets. Basically, this proposed scheme is the extension of AdapCode scheme with the difference of efficient neighbour discovery using network beacons for optimal segment creation to exploit coding opportunities. The performance of AdapCode scheme is dependent on efficient calculation of neighbouring node or segment density as it will create more coding opportunities. However, AdapCode scheme only discovers the active neighbours and cannot detect the neighbours which are not transmitting any packets. This problem is efficiently solved in [76] in which efficient neighbour discovery scheme by using beacons is proposed which will create more coding opportunities that will result in fewer transmission of packets and less delay in the network. In this scheme the neighbours are discovered, and then it follows all the phases of AdapCode for packet dissemination to the network.

In [96], the authors have proposed R-Code scheme based on minimum spanning tree approach for efficient data dissemination in multi-hop wireless mesh network with unreliable links. Basically, they have created a minimum spanning tree in which the weight of the link is equivalent to the expected transmission values. In this scheme, the parent nodes are responsible for delivering the packets to their child nodes, and relay nodes are the non-leaf nodes in the network. When the parent node sends the packets to child nodes, they will send the acknowledgement about the packet reception as well. This technique forced the non-leaf nodes in the tree to migrate to the new tree as soon as possible to reduce the broadcast delay in the network. Moreover, they have also utilized the opportunistic overhearing to acquire the information about the nodes for the reduction of transmission in the network.

In [12], the authors have proposed DutyCode technique which utilizes the NC along with duty-cycling technique to save the energy, transmission reduction and efficient utilization of spectrum resources. Basically, the combination of NC with duty-cycling enables the energy efficiency as NC exploits overhearing of wireless medium, while duty-cycling minimizes the idle listening which reduces overhearing for energy saving in WSNs. Moreover, they have also presented the solution of reducing NC redundancy in the network by identifying the periods of time in which the mobile nodes do not get benefit from overhearing of coded packets and have to switch to sleep mode. Basically in this cross-layer scheme, the MAC layer provides synchronization, random sleeping and streaming facilities while application layer determines the sleeping time and its duration. Basically, the request for switching the node to sleeping mode is placed by application layer, and MAC layer performs the task to put the sensor node to sleep mode or shut it down at user request if there is no pending transmission.

In [13], the authors have presented a scheme based on NC for broadcasting in multi-hop wireless network that prevents the deadlock. Basically, this scheme used NC to solve problem of many-to-all minimum transmission broadcast that calculates the lower bound for transmission. Moreover, they have proposed a distributed NC-based broadcast protocol which creates an efficient broadcast tree and guides each node for transmitting the NC packet. They also presented the algorithm which prevents the deadlock situation during transmission of packets.

References

1. Abdel-Rahman MJ, Rahbari H, Krunz M, Nain P (2013) Fast and secure rendezvous protocols for mitigating control channel DoS attacks. In: INFOCOM, 2013 proceedings IEEE, 2013. IEEE, pp 370–374
2. Ahlswede R, Cai N, Li S-Y, Yeung RW (2000) Network information flow. IEEE Trans Inf Theory 46(4):1204–1216
3. Akhtar F, Rehmani MH, Reisslein M (2016) White space: definitional perspectives and their role in exploiting spectrum opportunities. Telecommun Policy 40(4):319–331
4. Akyildiz IF, Lee W-Y, Vuran MC, Mohanty S (2006) NeXt generation/dynamic spectrum access/cognitive radio wireless networks: a survey. Comput Netw 50(13):2127–2159
5. Ali A, Panda M, Chahed T, Altman E (2013) Improving the transport performance in delay tolerant networks by random linear network coding and global acknowledgments. Ad Hoc Netw 11(8):2567–2587
6. Al-Mathehaji, Y., Boussakta, S., Johnston, M. and Hussein, J. (2016) Primary receiver-aware opportunistic broadcasting in cognitive radio ad hoc networks. In: Ubiquitous and Future Networks (ICUFN), 2016 Eighth international conference. IEEE, pp 30–35
7. Arachchige CJL, Venkatesan S, Chandrasekaran R, Mittal N (2011) Minimal time broadcasting in cognitive radio networks. In: International conference on distributed computing and networking, 2011. Springer, pp 364–375
8. Asterjadhi A, Zorzi M (2010) JENNA: a jamming evasive network-coding neighbor-discovery algorithm for cognitive radio networks [Dynamic Spectrum management]. IEEE Wirel Commun 17 (4), 24
9. Asterjadhi A, Baldo N, Zorzi M (2009) A distributed network coded control channel for multihop cognitive radio networks. IEEE Netw 23(4):26
10. Baldo N, Asterjadhi A, Zorzi M (2010) Dynamic spectrum access using a network coded cognitive control channel. IEEE Trans Wirel Commun 9(8):2575–2587
11. Bian K, Park J-M, Chen R A (2009) Quorum-based framework for establishing control channels in dynamic spectrum access networks. In: Proceedings of the 15th annual international conference on mobile computing and networking, ACM, pp 25–36
12. Chandanala R, Stoleru R (2010) Network coding in duty-cycled sensor networks. In: Networked Sensing Systems (INSS), 2010 seventh international conference on, IEEE, pp 203–210

13. Chang C-H, Kao J-C, Chen F-W Cheng SH (2014) Many-to-all priority-based network-coding broadcast in wireless multihop networks. In: Wireless Telecommunications Symposium (WTS) IEEE, pp 1–6
14. Chaoub A, Ibn-Elhaj E (2012) Multiple description coding for cognitive radio networks under secondary collision errors. In: Electrotechnical conference (MELECON), 2012 16th IEEE Mediterranean, 2012 IEEE, pp 27–30
15. Chatzigeorgiou I, Tassi A (2017) Decoding delay performance of random linear network coding for broadcast. IEEE Trans Veh Technol 66(8):7050–7060
16. Chen T, Zhang H, Maggio GM, Chlamtac I (2007) CogMesh: a cluster-based cognitive radio network. In: New frontiers in dynamic spectrum access networks, 2007. DySPAN 2007. 2nd IEEE international symposium on. IEEE, pp 168–178
17. Chen J, Lee VC, Liu K, Ali GMN, Chan E (2013) Efficient processing of requests with network coding in on-demand data broadcast environments. Inf Sci 232:27–43
18. Chun YJ, Hasna MO, Ghrayeb A (2015) Adaptive network coding for spectrum sharing systems. IEEE Trans Wirel Commun 14(2):639–654
19. Dong Z, Zhan C, Xu Y (2010) Delay aware broadcast scheduling in wireless networks using network coding. In: Networks Security Wireless Communications and Trusted Computing (NSWCTC) 2010 second international conference on IEEE, pp 214–217
20. Fang W, Liu F, Liu Z, Shu L, Nishio S (2011) Reliable broadcast transmission in wireless networks based on network coding. In: Computer Communications Workshops (INFOCOM WKSHPS), 2011 I.E. conference on IEEE, pp 555–559
21. Fanous A, Sagduyu YE, Ephremides A (2014) Reliable spectrum sensing and opportunistic access in network-coded communications. IEEE J Sel Areas Commun 32(3):400–410
22. Farooqi MZ, Tabassum SM, Rehmani MH, Saleem Y (2014) A survey on network coding: from traditional wireless networks to emerging cognitive radio networks. J Netw Comput Appl 46:166–181
23. Gheorghiu S, Lima L, Toledo AL, Barros J, Medard M (2010) On the performance of network coding in multi-resolution wireless video streaming. In: Network Coding (NetCod), 2010 I.E. international symposium on. IEEE, pp 1–6
24. Guirguis A, Guirguis R, Youssef M (2014) Primary user-aware network coding for multi-hop cognitive radio networks
25. Halloush M, Radha H (2009) Practical Network Coding for scalable video in error prone networks. In: Picture Coding Symposium. PCS 2009, 2009. IEEE, pp 1–4
26. Hatamnia S, Vahidian S, Aïssa S, Champagne B, Ahmadian-Attari M (2017) Network-coded two-way relaying in spectrum-sharing systems with quality-of-service requirements. IEEE Trans Veh Technol 66(2):1299–1312
27. Haykin S (2005) Cognitive radio: brain-empowered wireless communications. IEEE J Sel Areas Commun 23(2):201–220
28. Ho T, Lun D (2008) Network coding: an introduction. Cambridge University Press, Cambridge
29. Holland O (2016) Some are born with white space, some achieve white space, and some have white space thrust upon them. IEEE Trans Cogn Commun Netw 2(2):178–193
30. Hou I-H, Tsai Y-E, Abdelzaher TF, Gupta I (2008) Adapcode: adaptive network coding for code updates in wireless sensor networks. In: INFOCOM 2008. The 27th conference on computer communications. IEEE, 2008. IEEE, pp 1517–1525
31. Htike Z, Hong CS (2013) Broadcasting in multichannel cognitive radio ad hoc networks. In: Wireless Communications and Networking Conference (WCNC) IEEE, pp 733–737
32. Huang X-L, Wang G, Hu F, Kumar S (2011) Stability-capacity-adaptive routing for high-mobility multihop cognitive radio networks. IEEE Trans Veh Technol 60(6):2714–2729
33. Jamil F, Javaid A, Umer T, Rehmani MH (2017) A comprehensive survey of network coding in vehicular ad-hoc networks. Wirel Netw 23(8):2395–2414

34. Ji S, Beyah R, Cai Z (2013) Minimum-latency broadcast scheduling for cognitive radio networks. In: Sensor, Mesh and ad hoc Communications and Networks (SECON),10th annual IEEE communications society conference on IEEE, pp 389–397
35. Ji S, Beyah R, Cai Z (2015) Broadcast scheduling with latency and redundancy analysis for cognitive radio networks. IEEE Trans Veh Technol 64(7):3090–3097
36. Jia J, Zhang Q, Shen XS (2008) HC-MAC: a hardware-constrained cognitive MAC for efficient spectrum management. IEEE J Sel Areas Commun 26(1):106
37. Jiang J, Marić I, Goldsmith A, Cui S (2009) Achievable rate regions for broadcast channels with cognitive relays. In: Information Theory Workshop. ITW 2009. IEEE, pp 500–504
38. Katti S, Gollakota S, Katabi D (2007) Embracing wireless interference: analog network coding. ACM SIGCOMM Comput Commun Rev 37(4):397–408
39. Ke K, Xie Y, Liu Y, Hua C (2015) A cognitive two-way relay transmission scheme based on adaptive QAM and wireless network coding. In: New Technologies, Mobility and Security (NTMS), 7th international conference on IEEE, pp 1–4
40. Kondareddy YR (2010) Protocol design and performance issues in cognitive radio networks. Auburn University
41. Kondareddy YR, Agrawal P (2008) Selective broadcasting in multi-hop cognitive radio networks. In: Sarnoff Symposium IEEE, pp 1–5
42. Koutsonikolas D, Hu YC, Wang C-C, Comer M, Mohamed AMS (2011) Efficient online WiFi delivery of layered-coding media using inter-layer network coding. Distributed Computing Systems (ICDCS), 2011 31st international conference on IEEE:237–247
43. Kushwaha H, Xing Y, Chandramouli R, Heffes H (2008) Reliable multimedia transmission over cognitive radio networks using fountain codes. Proc IEEE 96(1):155–165
44. Lai L, El Gamal H, Jiang H, Poor HV (2011) Cognitive medium access: exploration, exploitation, and competition. IEEE Trans Mob Comput 10(2):239–253
45. Lazos L, Liu S, Krunz M (2009) Spectrum opportunity-based control channel assignment in cognitive radio networks. In: Sensor, mesh and ad hoc communications and networks, SECON'09. 6th annual IEEE communications society conference. IEEE, pp 1–9
46. Li L, Ramjee R, Buddhikot M, Miller S (2007) Network coding-based broadcast in mobile ad-hoc networks. In: INFOCOM 2007. 26th IEEE international conference on computer communications. IEEE, pp 1739–1747
47. Li X, Jiang T, Zhang Q, Wang L (2009) Binary linear multicast network coding on acyclic networks: principles and applications in wireless communication networks. IEEE J Sel Areas Commun 27(5):738
48. Li H, Dai H, Li C (2010) Collaborative quickest spectrum sensing via random broadcast in cognitive radio systems. IEEE Trans Wirel Commun 9(7):2338–2348
49. Li Y, Long H, Peng M, Wang W (2014) Spectrum sharing with analog network coding. IEEE Trans Veh Technol 63(4):1703–1716
50. Liang W, Nguyen HV, Ng SX, Hanzo L (2015) Network coding aided cooperative cognitive radio for uplink transmission. In: Global Communications Conference (GLOBECOM), 2015. IEEE, pp 1–6
51. Limmanee A, Dey S (2012) Optimal power policy and throughput analysis in cognitive broadcast networks under primary's outage constraint. In: Modeling and optimization in mobile, ad hoc and wireless networks (WiOpt), 10th international symposium on, IEEE, pp 391–397
52. Liu Z, Xu W, Li S, Lin J (2015) Network-coded primary-secondary cooperation in OFDM-based cognitive multicast networks. EURASIP J Wirel Commun Netw 2015(1):144
53. Lo BF (2011) A survey of common control channel design in cognitive radio networks. Phys Commun 4(1):26–39
54. Lou W, Wu J, (2003) On reducing broadcast redundancy in ad hoc wireless networks. In: System Sciences. Proceedings of the 36th annual Hawaii international conference on, IEEE, pp 10

55. Lu L, Xiao M, Skoglund M, Rasmussen L Wu G, Li S (2010) Efficient network coding for wireless broadcasting. In: Wireless Communications and Networking Conference (WCNC). IEEE, pp 1–6
56. Lu L, Xiao M, Rasmussen LK (2011) Relay-aided broadcasting with instantaneously decodable binary network codes. In: Computer Communications and Networks (ICCCN), proceedings of 20th international conference IEEE, pp 1–5
57. Manzoor MF, Qadir J, Baig A (2010) Broadcasting in cognitive wireless mesh networks with dynamic channel conditions. In: Emerging Technologies (ICET), 6th international conference on IEEE, pp 400–404
58. Mitola J, Maguire GQ (1999) Cognitive radio: making software radios more personal. IEEE Pers Commun 6(4):13–18
59. Mittal N, Krishnamurthy S, Chandrasekaran R, Venkatesan S, Zeng Y (2009) On neighbor discovery in cognitive radio networks. J Parallel Distrib Comput 69(7):623–637
60. Naeem A, Rehmani MH, Saleem Y, Rashid I, Crespi N (2017) Network coding in cognitive radio networks: a comprehensive survey. IEEE Commun Surv Tutorials 19(3):1945–1973
61. Nguyen D, Tran T, Nguyen T, Bose B (2009) Wireless broadcast using network coding. IEEE Trans Veh Technol 58(2):914–925
62. Nguyen D, Nguyen C, Duong-Ba T, Nguyen H, Nguyen A, Tran T (2017) Joint network coding and machine learning for error-prone wireless broadcast. In: Computing and Communication Workshop and Conference (CCWC). IEEE 7th annual, pp 1–7
63. Ostovari P, Khreishah A, Wu J (2012) Deadline-aware broadcasting in wireless networks with network coding. In: Global Communications Conference (GLOBECOM), 2012 IEEE. IEEE, pp 4435–4440
64. Ostovari P, Wu J, Khreishah A (2014) Network coding techniques for wireless and sensor networks. In: The art of wireless sensor networks. Springer, pp 129–162
65. Popovski P (2012) Communication-theoretic aspects of cognitive radio
66. Qadir J, Chou CT, Misra A, Lim JG (2009) Minimum latency broadcasting in multiradio, multichannel, multirate wireless meshes. IEEE Trans Mob Comput 8(11):1510–1523
67. Qadir J, Baig A, Ali A, Shafi Q (2014) Multicasting in cognitive radio networks: algorithms, techniques and protocols. J Netw Comput Appl 45:44–61
68. Qin C, Xian Y, Gray C, Santhapuri N, Nelakuditi S (2008) I²MIX: Integration of intra-flow and inter-flow wireless network coding. In: Sensor, Mesh and Ad Hoc Communications and Networks Workshops, SECON Workshops' 08. 5th IEEE Annual Communications Society Conference on. IEEE, pp 1–6
69. Qin Y, Zhong X, Yang Y, Li L, Wu F (2016) TCPJGNC: a transport control protocol based on network coding for multi-hop cognitive radio networks. Comput Commun 79:9–21
70. Rashid B, Rehmani MH, Ahmad A (2016) Broadcasting strategies for cognitive radio networks: taxonomy, issues, and open challenges. Comput Electr Eng 52:349–361
71. Rehmani MH, Viana AC, Khalife H, Fdida S (2013) Surf: a distributed channel selection strategy for data dissemination in multi-hop cognitive radio networks. Comput Commun 36 (10–11):1172–1185
72. Salameh HAB, El-Attar MF (2015) Cooperative OFDM-based virtual clustering scheme for distributed coordination in cognitive radio networks. IEEE Trans Veh Technol 64 (8):3624–3632
73. Saleem Y, Yau K-LA, Mohamad H, Ramli N, Rehmani MH (2015) SMART: a SpectruM-Aware ClusteR-based rouTing scheme for distributed cognitive radio networks. Comput Netw 91:196–224
74. Shahrasbi B, Rahnavard N (2011) Rateless-coding-based cooperative cognitive radio networks: design and analysis. In: Sensor, Mesh and Ad Hoc Communications and Networks (SECON), 8th annual IEEE communications society conference on IEEE, pp 224–232
75. Shu Z, Zhou J, Yang Y, Sharif H, Qian Y (2012) Network coding-aware channel allocation and routing in cognitive radio networks. Global Communications Conference (GLOBECOM). IEEE:5590–5595

76. Shwe HY, Adachi F (2011) Power efficient adaptive network coding in wireless sensor networks. In: Communications (ICC), IEEE international conference on IEEE, pp 1–5
77. Skevakis E, Lambadaris I, Halabian H (2017) Delay optimal scheduling for chunked random linear network coding broadcast. arXiv preprint arXiv:170602328
78. Song Y, Xie J (2014) QoS-based broadcast protocol under blind information in cognitive radio ad hoc networks. In: Broadcast design in cognitive radio ad hoc networks. Springer, Cham, pp 13–36
79. Song Y, Xie J (2015) BRACER: a distributed broadcast protocol in multi-hop cognitive radio ad hoc networks with collision avoidance. IEEE Trans Mob Comput 14(3):509–524
80. Song M, Wang J, Hao Q (2007) Broadcasting protocols for multi-radio multi-channel and multi-rate mesh networks. In: Communications. ICC'07. IEEE international conference. IEEE, pp 3604–3609
81. Stupia I, Vandendorpe L, Andreotti R, Lottici V (2012) A game theoretical approach for coded cooperation in cognitive radio networks. In: Communications Control and Signal Processing (ISCCSP), 5th international symposium IEEE, pp 1–6
82. Talay AC, Altilar DT (2011) United nodes: cluster-based routing protocol for mobile cognitive radio networks. IET Commun 5(15):2097–2105
83. Tessema WB, Kim B, Han K (2015) An asynchronous channel hopping sequence for rendezvous establishment in self organized cognitive radio networks. Wirel Pers Commun 81(2):649–659
84. Unnikrishnan J, Veeravalli VV (2008) Cooperative sensing for primary detection in cognitive radio. IEEE J Sel Top Sign Proces 2(1):18–27
85. Velmurugan P, Nandhini M, Thiruvengadam S (2015) Full duplex relay based cognitive radio system with physical layer network coding. Wirel Pers Commun 80(3):1113–1130
86. Wang F, Krunz M, Cui S (2008) Price-based spectrum management in cognitive radio networks. IEEE J Sel Top Sign Proces 2(1):74–87
87. Wang X, Chen W, Cao ZA (2009) Rateless coding based multi-relay cooperative transmission scheme for cognitive radio networks. GLOBECOM Glob Telecommun Conf, In, pp 1–6. https://doi.org/10.1109/GLOCOM.2009.5425269
88. Wang Z, Sagduyu YE, Jason HL, Zhang J (2010) Capacity and delay scaling laws for cognitive radio networks with routing and network coding. In: Military communication conference, Milcom IEEE, pp 1375–1380
89. Welsh DJ, Powell MB (1967) An upper bound for the chromatic number of a graph and its application to timetabling problems. Comput J 10(1):85–86
90. Wu T-Y, Liao W (2013) Time-efficient broadcasting in cognitive radio networks. In: Global Communications Conference (GLOBECOM) IEEE, pp 1191–1196
91. Xie X, Wang W (2013) Detecting primary user emulation attacks in cognitive radio networks via physical layer network coding. Proc Compt Sci 21:430–435
92. Xu Z, Hou X, Wei H (2012) Cooperative communication with physical-layer network coding in cognitive radio network. In: Instrumentation, Measurement, Computer, Communication and Control (IMCCC), 2012 Second international conference on IEEE, pp 1215–1219
93. Yang Y, Aïssa S (2014) Spectrum-sharing broadcast channels using fountain codes: energy, delay and throughput. IET Commun 8(14):2574–2583
94. Yang S, Wu J (2010) Efficient broadcasting using network coding and directional antennas in MANETs. IEEE Trans Parallel Distrib Syst 21(2):148–161
95. Yang S, Wu J, Dai F (2007) Efficient backbone construction methods in manets using directional antennas. In: Distributed computing systems, 2007. ICDCS'07. 27th international conference on. IEEE, pp 45–45
96. Yang Z, Li M, Lou W (2009) R-code: network coding based reliable broadcast in wireless mesh networks with unreliable links. In: Global telecommunications conference, 2009. GLOBECOM 2009. IEEE, pp 1–6
97. Yang Z, Li M, Lou W (2011) R-code: network coding-based reliable broadcast in wireless mesh networks. Ad Hoc Netw 9(5):788–798

98. Yang L, Z-y F, Zhang P (2012) Optimized in-band control channel with channel selection scheduling and network coding in distributed cognitive radio networks. J China Univ Posts Telecommun 19(2):48–56

99. Yang K, Xu W, Li S, Lin J (2013) A distributed multiple description coding multicast resource allocation scheme in OFDM-based cognitive radio networks. In: Wireless Communications and Networking Conference (WCNC), 2013 IEEE, pp 210–215

100. Yau KLA, Ramli N, Hashim W, Mohamad H (2014) Clustering algorithms for cognitive radio networks: a survey. J Netw Compt Appl 45:79–95

101. Zhan C, Xu Y (2010) Broadcast scheduling based on network coding in time critical wireless networks. In: Network Coding (NetCod), IEEE International symposium on, 2010. IEEE, pp 1–6

102. Zhang S, Liew S-C (2009) Channel coding and decoding in a relay system operated with physical-layer network coding. IEEE J Sel Areas Commun 27(5):788

103. Zhang R, Cui S, Liang Y-C (2009) On ergodic sum capacity of fading cognitive multiple-access and broadcast channels. IEEE Trans Inf Theory 55(11):5161–5178

104. Zhang J, Chen YP, Marsic I (2010a) MAC-layer proactive mixing for network coding in multi-hop wireless networks. Comput Netw 54(2):196–207

105. Zhang Q, Feng Z, Zhang P (2012) Efficient coding scheme for broadcast cognitive pilot channel in cognitive radio networks. In: Vehicular Technology Conference (VTC Spring), 2012 I.E. 75th. IEEE, pp 1–5

106. Zhang L, Ding G, Wu Q, Zou Y, Han Z, Wang J (2015) Byzantine attack and defense in cognitive radio networks: a survey. IEEE Commun Surv Tutorials 17(3):1342–1363

107. Zhao J, Zheng H, Yang GH (2007) Spectrum sharing through distributed coordination in dynamic spectrum access networks. Wirel Commun Mobile Compt 7(9):1061–1075

108. Zhao Z, Ding Z, Peng M, Wang W, Thompson JS (2015) On the design of cognitive-radio-inspired asymmetric network coding transmissions in MIMO systems. IEEE Trans Veh Technol 64(3):1014–1025

109. Zheng C, Dutkiewicz E, Liu RP, Vesilo R, Zhou Z (2012) Efficient network coding transmission in 2-hop multi-channel cognitive radio networks. In: Communications and Information Technologies (ISCIT), 2012 international symposium IEEE, pp 574–579

110. Zhong X, Qin Y, Yang Y, Li L (2014) CROR: coding-aware opportunistic routing in multi-channel cognitive radio networks. In: Global Communications Conference (GLOBECOM), 2014 IEEE, pp 100–105

111. Zhong X, Qin Y, Li L (2015) TCPNC-DGSA: efficient network coding scheme for TCP in multi-hop cognitive radio networks. Wirel Pers Commun 84(2):1243–1263

112. Zhong X, Lu R, Li L (2016) Social-based broadcast in cognitive radio networks: a network coding perspective. In: Internet of Things (iThings) and IEEE Green Computing and Communications (GreenCom) and IEEE Cyber, Physical and Social Computing (CPSCom) and IEEE Smart Data (SmartData), IEEE International conference, 2016 IEEE, pp 441–444

Chapter 5
Cooperative and Cognitive Hybrid Satellite-Terrestrial Networks

Vibhum Singh, Prabhat K. Upadhyay, Kyoung-Jae Lee, and Daniel Benevides da Costa

5.1 Introduction

The explosion of mobile applications and their integration in various aspects of everyday life necessitates the deployment of modern wireless systems that can handle such exponentially rising data traffic. We have already witnessed the evolution of communication technology from first generation (1G) to fourth generation (4G). High-speed broadband access, high capacity, low signal latency, long battery life time, wide coverage, etc. are the most important requirements to be considered for the deployment of the next-generation communication networks. Recently, satellite communication has gained significant attention owing to its numerous advantages. It has potential to offer services over a wide coverage area especially in the fields of broadcasting, navigation, and disaster relief, where the terrestrial network becomes unreachable or its deployment is economically not possible [1]. Moreover, satellite and terrestrial networks can be coalesced to harvest the benefits of both the systems [2]. For such a hybrid system, it is therefore important to find out an efficient way to share the resources of space-based networks with the terrestrial networks.

V. Singh (✉) · P. K. Upadhyay
Discipline of Electrical Engineering, Indian Institute of Technology Indore, Indore, Madhya Pradesh, India
e-mail: phd1701102002@iiti.ac.in; pkupadhyay@iiti.ac.in

K.-J. Lee
Department of Electronics and Control Engineering, Hanbat National University, Daejeon, South Korea
e-mail: kyoungjae@hanbat.ac.kr

D. B. da Costa
Department of Computer Engineering, Federal University of Ceará, Sobral, Ceará, Brazil
e-mail: danielbcosta@ieee.org

© Springer International Publishing AG, part of Springer Nature 2019
M. H. Rehmani, R. Dhaou (eds.), *Cognitive Radio, Mobile Communications and Wireless Networks*, EAI/Springer Innovations in Communication and Computing, https://doi.org/10.1007/978-3-319-91002-4_5

On another front, cooperative relaying has emerged as a prominent technique for terrestrial networks to provide wide coverage and high data rate services [3–6]. Recently, such communication techniques have found emerging applications in hybrid satellite-terrestrial networks (HSTNs) [7]. In fact, HSTNs could be impaired by the masking effect due to shadowing and obstacles, whereby the line-of-sight (LOS) path between satellite and end terrestrial users may be disrupted [8]. To redress this issue, a hybrid satellite-terrestrial relay system (HSTRN) has been proposed [2, 9] that employs terrestrial relay technique to aid the satellite transmission. The hybrid satellite-terrestrial architecture has been extensively used in Digital Video Broadcast-Satellite to a Handheld (DVB-SH) [10] standard using a geostationary earth orbit (GEO) satellite at S frequency band. However, the operational frequency band depends on various application scenarios and service requirements (like broadcasting, navigation, personal mobile communication, and disaster or emergency scenario) across different countries. With immense increase in wireless data traffic and limited availability of spectral resources, higher frequency bands (above 10 GHz), viz., Ku and Ka, are also needed to be allocated for mobile satellite services [8].

In a separate development, cognitive radio (CR) technology has been proposed for mitigating the spectrum scarcity problems in wireless communication systems. In CR networks, unlicensed secondary users (SUs) are benefited by sharing the spectrum with licensed primary users (PUs) network, provided that such spectrum sharing does not deteriorate the quality of service (QoS) of the primary network [11]. There are three main paradigms for spectrum sharing, namely, interweave, underlay, and overlay.

Interweave Spectrum Sharing In this paradigm, the SUs opportunistically access the unoccupied spectrum (also known as white spaces) of PUs without causing any interference to their transmission [12]. A lot of works have considered this paradigm of spectrum sharing for the performance assessment. For instance, in [13], the authors have discussed the dynamic spectrum access technique which allows the SUs to operate in the best available channel. Moreover, spectrum management process has been studied in CR to avoid interference in a better way [14]. In [15], the authors have investigated the spectrum opportunity (SOP) in terms of successful spectrum-aware communication between the communicating SUs in a dynamic radio environment under imperfections in the channel sensing for different network topologies. Recently, some works have considered utilization of CR technique into wireless sensor networks (WSNs). In addition, authors in [16] have considered a simulation model for cognitive radio sensor networks (CRSNs) with an intent to combine the useful properties of WSNs and CR networks. Authors have also tested their model by performing various experiments in different scenarios of CRSNs. Channel bonding (CB) technique has been studied to increase the bandwidth and reduce delays in wireless network [17], and hereby, a number of issues and challenges regarding CB in CRSN were highlighted. However, such spectrum sharing technique is highly sensitive to sensing errors and PU traffic patterns.

Underlay Spectrum Sharing In an underlay model [18], SUs can share the spectrum with PUs by satisfying the interference power criterion toward the PUs. In contrast to the interweave model, the underlay model has the advantage that the SUs can directly occupy licensed spectrum without considering the behavior of the PU's traffic patterns. Underlay CR network has been extensively studied in literature since it requires the least hardware complexity compared with the other two models for spectrum sharing [19, 20]. However, since the transmit power at the SU is restricted, the performance of the underlay cognitive network is severely affected in comparison with its noncognitive counterpart. Hence, to improve the performance of the SUs, relaying technique has been introduced to form a cognitive relay system. Several models have been proposed by using decode-and-forward (DF) and amplify-and-forward (AF) protocols [21–25]. However, the performance of secondary system may be very limited, especially when SUs lie in close proximity to the PUs.

Overlay Spectrum Sharing In this paradigm of spectrum sharing, SUs may be allowed to transmit over the spectrum owned by the PUs in exchange for the cooperation to the PU's signal transmission on a priority basis [26]. Thereby, in contrast to underlay model, overlay paradigm does not pose stringent transmit power restrictions to the SUs. Several existing works have considered the overlay systems for the performance analysis. For instance, in [27], the authors have proposed a two-phase protocol based on cooperative DF relaying for a secondary system to achieve spectrum access along with a primary system. Moreover, in [28], a scheme based on cooperative AF relaying with multiple antennas has been proposed which involves the design of antenna weights and power allocation to satisfy the QoS of both the PUs and SUs.

Related Works

HSTRN has emerged as a potential candidate for providing wide coverage and seamless data connectivity, especially when the LOS communication between satellite and end user is disrupted due to masking effect. Plenty of works have investigated the performance of such HSTRNs using AF [29] and DF [30, 31] relaying protocols in a single-user scenario. In contrast, multiuser relay network is the most widely accepted architecture for wireless communication systems wherein the communication between a source and multiple destinations/users is assisted through a relay [32]. Such architecture has been implemented in many standards like IEEE 802.11s and IEEE 802.16j [33]. Thereby, the HSTRN has been extended to a multiuser scenario in order to fulfill the high-throughput service requirements of a large number of terrestrial users [34, 35]. Specifically, in [34], a multiuser HSTRN has been studied where a user is selected opportunistically to exploit multiuser diversity. Further, to improve the performance, authors in [35] have considered a multiuser multi-relay architecture for HSTRN. However, these works assumed that the channel state information (CSI) has been acquired perfectly to facilitate the user selection process. But, in practice, the CSI for user selection may be outdated due to high feedback latency associated with satellite communication. Further, with dense frequency reuse in wireless networks, there is a significant degradation in the performance of the HSTRN due to co-channel interference (CCI). Although few works [31, 36] have analyzed the performance of HSTRN by considering the impact of CCI, they are restricted to the single-user scenarios. Different from these works, in this chapter, the performance of multiuser HSTRN is analyzed by considering both outdated CSI and CCI.

Additionally, integration of spectrum sharing models into HSTRN can provide significant performance improvements for the futuristic wireless networks in terms of spectral efficiency and transmission reliability [37, 38]. In few works [39, 40], the authors have considered interweave hybrid cognitive satellite-terrestrial network (HCSTN). In some more recent works, authors considered underlay HCSTN where a secondary satellite network shares the spectrum with a primary terrestrial network [41–44]. For instance, the authors in [41] have proposed a power allocation scheme for such networks and analyzed the effective capacity performance. The carrier, power, and bandwidth allocation schemes for these cognitive networks were elaborated in [42]. Under the constraints of interference power in underlay models, the authors in [43, 44] have investigated, respectively, the outage performance and optimal power control schemes. Different from these works, overlay models for hybrid satellite-terrestrial systems have been presented in [45, 46], where a secondary terrestrial wireless system gets an opportunity to access the spectrum with a licensed primary satellite user. In this chapter, the performance of an overlay cognitive radio model for HSTRNs is analyzed. The overlay paradigm of spectrum sharing has an absolute importance in satellite communication since it can bring network coverage even when the primary LOS satellite path is disrupted due to shadowing and obstacles. Further, performing the radio resource and interference management remotely in space segment would be a tedious task. Thereby, the overlay model becomes a promising candidate, essentially, where the secondary system can directly cooperate with existing primary network.

Organization of Book Chapter The rest of the chapter is composed of the following sections. In Sect. 5.2, a multiuser HSTRN model will be discussed and its performance will be analyzed over pertinent channel fading scenarios. Then, in Sect. 5.3, an overlay model of multiuser HCSTN will be comprehensively investigated. Some critical issues and future challenges will be discussed in Sect. 5.4, especially for the implementation of HCSTNs in practice. Finally, in Sect. 5.5, various important conclusions are drawn.

Modeling and Performance Analysis The main focus of this chapter will be on illuminating the system performance in terms of outage probability (OP). Specifically, in Sect. 5.2, the performance of a multiuser HSTRN system is analyzed by considering two inevitable imperfections, viz., outdated CSI and CCI. Further, in Sect. 5.3, an overlay HCSTN is investigated in which multiple terrestrial secondary networks compete for spectrum access opportunity with a primary satellite network. Note that all the analytical results in the numerical results sections are obtained by using symbolic software Mathematica. In addition, for the veracity of analytical results, the simulation results are obtained using the MATLAB.

5.2 Multiuser Hybrid Satellite-Terrestrial Relay Network

In this section, various aspects of a multiuser HSTRN are discussed for its performance assessment and possible implementation in realistic scenarios.

5.2.1 System Model

As depicted in Fig. 5.1, a multiuser HSTRN is considered, wherein a satellite transmitter exchanges the information with one out of other K terrestrial users $\{D_k\}_{k=1}^K$ by employing an AF-based terrestrial relay R. All the network nodes have a single antenna, and satellite and relay are assumed to be stationary while the users are mobile. There is no LOS transmission between satellite and terrestrial users due to heavy shadowing. Further, it is assumed that the signal at relay R is corrupted by M co-channel interferers $\{I_i\}_{i=1}^M$ and additive white Gaussian noise (AWGN), whereas each user is influenced by AWGN only. AWGN is assumed to have mean zero and variance σ^2.

The overall communication takes place in two orthogonal time slots by employing an opportunistic user selection scheme. The appropriate selection criteria will be discussed later in the section. During the first-time slot, the satellite broadcasts its signal x_s (having unit energy) to the relay node R. Consequently, the received signal at R can be given as

$$y_r = \sqrt{P_s}h_{sr}x_s + \sum_{i=1}^M \sqrt{P_i}h_{ir}x_i + n_r, \tag{5.1}$$

where P_s denotes the transmit power at the satellite, h_{sr} is the channel coefficient of the link between the satellite and relay, P_i is the transmit power of the ith interferer, h_{ir} is the channel coefficient of the link between i^{th} interferer and relay, x_i is the signal (having unit energy) of ith interferer, and n_r is the AWGN at the relay.

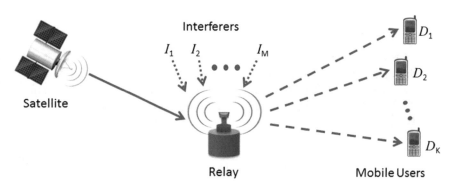

Fig. 5.1 HSTRN model. First phase (→), second phase (- - -▶), and interference (→)

During the second-time slot, the relay amplifies the received signal y_r with gain G given by

$$G = \sqrt{\frac{P_r}{P_s|h_{sr}|^2 + \sum_{i=1}^{M} P_i|h_{ir}|^2 + \sigma^2}}, \qquad (5.2)$$

and then forwards it to the selected user (say D_k). In Eq. (5.2), P_r denotes the transmit power at the relay node. After amplification, the received signal at the kth user can be written as

$$y_{d,k} = Gh_{rd,k}y_r + n_{d,k}, \qquad (5.3)$$

where $h_{rd,k}$ represents the channel coefficient of the link between relay and kth user and $n_{d,k}$ is the AWGN at the kth user. With the impact of dominant CCI, the noise power at the relay is assumed to be negligible as compared to the total interference power [19, 32]. With this, using Eqs. (5.1) and (5.2) into Eq. (5.3), the instantaneous end-to-end signal-to-interference-plus-noise ratio (SINR) at the kth user can be written as

$$\gamma_{sd,k} = \frac{\gamma_s\gamma_{d,k}}{\gamma_s + (\gamma_{d,k}+1)\gamma_c}, \qquad (5.4)$$

where $\gamma_s = \eta_s|h_{sr}|^2$, $\gamma_{d,k} = \eta_r|h_{rd,k}|^2$, and $\gamma_c = \sum_{i=1}^{M}\eta_i|h_{ir}|^2$ with $\eta_s = P_s/\sigma^2$, $\eta_r = P_r/\sigma^2$, and $\eta_i = P_i/\sigma^2$.

To exploit the benefits of multiuser diversity in the considered network, an opportunistic user selection scheme is adopted wherein the transmissions are scheduled on the basis of channel quality of multiple users. For this, the relay selects the user based on the best relay-user link and provides the index of selected user back to the satellite. The instantaneous signal-to-noise ratio (SNR) of the relay-user link can be expressed as

$$\gamma_d = \max_{k\in\{1,\cdots,K\}} \gamma_{d,k}. \qquad (5.5)$$

In practice, there may exist a delay in the feedback process, i.e., CSI obtained by the relay at the selection instant could be different from the one available during the data transmission and thereby the opportunistic scheduling is performed based on old CSI. Therefore, the actual end-to-end SINR pertaining to the scheduled user can be given by

$$\gamma_{sd} = \frac{\gamma_s\gamma_d^o}{\gamma_s + (\gamma_d^o+1)\gamma_c}, \qquad (5.6)$$

where γ_d^o is the delayed version of γ_d. Let $\gamma_{d,k}^o$ be the delayed version of $\gamma_{d,k}$ and is given by $\gamma_{d,k}^o = \eta_r |h_{rd,k}^o|^2$, where $h_{rd,k}^o$ is the delayed version of the CSI for the user-relay link. The relation between $h_{rd,k}^o$ and $h_{rd,k}$ can be given as [32, 47]

$$h_{rd,k} = \rho h_{rd,k}^o + \sqrt{1 - \rho^2}\varepsilon, \tag{5.7}$$

where ε is an independent and identically distributed with $|h_{rd,k}^o|$ and ρ is the correlation coefficient between $h_{rd,k}^o$ and $h_{rd,k}$ which is given by $\rho = J_0^2(2\Pi f_d \tau)$ with $J_0(.)$ being the zeroth order Bessel function of the first kind ([48], Eq. 8.411), f_d is the Doppler frequency, and τ is the time delay in feedback.

5.2.2 Channel Models

Considering Shadowed-Rician fading model for the satellite link, the probability density function (PDF) of $|h_{sr}|^2$ is given by [29]:

$$f_{|h_{sr}|^2}(x) = \alpha e^{-\beta x} {}_1F_1(m_s; 1; \delta x), x \geq 0, \tag{5.8}$$

where $\alpha = (2bm_s/(2bm_s + \Omega_s))^{m_s}/2b$, $\beta = 1/2b$, and $\delta = \Omega_s/(2b)(2bm_s + \Omega_s)$ with Ω_s and $2b$ the respective average power of the LOS and multipath components, m_s the fading severity parameter, and ${}_1F_1(.;.;.)$ the confluent hypergeometric function of first kind ([48], Eq. 9.210.1). Based on Eq. (5.8), the PDF and cumulative density function (CDF) of γ_s can be obtained by making a change of variables, $\gamma_s = \eta_s |h_{sr}|^2$, as

$$f_{\gamma_s}(x) = \alpha \sum_{k=0}^{m_s-1} \frac{\varsigma(k)}{(\eta_s)^{k+1}} x^k e^{-\left(\frac{\beta-\delta}{\eta_s}\right)x} \tag{5.9}$$

and

$$F_{\gamma_s}(x) = 1 - \alpha \sum_{k=0}^{m_s-1} \frac{\varsigma(k)}{(\eta_s)^{k+1}} \sum_{p=0}^{k} \frac{k!}{p!} \left(\frac{\beta-\delta}{\eta_s}\right)^{-(k+1-p)} x^p e^{-\left(\frac{\beta-\delta}{\eta_s}\right)x}, \tag{5.10}$$

where $\varsigma(k) = (-1)^k (1-m_s)_k \delta^k/(k!)^2$ and $(.)_k$ is the Pochhammer symbol.

For the terrestrial links with a cluster of K users, a Nakagami-m fading environment is considered for the analysis purpose. If the channel coefficients $h_{rd,k}$ follow Nakagami-m fading distribution with fading severity m_t and average power Ω_t, the PDF and CDF of SNR $\gamma_{d,k}$ are given, respectively, by

$$f_{\gamma_{d,k}}(x) = \left(\frac{m_t}{\Omega_t \eta_r}\right)^{m_t} \frac{x^{m_t-1}}{\Gamma(m_t)} e^{-\frac{m_t x}{\Omega_t \eta_r}} \tag{5.11}$$

and

$$F_{\gamma_{d,k}}(x) = \frac{1}{\Gamma(m_t)} \Upsilon\left(m_t, \frac{m_t x}{\Omega_t \eta_r}\right), \tag{5.12}$$

where $\Upsilon(\cdot, \cdot)$ and $\Gamma(\cdot)$ represent, respectively, the lower incomplete and the complete gamma functions ([48], Eqs. 8.310.1 and 8.350.1).

If there are M equal power interferers ($\eta_1 = \ldots = \eta_M = \eta_c$) with independent and identically distributed fading channels, the PDF of γ_c under Nakagami-m fading, with its parameters as m_c and Ω_c, can be given as

$$f_{\gamma_c}(\omega) = \left(\frac{m_c}{\Omega_c \eta_c}\right)^{m_c M} \frac{\omega^{m_c M - 1}}{\Gamma(m_c M)} e^{-\frac{m_c \omega}{\Omega_c \eta_c}}. \tag{5.13}$$

5.2.3   Statistical Characterizations

Here, PDF of γ_d^o is formulated. For this, we first obtain the PDF of γ_d. As such, by applying order statistics, $F_{\gamma_d}(x) = \left[F_{\gamma_{d,k}}(x)\right]^K$, the PDF of γ_d can be expressed as

$$f_{\gamma_d}(x) = K\left[F_{\gamma_{d,k}}(x)\right]^{K-1} f_{\gamma_{d,k}}(x), x \geq 0. \tag{5.14}$$

The PDF in Eq. (5.14) can be further simplified by using Eq. (5.11) and Eq. (5.12) as

$$f_{\gamma_d}(x) = K \sum_{j=0}^{K-1} \binom{K-1}{j} \frac{(-1)^j}{\Gamma(m_t)} \sum_{l=0}^{j(m_t-1)} \left(\frac{m_t}{\Omega_t \eta_r}\right)^{m_t+l} \omega_l^j \, x^{m_t+l-1} e^{-\frac{m_t(j+1)x}{\Omega_t \eta_r}}, \tag{5.15}$$

where the coefficients ω_l^j, for $0 \leq l \leq j(m_t - 1)$, can be evaluated recursively (with $\varepsilon_l = \frac{1}{l!}$) as $\omega_0^j = (\varepsilon_0)^j$, $\omega_1^j = j(\varepsilon_1)$, $\omega_{j(m_t-1)}^j = (\varepsilon_{m_t-1})^j$, $\omega_l^j = \frac{1}{l\varepsilon_0} \sum_{g=1}^{l} [gj - l + g] \, \varepsilon_g \omega_{l-g}^j$ for $2 \leq l \leq m_t - 1$, and $\omega_l^j = \frac{1}{l\varepsilon_0} \sum_{g=1}^{m_t-1} [gj - l + g] \, \varepsilon_g \omega_{l-g}^j$ for $m_t \leq l \leq j(m_t - 1)$.

Since γ_d^o and γ_d are correlated Gamma-distributed random variables, the PDF of γ_d^o can be obtained as

$$f_{\gamma_d^o}(x) = \int_0^\infty f_{\gamma_d^o|\gamma_d}(x|y) f_{\gamma_d}(y) dy, \tag{5.16}$$

where $f_{\gamma_d^o|\gamma_d}(x|y)$ is the conditional PDF of γ_d^o, conditioned on γ_d. It can be given by [49]:

$$f_{\gamma_d^o|\gamma_d}(x|y) = \frac{1}{1-\rho}\left(\frac{m_t}{\Omega_t \eta_r}\right)\left(\frac{x}{\rho y}\right)^{\frac{m_t-1}{2}} e^{-\frac{m_t(\rho y+x)}{(1-\rho)\Omega_t \eta_r}} I_{m_t-1}\left(\frac{2m_t\sqrt{\rho xy}}{(1-\rho)\Omega_t \eta_r}\right), \quad (5.17)$$

where $I_\nu(.)$ is the νth order modified Bessel function of the first kind ([48], Eq. 8.406.1). On substituting Eq. (5.15) and Eq. (5.17) in Eq. (5.16), and simplifying the result using [50], we obtain

$$f_{\gamma_d^o}(x) = K \sum_{j=0}^{K-1} \binom{K-1}{j}\frac{(-1)^j}{\Gamma(m_t)} \theta_j^{j(m_t-1)} \sum_{l=0}^{j} \omega_l^j \sum_{i=0}^{l} \binom{l}{i}\left(\frac{m_t}{\Omega_t \eta_r}\right)^{i+m_t} \xi_{i,l,j} x^{i+m_t-1} e^{-\frac{x}{\theta_j}}, \quad (5.18)$$

with $\xi_{i,l,j} = \frac{(l+m_t-1)!\rho^i(1-\rho)^{l-i}}{(i+m_t-1)![j(1-\rho)+1]^{l+i+m_t}}$ and $\theta_j = \frac{[j(1-\rho)+1]\Omega_t \eta_r}{m_t(j+1)}$.

5.2.4 Outage Performance Analysis

In this section, the OP and its asymptotic behavior at high SNR are examined for the considered multiuser HSTRN in the presence of CCI and feedback delay.

5.2.4.1 Exact Outage Probability

The OP can be defined as the probability that the instantaneous SINR at the receiving node falls below a certain threshold γ_{th}. For the considered HSTRN, it can be formulated as

$$P_{out}(\gamma_{th}) = \Pr[\gamma_{sd} < \gamma_{th}], \quad (5.19)$$

which can be further written using Eq. (5.6) as

$$P_{out}(\gamma_{th}) = \Pr\left[\frac{\gamma_s \gamma_d^o}{\gamma_s + (\gamma_d^o + 1)\gamma_c} < \gamma_{th}\right]. \quad (5.20)$$

The OP in Eq. (5.20) can be re-expressed in terms of expectation over γ_c (after some manipulations) as

$$P_{out}(\gamma_{th}) = E_{\gamma_c}\left[\int_0^{\gamma_{th}} f_{\gamma_d^o}(x)dx + \int_0^{\infty} F_{\gamma_s}\left(\frac{\gamma_{th}(x+1)\gamma_c}{x-\gamma_{th}}\right)f_{\gamma_d^o}(x)dx\right], \quad (5.21)$$

which can be further simplified to represent

$$P_{\text{out}}(\gamma_{\text{th}}) = 1 - E_{\gamma_c}[\varphi(\gamma_{\text{th}}, \gamma_c)], \tag{5.22}$$

where $\varphi(\gamma_{\text{th}}, \gamma_c)$ is defined by an integral as

$$\varphi(\gamma_{\text{th}}, \gamma_c) = \int_{\gamma_{\text{th}}}^{\infty} \left[1 - F_{\gamma_s}\left(\frac{\gamma_{\text{th}}(x+1)\gamma_c}{x - \gamma_{\text{th}}}\right)\right] f_{\gamma_d^o}(x) dx. \tag{5.23}$$

Now, invoking the respective CDF and PDF expressions from Eq. (5.10) and Eq. (5.18) into Eq. (5.23), and simplifying the result using ([48], Eqs. 1.111 and 3.471.9), we obtain

$$
\begin{aligned}
\varphi(\gamma_{\text{th}}, \gamma_c) = \ & 2\alpha K \sum_{k=0}^{m_s-1} \frac{\varsigma(k)}{(\eta_s)^{k+1}} \sum_{p=0}^{k} \frac{k!}{p!} \left(\frac{\beta-\delta}{\eta_s}\right)^{-(k+1-p)} \sum_{j=0}^{K-1} \binom{K-1}{j} \\
& \times \frac{(-1)^{jj(m_t-1)}}{\Gamma(m_t)} \sum_{l=0}^{l} \omega_l^j \sum_{i=0}^{l} \binom{l}{i} \left(\frac{m_t}{\Omega_t \eta_r}\right)^{i+m_t} \xi_{i,l,j} \ e^{-\frac{\gamma_{\text{th}}}{\theta_j}} \sum_{q=0}^{p} \binom{p}{q} \sum_{v=0}^{m_t+i-1} \binom{m_t+i-1}{v} \\
& \times \left(\left(\frac{\beta-\delta}{\eta_s}\right)\theta_j\right)^{\frac{v-q+1}{2}} \gamma_{\text{th}}^{m_t+p+i-v-1+\left(\frac{v-q+1}{2}\right)} (\gamma_{\text{th}}+1)^{q+\frac{v-q+1}{2}} \\
& \times \gamma_c^{p+\frac{v-q+1}{2}} e^{-\left(\frac{\beta-\delta}{\eta_s}\right)\gamma_{\text{th}} \gamma_c} \mathcal{K}_{v-q+1}\left(2\sqrt{\left(\frac{\beta-\delta}{\eta_s}\right)\frac{\gamma_{\text{th}} \gamma_c (\gamma_{\text{th}}+1)}{\theta_j}}\right),
\end{aligned}
\tag{5.24}
$$

where $\mathcal{K}_v(.)$ is the modified Bessel function of second kind ([48], Eq. 8.432.6).

Finally, on inserting Eq. (5.24) into Eq. (5.22) and then taking the expectation over γ_c, using the PDF from Eq. (5.13), with the aid of ([48], Eq. 6.631.3), expression of $P_{\text{out}}(\gamma_{\text{th}})$ can be obtained as

$$
\begin{aligned}
P_{\text{out}}(\gamma_{\text{th}}) = \ & 1 - \alpha K \sum_{k=0}^{m_s-1} \frac{\varsigma(k)}{(\eta_s)^{k+1}} \sum_{p=0}^{k} \frac{k!}{p!} \left(\frac{\beta-\delta}{\eta_s}\right)^{-(k+1-p)} \sum_{j=0}^{K-1} \binom{K-1}{j} \\
& \times \frac{(-1)^{jj(m_t-1)}}{\Gamma(m_t)} \sum_{l=0}^{l} \omega_l^j \sum_{i=0}^{l} \binom{l}{i} \left(\frac{m_t}{\Omega_t \eta_r}\right)^{i+m_t} \xi_{i,l,j} \ e^{-\frac{\gamma_{\text{th}}}{\theta_j}} \sum_{q=0}^{p} \binom{p}{q} \sum_{v=0}^{m_t+i-1} \binom{m_t+i-1}{v} \\
& \times \left(\frac{\beta-\delta}{\eta_s}\right)^{-p-m_cM} \left(\frac{m_c}{\Omega_c \eta_c}\right)^{m_cM} \frac{1}{\Gamma(m_cM)} (\theta_j)^{p+m_cM+v-q+1} \\
& \times \gamma_{\text{th}}^{m_t+i-v-m_cM-1} (\gamma_{\text{th}}+1)^{q-p-m_cM} \Gamma(1+v-q+p+m_cM)\Gamma(p+m_cM) \\
& \times e^{\left(\frac{\beta-\delta}{\eta_s}\right)\frac{\gamma_{\text{th}}(\gamma_{\text{th}}+1)}{2\theta_j \vartheta}} \times \left(\left(\frac{\beta-\delta}{\eta_s}\right)\frac{\gamma_{\text{th}}(\gamma_{\text{th}}+1)}{\theta_j \vartheta}\right)^{\frac{1}{2}[2(p+m_cM)+v-q]} \\
& \times \mathcal{W}_{-\frac{1}{2}[2(p+m_cM)+v-q], \frac{1}{2}(v-q+1)}\left(\left(\frac{\beta-\delta}{\eta_s}\right)\frac{\gamma_{\text{th}}(\gamma_{\text{th}}+1)}{\theta_j \vartheta}\right),
\end{aligned}
\tag{5.25}
$$

where $\vartheta = \left(\frac{\beta-\delta}{\eta_s}\right)\gamma_{\text{th}} + \frac{m_c}{\Omega_c \eta_c}$ and $\mathcal{W}_{u,v}(.)$ are the Whittaker functions ([48], Eq. 9.221.1).

5.2.4.2 Asymptotic Outage Probability

To further delve into the system performance, Eq. (5.25) can be approximated at high SNR ($\eta_s, \eta_r \rightarrow \infty$) as

$$
P_{out}(\gamma_{th}) \approx \alpha \gamma_{th} M \Omega_c \left(\frac{\eta_c}{\eta_s} \right)
$$
$$
+ K \sum_{j=0}^{K-1} \binom{K-1}{j} \frac{(-1)^j}{\Gamma(m_t)} \left(\frac{m_t}{\Omega_t \eta_r} \right)^{m_t} \frac{\gamma_{th}^{m_t}}{[j(1-\rho)+1]^{m_t}} \quad . \tag{5.26}
$$

From Eq. (5.26), one can obtain the diversity order for the considered system with ($\rho < 1$) under the following cases:

(i) When $\eta_c \ll \eta_s$ and $m_t = 1$, the achievable diversity order is one.
(ii) When $\eta_c \ll \eta_s$ and $m_t > 1$, the achievable diversity order is still one.
(iii) When $\eta_c = \eta_s$, the diversity order becomes zero.

5.2.5 Numerical Results

In this section, numerical and simulation results are presented to demonstrate the effect of key parameters on the performance of considered system. For this, various parameters are set as $m_t = 1$, $\Omega_t = 1$, $m_c = 1$, $\Omega_c = 0.01$, and $\gamma_{th} = 0$ dB with $\eta_s = \eta_r$ as transmit SNR. The satellite link is subject to two different fading extents, as shown in Table 5.1.

In Fig. 5.2, OP versus SNR curves are drawn for various parameters (K, ρ, M, η_c) under both average and heavy shadowing scenarios. The analytical and asymptotic curves are drawn using the derived expression in Eqs. (5.25 and 5.26), respectively. From these curves, it can be observed that the system has a unity diversity order if the interference power η_c satisfies the inequality constraint $\eta_c \ll \eta_s$. It can also be witnessed that the system performance is badly affected if either the interference power or the number of interferers increases. Apparently, the performance is further deteriorated for the case of outdated CSI ($\rho = 0.6$ or 0.8). Expectedly, the overall system performance degrades severely when the satellite link undergoes heavy shadowing. Moreover, the effect on the coding gain of the system, by various parameters, can be clearly visualized by the relative shift in the curves.

Figure 5.3 illustrates the OP versus SNR curves for the considered HSTRN to obtain some useful design insights. Herein, the number of users is kept as $M = 2$,

Table 5.1 Corresponding parameters of shadowed Rician fading model (N. I. [51])

Fading scenario	m	b	Ω
Heavy shadowing	1	0.063	0.0007
Average shadowing	5	0.251	0.279

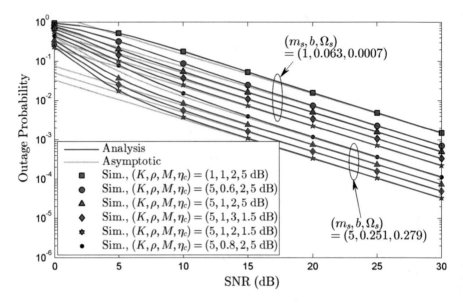

Fig. 5.2 Outage performance for considered HSTRN under low CCI level

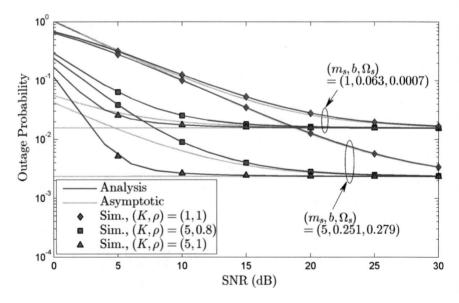

Fig. 5.3 Outage performance for considered HSTRN under high CCI level

while a scenario of high CCI level such that $\frac{\eta_s}{\eta_c} = 10$ dB is considered. It can be observed that high interference level induces zero diversity outage floors, irrespective of the value of the parameter ρ. More importantly, when CSI is outdated for a fixed value of K, the performance deteriorates significantly. Further, the outage performance improves dramatically as the number of users K increases from one to five. One can realize this by observing an increment in the slope of OP curves.

5.3   Multiuser Hybrid Cognitive Satellite-Terrestrial Network

An overlay cognitive radio paradigm is of vital importance in satellite communication since it can bring network coverage even when the primary direct satellite (DS) link is disrupted due to shadowing and obstacles. In this section, an overlay HCSTN is described as shown in Fig. 5.4, wherein a satellite transmitter communicates with its intended receiver via multiple secondary relays lying inside the satellite footprint [52, 53]. Hereby, the comprehensive investigation of the performance for the considered overlay HCSTN is carried out by adopting Shadowed-Rician fading for satellite links and Nakagami-m fading for terrestrial links.

5.3.1   System Model

As depicted in Fig. 5.4, an overlay HCSTN is considered, wherein the primary and secondary networks coexist together. The primary network comprises a satellite

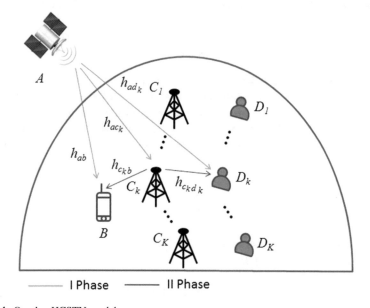

Fig. 5.4 Overlay HCSTN model

transmitter A and its intended terrestrial receiver B, whereas the secondary network consists of multiple terrestrial transmitter-receiver pairs:

$$C_k - D_k, k = 1, \ldots, K.$$

The node A is considered to be a GEO satellite, while nodes B, C_k, and D_k are fixed terrestrial service terminals. These terminals are equipped with the wideband omnidirectional antennas for the signal transmission and/or reception. The Doppler spread at all the terrestrial nodes is assumed to be negligible owing to the static nature of these nodes. Herein, all the secondary transmitters (STs) seek the spectrum access opportunity to communicate with their intended receivers. As such, these STs may be allowed to transmit over the spectrum owned by the primary network in exchange for the cooperation to the primary satellite transmission on a priority basis. Further, it is assumed that the satellite links experience Shadowed-Rician fading, while the terrestrial links undergo Nakagami-m fading. The channel coefficients for the links $A{\rightarrow}B$, $A{\rightarrow}C_k$, $A{\rightarrow}D_k$, $C_k{\rightarrow}B$, and $C_k{\rightarrow}D_k$ are represented by h_{ab}, h_{ac_k}, h_{ad_k}, $h_{c_k b}$, and $h_{c_k d_k}$, respectively. All the receiving nodes are influenced by AWGN with mean zero and variance σ^2.

Hereby, the overall communication takes place in two-time phases by using an AF-based relay cooperation. For this, a secondary network pair is appropriately selected (say $C_k - D_k$). The selection criteria will be discussed later in the next subsection. From this selected pair, the ST node C_k serves as an AF relay for the primary transmission while at the same time being the transmitter for its own intended secondary receiver D_k. In the first phase, satellite A transmits a unit energy signal x_a to B, which is also received by secondary users C_k and D_k. Consequently, the received signals y_{ab}, y_{ac_k}, and y_{ad_k} at B, C_k, and D_k, respectively, can be obtained by

$$y_{ai} = \sqrt{P_a} h_{ai} x_a + n_{ai}, \tag{5.27}$$

where $i \in \{b, c_k, d_k\}$, P_a is the transmit power at A, and n_{ai} is the AWGN. In the second phase, the node C_k amplifies and forwards the received primary signal y_{ac_k} to node B while simultaneously transmitting its own information signal x_{c_k} to node D_k. For this concurrent transmission, C_k splits its transmit power P_c in order to superimpose its signal x_{c_k} with the primary signal y_{ac_k} to generate a network-coded signal given as

$$z_{c_k} = \sqrt{\mu P_c} \frac{y_{ac_k}}{\sqrt{|y_{ac_k}|^2}} + \sqrt{(1 - \mu)P_c} x_{c_k}, \tag{5.28}$$

where $\mu \in (0, 1)$ is the power allocation factor. Thus, the received signals $y_{c_k b}$ and $y_{c_k d_k}$ at nodes B and D_k, respectively, from C_k can be given by

$$y_{c_k j} = h_{c_k j} z_{c_k} + n_{c_k j}, \tag{5.29}$$

where $j \in \{b, d_k\}$ and $n_{c_k j}$ are the AWGN. As such, the resulting SNRs at B via direct and relay links can be given, respectively, as

$$\Lambda_{ab} = \eta_a |h_{ab}|^2 \tag{5.30}$$

and

$$\Lambda_{ac_k b} = \frac{\mu \Lambda_{ac_k} \Lambda_{c_k b}}{(1 - \mu) \Lambda_{ac_k} \Lambda_{c_k b} + \Lambda_{ac_k} + \Lambda_{c_k b} + 1}, \tag{5.31}$$

where $\Lambda_{ac_k} = \eta_a |h_{ac_k}|^2$, $\Lambda_{c_k b} = \eta_c |h_{c_k b}|^2$ with $\eta_a = P_a/\sigma^2$, and $\eta_c = P_c/\sigma^2$. From Eq. (5.29), it can be observed that the received signal at D_k involves a primary signal component x_a as interference which can be successfully removed [27] using the primary signal decoded in the first transmission phase. Thus, the resultant SNR at D_k can be expressed as

$$\Lambda_{ac_k d_k} = \frac{(1 - \mu) \Lambda_{c_k d_k} (\Lambda_{ac_k} + 1)}{\mu \Lambda_{c_k d_k} + \Lambda_{ac_k} + 1}, \tag{5.32}$$

where $\Lambda_{c_k d_k} = \eta_c |h_{c_k d_k}|^2$. It is worth mentioning that on substituting $\mu = 1$, the SNR expression in Eq. (5.31) reduces to the case of conventional hybrid satellite-terrestrial AF relay system without spectrum sharing [34, 35] and thereby the end-to-end SNR at node D_k for secondary communication in Eq. (5.32) vanishes completely.

5.3.2 Criteria for Secondary Network Selection

In this subsection, we discuss the criteria for the selection of the best secondary network ($C_{k^*} - D_{k^*}$ Pair). Essentially, the best secondary pair should be selected in such a way that it improves the QoS of primary network. Herein, based on the availability of the CSI, two selection schemes can be employed, i.e., partial secondary network selection (PSNS) and opportunistic secondary network selection (OSNS). For instance, when the CSI of only $A - C_k$ links are available, PSNS scheme can be designed as

$$k^* = \arg \max_{k \in \{1, \cdots, K\}} \Lambda_{ac_k}. \tag{5.33}$$

Thus, when CSI of all the $A - C_k$ and $C_k - B$ links are available, the OSNS scheme can be formulated as

$$k^* = \arg \max_{k \in \{1, \cdots, K\}} \Lambda_{ac_k b}. \tag{5.34}$$

Note that the PSNS scheme maximizes only the first hop SNR, while OSNS maximizes the end-to-end SNR. Nevertheless, both these schemes are designed to minimize the OP of primary network. Herein, it is assumed that the required CSI is perfectly available for employing these schemes in the considered HCSTN.

5.3.3 Channel Models

Under Shadowed-Rician fading model for the satellite links, the PDF of $|h_{ai}|^2$, for $i \in \{b, c_k, d_k\}$, is given by [29]:

$$f_{|h_{ai}|^2}(x) = \alpha_i e^{-\beta_i x} {}_1F_1(m_{ai}; 1; \delta_i x), x \geq 0, \tag{5.35}$$

where $\alpha_i = (2b_{ai}m_{ai}/(2b_{ai}m_{ai} + \Omega_{ai}))^{m_{ai}}/2b_{ai}$, $\beta_i = 1/2b_{ai}$, and $\delta_i = \Omega_{ai}/(2b_{ai})$ $(2b_{ai}m_{ai} + \Omega_{ai})$, with Ω_{ai} and $2b_{ai}$ being the respective average powers of the LOS and multipath components, m_{ai} the fading severity parameter, and ${}_1F_1(.; .; .)$ the confluent hypergeometric function of first kind. After simplifying and making a change of variables, the PDF and CDF of $\Lambda_{ai} = \eta_a |h_{ai}|^2$ are given, respectively, by

$$f_{\Lambda_{ai}}(x) = \alpha_i \sum_{k=0}^{m_{ai}-1} \frac{\varsigma(k)}{(\eta_a)^{k+1}} x^k e^{-\left(\frac{\beta_i - \delta_i}{\eta_a}\right)x} \tag{5.36}$$

and

$$F_{\Lambda_{ai}}(x) = 1 - \alpha_i \sum_{k=0}^{m_{ai}-1} \frac{\varsigma(k)}{(\eta_a)^{k+1}} \sum_{p=0}^{k} \frac{k!}{p!} \left(\frac{\beta_i - \delta_i}{\eta_a}\right)^{-(k+1-p)} x^p e^{-\left(\frac{\beta_i - \delta_i}{\eta_a}\right)x}, \tag{5.37}$$

where $\varsigma(k) = (-1)^k (1 - m_{ai})_k \delta^k/(k!)^2$ and $(.)_k$ are the Pochhammer symbols.

For terrestrial links, the Nakagami-m fading channel for analyzing the performance of HCSTN is considered. If the channel coefficient $h_{c_k j}$, for $j \in \{b, d_k\}$, follows Nakagami-m distribution with fading severity $m_{c_k j}$ and average power $\Omega_{c_k j}$, the PDF and CDF of $\Lambda_{c_k j}$ are given, respectively, by

$$f_{\Lambda_{c_k j}}(x) = \left(\frac{m_{c_k j}}{\Omega_{c_k j}\eta_c}\right)^{m_{c_k j}} \frac{x^{m_{c_k j}-1}}{\Gamma(m_{c_k j})} e^{-\frac{m_{c_k j}x}{\Omega_{c_k j}\eta_c}} \tag{5.38}$$

and

$$F_{\Lambda_{c_k j}}(x) = \frac{1}{\Gamma(m_{c_k j})} \Upsilon\left(m_{c_k j}, \frac{m_{c_k j}x}{\Omega_{c_k j}\eta_c}\right). \tag{5.39}$$

5.3.4 *Performance Analysis of the Primary Network*

In this subsection, the performance of the primary network in terms of OP is investigated. Further, asymptotic OP expression is derived to assess the diversity order. Furthermore, the power allocation policy for spectrum sharing between primary and secondary network is also discussed.

5.3.4.1 For Direct Satellite (DS) Transmission Only

For a target rate R_p, the OP of primary network with only DS transmission is given by

$$P_{out}^{DS}(R_p) = Pr\left[\log_2(1 + \Lambda_{ab}) < R_p\right] = F_{\Lambda_{ab}}\left(\gamma_p'\right), \tag{5.40}$$

where $\gamma_p' = 2^{R_p} - 1$. The CDF $F_{\Lambda_{ab}}(x)$ in Eq. (5.40) can be expressed using Eq. (5.37) as

$$F_{\Lambda_{ab}}(x) = 1 - \alpha_b \sum_{k=0}^{m_{ab}-1} \frac{\varsigma(k)}{(\eta_a)^{k+1}} \sum_{p=0}^{k} \frac{k!}{p!} \left(\frac{\beta_b - \delta_b}{\eta_a}\right)^{-(k+1-p)} x^p e^{-\left(\frac{\beta_b - \delta_b}{\eta_a}\right)x}, \tag{5.41}$$

which can be further approximated at high SNR as

$$F_{\Lambda_{ab}}(x) \approx \frac{\alpha_b}{\eta_a} x. \tag{5.42}$$

It can be clearly seen from Eq. (5.42) that achievable diversity order is one irrespective of the fading parameter m_{ab}.

5.3.4.2 For Spectrum Sharing with DS Transmission

With the application of maximal-ratio combining (MRC), using Eqs. (5.30) and (5.31), the OP of primary network in the considered HCSTN with the best selected pair $(C_{k*} - D_{k*})$ is given by

$$P_{out}(R_p) = Pr\left[\frac{1}{2}\log_2(\Lambda_{ab} + \Lambda_{ac_k*b}) < R_p\right]$$
$$= Pr\left[\Lambda_{ab} + \Lambda_{ac_k*b} < \gamma_p\right], \tag{5.43}$$

where $\gamma_p = 2^{2R_p} - 1$. Further, Eq. (5.43) can be evaluated as

$$P_{out}(R_p) = \int_0^{\gamma_p} \int_0^{\gamma_p - v} f_{\Lambda_{ac_k * b}}(u) f_{\Lambda_{ab}}(v) du dv. \tag{5.44}$$

As such, it is rather difficult to obtain an exact closed-form solution of Eq. (5.44). Therefore, the I-step staircase approximation to the actual triangular integral region is used [54] in Eq. (5.44) to express

$$P_{out}(R_p) \approx \sum_{i=0}^{I-1} \left\{ F_{\Lambda_{ab}}\left(\frac{i+1}{I}\gamma_p\right) - F_{\Lambda_{ab}}\left(\frac{i}{I}\gamma_p\right) \right\} F_{\Lambda_{ac_k * b}}\left(\frac{I-i}{I}\gamma_p\right). \tag{5.45}$$

Evaluation of Eq. (5.45) requires only the CDFs of Λ_{ab} and $\Lambda_{ac_k * b}$. The respective CDFs of $\Lambda_{ac_k * b}$ for the PSNS and OSNS scheme can be expressed [53] as shown in Eqs. (5.46) and (5.48):

$$F_{\Lambda_{ac_k * b}}(x) = \begin{cases} \Phi_1(x), & \text{if } x < \mu' \\ 1, & \text{if } x \geq \mu' \end{cases}, \tag{5.46}$$

where $\Phi_1(x)$ is given by

$$
\begin{aligned}
\Phi_1(x) = 1 &- \frac{2}{\Gamma(m_{cb})}\left(\frac{m_{cb}}{\Omega_{cb}\eta_c}\right)^{m_{cb}} \sum_{n=1}^{K} C_n^K (-1)^{n-1} \\
&\times (\alpha_c)^n e^{-\frac{x}{\theta_x}\left(\frac{n(\beta_c - \delta_c)}{\eta_a} + \frac{m_{cb}}{\Omega_{cb}\eta_c}\right)} \sum_{S_m \in S} \frac{n!}{\prod_{m=0}^{m_{ac}-1} s_m!} \\
&\times \prod_{0 \leq m \leq m_{ac}-1} (A_m)^{s_m} \sum_{q=0}^{\Delta_{ac}} C_q^{\Delta_{ac}} \sum_{r=0}^{m_{cb}-1} C_r^{m_{cb}-1} \left(\frac{1}{\theta_x}\right)^{m_{cb}+\Delta_{ac}} \\
&\times x^{m_{cb}+\Delta_{ac}-\left(\frac{r+q+1}{2}\right)}(x+\theta_x)^{\frac{r+q+1}{2}}\left(\frac{n(\beta_c - \delta_c)\Omega_{cb}\eta_c}{\eta_a m_{cb}}\right)^{\frac{r-q+1}{2}} \\
&\times \mathcal{K}_{r-q+1}\left(2\sqrt{n\left(\frac{x^2}{\theta_x}+x\right)\frac{(\beta_c - \delta_c)m_{cb}}{\Omega_{cb}\eta_c\eta_a\theta_x}}\right),
\end{aligned}
\tag{5.47}
$$

with $\mu' = \frac{\mu}{1-\mu}, \theta_x = \mu - (1-\mu)x, -$

$S = \left\{ S_m \middle| \sum_{m=0}^{m_{ac}-1} s_m = n \right\}$, containing non-negative integers

$\{s_m\}, A_m = \sum_{k=m}^{m_{ac}-1} \frac{\varsigma(k)}{(\eta_a)^{k+1}} \frac{k!}{m!}\left(\frac{\beta_c - \delta_c}{\eta_a}\right)^{-(k+1-m)}, \Delta_{ac} = \sum_{m=0}^{m_{ac}-1} m s_m, C_i^l = \frac{l!}{(l-i)!i!}$ is the binomial coefficient, and $\mathcal{K}_v(.)$ is the modified Bessel function of the second kind ([48], Eq. 432.6):

$$F_{\Lambda_{ac_{k*b}}}(x) = [\Phi_2(x)]^K, \text{ for } x < \mu' \tag{5.48}$$

where $\Phi_2(x)$ is given by

$$
\begin{aligned}
\Phi_2(x) = {} & 1 - 2\alpha_c e^{-\frac{x}{\theta_x}\left(\frac{\beta_c - \delta_c}{\eta_a} + \frac{m_{cb}}{\Omega_{cb}\eta_c}\right)} \sum_{k=0}^{m_{ac}-1} \frac{\varsigma(k)}{(\eta_a)^{k+1}} \\
& \times \sum_{p=0}^{k} \frac{k!}{p!}\left(\frac{\beta_c - \delta_c}{\eta_a}\right)^{-(k+1-p)} \frac{1}{\Gamma(m_{cb})}\left(\frac{m_{cb}}{\Omega_{cb}\eta_c}\right)^{m_{cb}} \\
& \times \sum_{q=0}^{p} C_q^p \sum_{r=0}^{m_{cb}-1} C_r^{m_{cb}-1}\left(\frac{1}{\theta_x}\right)^{m_{cb}+p} x^{m_{cb}+p-\left(\frac{r-q+1}{2}\right)} \\
& \times (x + \theta_x)^{\frac{r+q+1}{2}}\left(\frac{(\beta_c-\delta_c)\Omega_{cb}\eta_c}{\eta_a m_{cb}}\right)^{\frac{r-q+1}{2}} \mathcal{K}_{r-q+1}\left(2\sqrt{\left(\frac{x^2}{\theta_x}+x\right)\frac{(\beta_c-\delta_c)m_{cb}}{\Omega_{cb}\eta_c\eta_a\theta_x}}\right).
\end{aligned}
\tag{5.49}
$$

Finally, on invoking the respective CDF expressions from Eqs. (5.46) and (5.48) along with Eq. (5.41) into Eq. (5.45), respective OPs can be evaluated for both PSNS and OSNS schemes.

At high SNR, CDFs for $F_{\Lambda_{ai}}(x)$ and $F_{\Lambda_{c_k j}}(x)$ can be approximated, respectively, as

$$F_{\Lambda_{ai}}(x) = \frac{\alpha_i}{\eta_a}x \tag{5.50}$$

and

$$F_{\Lambda_{c_k j}}(x) = \frac{1}{\Gamma(m_{c_k j}+1)}\left(\frac{m_{c_k j}x}{\Omega_{c_k j}\eta_c}\right)^{m_{c_k j}}. \tag{5.51}$$

Now, one can utilize these approximated CDF expressions to evaluate the asymptotic OP for PSNS and OSNS, respectively, as

$$
P_{\text{out}}(R_p) \approx \sum_{i=0}^{I-1}\left(\frac{\alpha_b\gamma_p}{I\eta_a}\right)\left[\left(\frac{\alpha_c(I-i)\gamma_p}{I\theta_{\frac{L-i}{I}\gamma_p}\eta_a}\right)^K + \frac{1}{\Gamma(m_{cb}+1)}\left(\frac{m_{cb}(I-i)\gamma_p}{\Omega_{cb}I\theta_{\frac{L-i}{I}\gamma_p}\eta_c}\right)^{m_{cb}}\right]
\tag{5.52}
$$

and

$$
P_{\text{out}}(R_p) \approx \sum_{i=0}^{I-1}\left(\frac{\alpha_b\gamma_p}{I\eta_a}\right)\left[\frac{\alpha_c(I-i)\gamma_p}{I\theta_{\frac{L-i}{I}\gamma_p}\eta_a} + \frac{1}{\Gamma(m_{cb}+1)}\left(\frac{m_{cb}(I-i)\gamma_p}{\Omega_{cb}I\theta_{\frac{L-i}{I}\gamma_p}\eta_c}\right)^{m_{cb}}\right]^K, \tag{5.53}
$$

where $\theta_{(.)} = \mu - (1 - \mu)(.)$. From Eqs. (5.52) and (5.53), it can be observed that the achievable diversity order for the primary network with PSNS and OSNS is 1 + min (K, m_{cb}) and 1 + K, respectively.

5.3.4.3 Constrained Power Allocation Policy for Spectrum Sharing

Herein is the power allocation policy for ST to obtain the appropriate value of μ. Recalling the condition $x < \mu'$ from [53], the dynamic range of power allocation factor for certain threshold γ_p of the primary network can be calculated as $\frac{\gamma_p}{1+\gamma_p} \leq \mu \leq 1.$. Further, based on the QoS constraint, the ST could determine the effective value of μ such that the primary OP for HCSTN lies below or equal to that for DS transmission only scheme. Consequently, the value of μ can be numerically obtained based on the following condition:

$$P_{out}(R_p) \leq P_{out}^{DS}(R_p). \tag{5.54}$$

As such, a smaller value of μ can provide more spectrum sharing opportunities for the secondary network.

5.3.4.4 Numerical Results

In this section, numerical and simulation results are presented to demonstrate the effect of key parameters on the performance of HCSTN. For this, various parameters are set as $R_p = 0.5$ bps/Hz, so that $\gamma'_p = 0.414$, $\gamma_p = 1$, $\Omega_{cb} = 1$, and $\eta_a = \eta_c$ as the SNR. Link A–B is subject to heavy shadowing with its parameters as $\{m_{ab}, b_{ab}, \Omega_{ab}\} = \{2, 0.063, 0.0005\}$, whereas the link A–C_k experiences average shadowing with its parameters as $\{m_{ac}, b_{ac}, \Omega_{ac}\} = \{5, 0.251, 0.279\}$, as provided in Table 5.1 in Sect. 5.2.5. Further, the number of steps is kept as $I = 50$ to make relative approximation error negligible.

In Fig. 5.5, the OP versus SNR curves are plotted, by setting $\mu = 0.75$, under Nakagami-m faded terrestrial links. Specifically, analytical and asymptotic OP curves are drawn for both PSNS and OSNS schemes. For the comparison purpose, the OP curves for the benchmark DS transmission only scheme are also plotted. It can be observed that there is a significant performance improvement due to exploitation of diversity. Further, the diversity order of 1 + min (K, m_{cb}) and 1 + K, respectively, for PSNS and OSNS can be verified from the pertinent curves.

Figure 5.6 demonstrates the feasible values of power allocation factor μ for the spectrum sharing in HCSTN. For this, the SNR is kept as 15 dB. In accordance with the spectrum sharing condition given in Eq. (5.53), one can obtain the value of μ^* at the intersection points between the curves of considered HCSTN (for PSNS and OSNS) and DS transmission only. For $\gamma_p = 1$, the acceptable limit of μ^* is $0.5 < \mu^* < 1$ based on $\frac{\gamma_p}{1+\gamma_p} \leq \mu \leq 1$. Therefore, all the OP curves exhibit unity value below

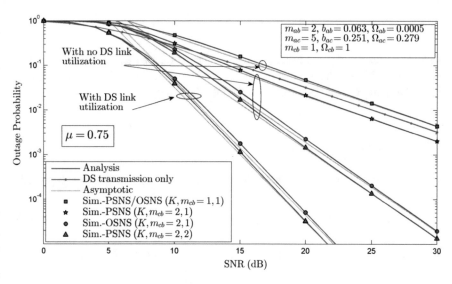

Fig. 5.5 OP of primary network against SNR

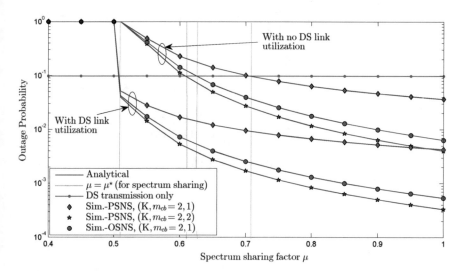

Fig. 5.6 OP of primary network against μ

0.5. It can be further observed that without DS link, for a given set of parameters, the μ^* takes lower values with OSNS as compared to PSNS. Moreover, with DS link utilization μ^* quickly attains its minimum possible value for both PSNS and OSNS. This can be attributed to the harvested diversity gain due to exploitation of DS link.

5.3.5 Performance Analysis of Secondary Network

In this section, numerical investigation is conducted for the secondary network of the considered HCSTN. The OP expression is evaluated and then it is further approximated at high SNR to gain more insight into the system behavior.

5.3.5.1 Outage Performance

For a given target rate R_s, the OP of the selected secondary network can be given by

$$P_{out}(R_s) = Pr\left[\frac{1}{2}\log_2(1 + \Lambda_{ac_{k*}d_{k*}}) < R_s\right], \tag{5.55}$$

which can be expressed using Eq. (5.32), after some manipulations, as

$$P_{out}(R_s) = Pr\left[\frac{\mu\Lambda_{c_{k*}d_{k*}}(1 + \Lambda_{ac_{k*}})}{\mu\Lambda_{c_{k*}d_{k*}} + \Lambda_{ac_{k*}} + 1} < \mu'\gamma_s\right], \tag{5.56}$$

where $\gamma_s = 2^{2R_s} - 1$. To evaluate the $P_{out}(R_s)$ in Eq. (5.56), the bound $\frac{XY}{X+Y} < \min(X, Y)$ [55] is used to approximate Eq. (5.56) as

$$P_{out}(R_s) \approx Pr[\min(\mu\Lambda_{c_{k*}d_{k*}}, \Lambda_{ac_{k*}} + 1) < \mu'\gamma_s]. \tag{5.57}$$

Hereby, the CDF of $\Lambda_{ac_{k*}} + 1$ can be expressed by a change in variables as

$$F_{\Lambda_{ac_{k*}}+1}(x) = \begin{cases} 0, & \text{if } x < 1 \\ F_{\Lambda_{ac_{k*}}}(x-1), & \text{if } x \geq 1 \end{cases}. \tag{5.58}$$

Since the CDF in Eq. (5.58) is discontinuous for $x \in [0, \infty)$, the OP in Eq. (5.57) can be represented as

$P_{out}(R_s)$

$$\approx \begin{cases} F_{\mu\Lambda_{c_{k*}d_{k*}}}(\mu'\gamma_s), & \text{if } \gamma_s < \frac{1}{\mu'} \\ F_{\mu\Lambda_{c_{k*}d_{k*}}}(\mu'\gamma_s) + F_{\Lambda_{ac_{k*}}}(\mu'\gamma_s - 1) - \left[F_{\mu\Lambda_{c_{k*}d_{k*}}}(\mu'\gamma_s)F_{\Lambda_{ac_{k*}}}(\mu'\gamma_s - 1)\right], & \text{if } \gamma_s \geq \frac{1}{\mu'} \end{cases}, \tag{5.59}$$

where $\mu' = \frac{\mu}{1-\mu}$.

Further, based on PSNS in Eq. (5.33), we have $F_{\Lambda_{ac_{k*}}}(x) = \left[F_{\Lambda_{ac_k}}(x)\right]^K$ as derived in Eq. (63) [53]. However, since the PSNS criterion does not follow order statistics, the CDF of $\mu\Lambda_{c_{k*}d_{k*}}$ can be expressed using Eq. (5.39) as

$$F_{\mu\Lambda_{c_{k*}d_{k*}}}(x) = F_{\Lambda_{c_k d_k}}\left(\frac{x}{\mu}\right) = \frac{1}{\Gamma(m_{cd})}\Upsilon\left(m_{cd},\frac{m_{cd}x}{\Omega_{cd}\eta_c\mu}\right). \tag{5.60}$$

Thus, making use of CDFs from Eq. (5.60) and Eq. (63) [53] into Eq. (5.59), one can obtain the OP of the secondary network with PSNS scheme.

For OSNS case, the required statistics of $\Lambda_{ac_{k*}}$ depends upon the channel conditions of dual hops of primary network as derived in Eq. (5.50) [53]. Further, the CDF of $\mu\Lambda_{c_{k*}d_{k*}}$ for OSNS does not follow order statistics, and it remains same as obtained in (Eq. (5.60) for PSNS. Hence, on invoking the respective CDFs from Eq. (5.60) and Eq. (5.50) [53] into Eq. (5.59), one can obtain the OP of the secondary network with OSNS scheme.

To further delve into the system performance, the OP can be approximated for PSNS and OSNS at high SNR, respectively, as

$$P_{out}(R_s) \approx \begin{cases} \dfrac{1}{\Gamma(m_{cd}+1)}\left[\dfrac{m_{cd}}{\Omega_{cd}\eta_c}\left(\dfrac{\gamma_s}{1-\mu}\right)\right]^{m_{cd}}, \text{if}\,\gamma_s < \dfrac{1}{\mu'} \\[3mm] \left[\dfrac{\alpha_c}{\eta_c}(\mu'\gamma_s-1)\right]^K + \dfrac{1}{\Gamma(m_{cd}+1)}\left[\dfrac{m_{cd}}{\Omega_{cd}\eta_c}\left(\dfrac{\gamma_s}{1-\mu}\right)\right]^{m_{cd}}, \text{if}\,\gamma_s \geq \dfrac{1}{\mu'} \end{cases}. \tag{5.61}$$

$$P_{out}(R_s)$$

$$\times \approx \begin{cases} \dfrac{1}{\Gamma(m_{cd}+1)}\left[\dfrac{m_{cd}}{\Omega_{cd}\eta_c}\left(\dfrac{\gamma_s}{1-\mu}\right)\right]^{m_{cd}}, \text{if}\,\gamma_s < \dfrac{1}{\mu'}\,\text{and}\,m_{cd} \geq 1 \\[3mm] \left(1+\dfrac{\eta_a}{\alpha_c\Omega_{cd}\eta_c}\right)^{K-1}\left[\dfrac{\alpha_c}{\eta_a}(\mu'\gamma_s-1)\right]^K + \dfrac{1}{\Gamma(m_{cd}+1)}\left[\dfrac{m_{cd}}{\Omega_{cd}\eta_c}\left(\dfrac{\gamma_s}{1-\mu}\right)\right]^{m_{cd}}, \text{if}\,\gamma_s \geq \dfrac{1}{\mu'}\,\text{and}\,m_{cd} = 1 \\[3mm] \left[\dfrac{\alpha_c}{\eta_a}(\mu'\gamma_s-1)\right]^K + \dfrac{1}{\Gamma(m_{cd}+1)}\left[\dfrac{m_{cd}}{\Omega_{cd}\eta_c}\left(\dfrac{\gamma_s}{1-\mu}\right)\right]^{m_{cd}}, \text{if}\,\gamma_s \geq \dfrac{1}{\mu'}\,\text{and}\,m_{cd} > 1 \end{cases}. \tag{5.62}$$

From Eqs. (5.61) and (5.62), it can be deduced that the achievable diversity order for a high-rate requirement is min (K, m_{cd}), whereas for the low-rate requirement, the system's diversity solely depends upon the fading parameter m_{cd}.

5.3.5.2 Numerical Results

In Fig. 5.7, the OP curves of the secondary network are plotted for the considered HCSTN under the setting $\mu = 0.75$, $\Omega_{cd} = 1$. It is evident from the figure that for a fixed value of μ, the OP of the secondary network is almost the same for both the PSNS and OSNS schemes. From the respective curves, the diversity order of the secondary network can be verified. For example, in the case of high-rate require-ment, i.e., $\gamma_s = 1$, one can manifestly observe the increase in slope of OP curves, when (K, m_{cd}) changes from (2,1) to (2,2). On the contrary, for low-rate requirement, i.e., $\gamma_s = 0.3$, the diversity order becomes m_{cd} only, irrespective of the other parameters.

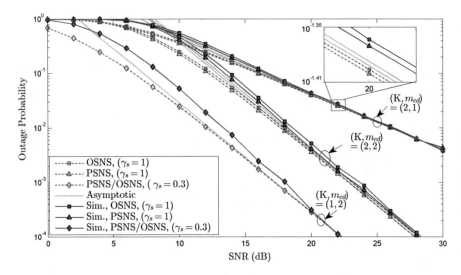

Fig. 5.7 OP of secondary network against SNR

5.4 Issues and Challenges

The major issues and challenges for the implementation and design of the considered system can be summarized as follows:

- One of the main issues with satellite communication is high latency. This latency depends on the distance between satellite and terrestrial nodes and the speed of light. For instance, the latency of a GEO satellite communication network is almost 20 times that of terrestrial link-based network. Moreover, the feedback delay in the signal reception would further increase the latency to a great extent. Thus, modeling and designing of low-latency satellite communication system is quite challenging.
- Satellite links are prone to fast variations due to atmospheric fluctuations and tropospheric scattering. As a consequence, it becomes very hard to acquire the perfect CSI. The main aspect of designing the considered system relies on the atmosphere attenuation and the fading. Thereby, for the analysis purpose, the presence of perfect CSI has been assumed.
- For the presented analysis, it is assumed that the transceiver hardware of the network nodes is perfect. However, in practice, the hardware suffers from various types of impairments such as I/Q imbalances, phase noises, and high-power amplifier nonlinearities. Incorporating the hardware impairments for system performance assessment is interesting and challenging as well.
- For offloading the data traffic from the existing bands, the researchers are also dwelling over the possibility of exploiting higher frequency bands, viz., Ku and Ka. Nevertheless, it is quite challenging to transmit over such high frequencies due to severe effects of atmospheric turbulence and scattering.

5.5   Conclusion

HSTN has evolved in recent years as a prominent candidate to provide the wide coverage, higher throughput, and seamless data connectivity to cope with the demands of the futuristic wireless networks. In this context, this chapter dealt with multiuser configurations for a HSTRN and for a HCSTN to enhance the radio spectrum utilization and coverage by integrating cooperative spectrum sharing techniques into satellite communication systems. The chapter demonstrated the outage behavior of a multiuser HSTRN employing opportunistic user selection scheme with the help of an AF-based relay. Moreover, the outage performance is analyzed under two inevitable imperfections, viz., outdated CSI and CCI. It is revealed that the achievable diversity order is either one or zero depending upon the interference level at the relay, even when the CSI is outdated. Further, an overlay HCSTN is investigated in which multiple terrestrial secondary networks compete for spectrum access with a primary satellite network. For this system, two complexity-aware schemes, viz., PSNS and OSNS, are discussed to facilitate dynamic spectrum sharing through an AF-based relay cooperation. By analyzing the OP performance, it is highlighted that the primary network can achieve full diversity with secondary network selection under the bottleneck effects of Shadowed Rician channels of pertinent links. It is observed that, in contrast to PSNS, OSNS can explore more spectrum sharing opportunities with a broader coverage for secondary network. Finally, the impact of key parameters is illustrated which eventually may be useful for designing the practical systems.

References

1. Vanelli-Coralli A et al (2007) Satellite communications: research trends and open issues. Paper presented at proceedings international workshop on satellite and space communications, University of Salzburg, Salzburg, Austria, 13–14 September 2007
2. Evans B et al (2005) Integration of satellite and terrestrial systems in future media communications. IEEE Trans Wirel Commun 12(5):72–80. https://doi.org/10.1109/MWC.2005.1522108
3. Laneman JN et al (2004) Cooperative diversity in wireless networks: efficient protocols and outage behavior. IEEE Trans Inf Theory 50(12):3062–3080. https://doi.org/10.1109/TIT.2004.838089
4. Bletsas A et al (2006) A simple cooperative diversity method based on network path selection. IEEE J Sel Areas Commun 24(3):659–672. https://doi.org/10.1109/JSAC.2005.862417
5. Krikidis I et al (2008) Amplify-and-forward with partial relay selection. IEEE Commun Lett 12(4):235–237. https://doi.org/10.1109/LCOMM.2008.071987
6. Yang Y et al (2009) Relay technologies for WiMax and LTE-advanced mobile systems. IEEE Commun Mag 47(10):100–105. https://doi.org/10.1109/MCOM.2009.5273815
7. Paillassa B et al (2011) Improving satellite services with cooperative communications. Int J Satell Commun Netw 29(6):479–500. https://doi.org/10.1002/sat.989
8. Chini P et al (2009) A survey on mobile satellite systems. Int J Satell Commun Netw 28(1):29–57. https://doi.org/10.1002/sat.941

9. Sakarellos V et al (2014) Cooperative hybrid land mobile satellite–terrestrial broadcasting systems: outage probability evaluation and accurate simulation. Wirel Pers Commun 79 (2):1471–1481. https://doi.org/10.1007/s11277-014-1941-6

10. Digital Video Broadcasting (DVB) (2007) Framing structure, channel coding and modulation for satellite transmission to handheld (DVB-SH) document ETSI EN 302583 V1.2.1

11. Haykin S (2005) Cognitive radio: brain-empowered wireless communications. IEEE J Sel Areas Commun 23(2):201–220. https://doi.org/10.1109/JSAC.2004.839380

12. Huang K et al (2009) Spectrum sharing between cellular and mobile ad hoc networks: transmission-capacity trade-off. IEEE J Sel Areas Commun 27(7):1256–1267. https://doi.org/10.1109/JSAC.2009.090921

13. Akyildiz IF et al (2006) Next generation/dynamic spectrum access/cognitive radio wireless networks: A survey. Comput Netw 50(13):2127–2159. https://doi.org/10.1016/j.comnet.2006.05.001

14. Akyildiz IF et al (2008) A survey on spectrum management in cognitive radio networks. IEEE Commun Mag 46(4). https://doi.org/10.1109/MCOM.2008.4481339

15. Ozger M, Akan OB (2016) On the utilization of spectrum opportunity in cognitive radio networks. IEEE Commun Lett 20(1):157–160. https://doi.org/10.1109/LCOMM.2015.2504103

16. Bukhari SHR et al (2016a) NS-2 based simulation Framework for cognitive radio sensor networks. http://eprints.whiterose.ac.uk/108661

17. Bukhari SHR et al (2016b) A survey of channel bonding for wireless networks and guidelines of channel bonding for futuristic cognitive radio sensor networks. IEEE Commun Surv Tutorials 18(2):924–948. https://doi.org/10.1109/COMST.2015.2504408

18. Zou Y, Zhu J et al (2010) An adaptive cooperation diversity scheme with best-relay selection in cognitive radio networks. IEEE Trans Signal Process 58(10):5438–5445. https://doi.org/10.1109/TSP.2010.2053708

19. Suraweera HA et al (2010) Capacity limits and performance analysis of cognitive radio with imperfect channel knowledge. IEEE Trans Veh Technol 59(4):1811–1822. https://doi.org/10.1109/TVT.2010.2043454

20. Zhong C et al (2011) Outage analysis of decode and-forward cognitive dual-hop systems with the interference constraint in Nakagami-m fading channels. IEEE Trans Veh Technol 60 (6):2875–2879. https://doi.org/10.1109/TVT.2011.2159256

21. Ding H et al (2011) Asymptotic analysis of cooperative diversity systems with relay selection in a spectrum sharing scenario. IEEE Trans Veh Technol 60(2):457–472. https://doi.org/10.1109/TVT.2010.2100053

22. Sagong S et al (2011) Capacity of reactive DF scheme in cognitive relay networks. IEEE Trans Wirel Commun 10(10):3133–3138. https://doi.org/10.1109/TWC.2011.081011.101849

23. Lee J et al (2011) Outage probability of cognitive relay networks with interference constraints. IEEE Trans Wirel Commun 10(2):390–395. https://doi.org/10.1109/TWC.2010.120310.090852

24. da Costa DB et al. (2012) Dual-hop cooperative spectrum sharing systems with multi-primary users and multi-secondary destinations over Nakagami-m fading. Paper presented at proceedings IEEE international symposium on personal, indoor and mobile radio communications, Sydney, NSW, Australia, 9–12 September 2012

25. Duong TQ et al (2012) Cognitive relay networks with multiple primary transceivers under spectrum sharing. IEEE Signal Process Lett 19(11):741–744. https://doi.org/10.1109/LSP.2012.2217327

26. Shin E-H, Kim D (2011) Time and power allocation for collaborative primary–secondary transmission using superposition coding. IEEE Commun Lett 15(2):196–198. https://doi.org/10.1109/LCOMM.2011.122810.101486

27. Han Y et al (2009) Cooperative decode-and-forward relaying for secondary spectrum access. IEEE Trans Wirel Commun 8(10):4945–4950. https://doi.org/10.1109/TWC.2009.081484

28. Manna R et al (2011) Cooperative spectrum sharing in cognitive radio networks with multiple antennas. IEEE Trans Signal Process 59(11):5509–5522. https://doi.org/10.1109/TSP.2011. 2163068

29. Bhatnagar MR, Arti MK (2013) Performance analysis of AF based hybrid satellite-terrestrial cooperative network over generalized fading channels. IEEE Commun Lett 17(10):1912–1915. https://doi.org/10.1109/LCOMM.2013.090313.131079

30. Sreng S, Escrig B, Boucheret ML (2013) Exact outage probability of a hybrid satellite terrestrial cooperative system with best relay selection. Paper presented at proceedings IEEE international conference on communications, Budapest, Hungary, 9–13 June 2013

31. An K et al (2014) Symbol error analysis of hybrid satellite-terrestrial cooperative networks with cochannel interference. IEEE Commun Lett 18(11):1947–1950. https://doi.org/10.1109/ LCOMM.2014.2361517

32. Hemachandra KT, Beaulieu NC (2013) Outage analysis of opportunistic scheduling in dual-hop multiuser relay networks in the presence of interference. IEEE Trans Commun 61 (5):1786–1796. https://doi.org/10.1109/TCOMM.2013.031213.120686

33. Erwu L et al (2007) Performance evaluation of bandwidth allocation in 802.16j mobile multi-hop relay networks. In: Paper presented at proceedings IEEE vehicular technology conference-Spring, Dublin, Ireland, 22–25 April 2007

34. An K et al (2015) On the performance of multiuser hybrid satellite-terrestrial relay networks with opportunistic scheduling. IEEE Commun Lett 19(10):1722–1725. https://doi.org/10.1109/ LCOMM.2015.2466535

35. Upadhyay PK, Sharma PK (2016) Max-max user-relay selection scheme in multiuser and multirelay hybrid satellite terrestrial relay systems. IEEE Commun Lett 20(2):268–271. https://doi.org/10.1109/LCOMM.2015.2502599

36. An K et al (2015) Performance analysis of multi-antenna hybrid satellite-terrestrial relay networks in the presence of interference. IEEE Trans Commun 63(11):4390–4404. https:// doi.org/10.1109/TCOMM.2015.2474865

37. Kandeepan S et al (2010) Cognitive satellite terrestrial radios. In: Paper presented at proceedings IEEE global telecommunication conference, Miami, FL, USA, 6–10 December 2010

38. Sharma SK et al (2013) Cognitive radio techniques for satellite communication systems. Paper presented at proceedings IEEE vehicular technology conference-Fall, Las Vegas, NV, USA, 2–5 September 2013

39. Jia M et al (2016) Broadband hybrid satellite-terrestrial communication systems based on cognitive radio toward 5G. IEEE Wirel Commun 23(6):96–106. https://doi.org/10.1109/ MWC.2016.1500108WC

40. Chu TMC, Zepernick H-J (2018) Optimal power allocation for hybrid cognitive cooperative radio networks with imperfect spectrum sensing. IEEE Access PP 99:1–1. https://doi.org/10. 1109/ACCESS.2018.2792063

41. Vassaki S et al (2013) Power allocation in cognitive satellite terrestrial networks with QoS constraints. IEEE Commun Lett 17(7):1344–1347. https://doi.org/10.1109/LCOMM.2013. 051313.122923

42. Lagunas E et al (2015) Resource allocation for cognitive satellite communications with incumbent terrestrial networks. IEEE Trans Cogn Commun Netw 1(3):305–317. https://doi. org/10.1109/TCCN.2015.2503286

43. An K et al (2016) Outage performance of cognitive hybrid satellite–terrestrial networks with interference constraint. IEEE Trans Veh Technol 65(11):9397–9404. https://doi.org/10.1109/ TVT.2016.2519893

44. Shi S et al (2017) Optimal power control for real-time applications in cognitive satellite terrestrial networks. IEEE Commun Lett 21(8):1815–1818. https://doi.org/10.1109/LCOMM. 2017.2684798

45. Suffritti R et al (2011) Cognitive hybrid satellite-terrestrial systems. Paper presented at proceedings international conference on cognitive radio and advanced spectrum management, Barcelona, Spain, 26–29 October 2011

46. Liolis K et al (2013) Cognitive radio scenarios for satellite communications: the CoRaSat approach. Paper presented at proceedings future network and mobile summit, Lisboa, Portugal, 3–5 July 2013
47. Kim KJ, Tsiftsis TA (2010) Performance analysis of cyclically prefixed single-carrier transmissions with outdated opportunistic user selection. IEEE Signal Process Lett 17(10):847–850. https://doi.org/10.1109/LSP.2010.2060330
48. Gradshteyn IS, Ryzhik IM (2000) Tables of integrals, series and products, 6th edn. Academic Press, New York
49. Simon MK, Alouini MS (2000) Digital communications over fading channels: a unified approach to performance analysis. Wiley, New York
50. Tang J, Zhang X (2006) Transmit selection diversity with maximal- ratio combining for multicarrier DS-CDMA wireless networks over Nakagami-m fading channels. IEEE J Sel Areas Commun 24(1):104–112. https://doi.org/10.1109/JSAC.2005.858884
51. Miridakis NI et al (2015) Dual-hop communication over a satellite relay and shadowed Rician channels. IEEE Trans Veh Technol 64(9):4031–4040. https://doi.org/10.1109/TVT.2014.2361832
52. Sharma PK et al (2017) Hybrid satellite-terrestrial spectrum sharing system with opportunistic secondary network selection. Paper presented at proceedings IEEE international conference on communications, Paris, France, 21–25 May 2017
53. Sharma PK et al (2017) Performance analysis of overlay spectrum sharing in hybrid satellite-terrestrial system with secondary network selection. IEEE Trans Wirel Commun 16 (10):6586–6601. https://doi.org/10.1109/TWC.2017.2725950
54. Zhang C et al (2015) A unified approach for calculating the outage performance of two-way AF relaying over fading channels. IEEE Trans Veh Technol 64(3):1218–1229. https://doi.org/10.1109/TVT.2014.2329853
55. Suraweera HA et al (2009) Two hop amplify-and-forward transmission in mixed Rayleigh and Rician fading channels. IEEE Commun Lett 13(4):227–229. https://doi.org/10.1109/LCOMM.2009.081943

Chapter 6
Health Monitoring Using Wearable Technologies and Cognitive Radio for IoT

Raluca Maria Aileni, George Suciu, Victor Suciu, Sever Pasca, and Rodica Strungaru

6.1 Introduction

The human services division is experiencing significant change, because of the conceivable outcomes offered by the Internet of Things (IoT) and new innovations: the portable and wearable. The new model is arranged to the general wellbeing of the patient, fortified and executed through a solid star exercise of the patient and acknowledged utilizing cell phones and multichannel innovation. Because of the conceivable outcomes offered by the Internet of Things and new technologies, the mobile and wearable, the social insurance part is experiencing significant change [1]. So, Internet of Things has made a remarkable process, especially in e-health and medicine. IoT includes home health monitoring and wearable devices. In other words, the evolution of the medicine was boosted by IoT, an important factor being the environmental, wearable and implanted sensors spread everywhere in the purpose of monitoring the people's health [2].

Wearable technology means smart devices integrated with different types of accessories such as wristband, wristwatches, eyeglasses and smartphones. These devices can include wirelessly connected scales, glucometers, blood pressure and heart rate monitors, etc.

Nowadays, using technologies like IoT sensors, tablets, wearable devices, etc. became indispensable for monitoring your health. The data provided by these devices is useful not just for real-time self-monitoring of health but also for healthcare organizations, hospitals and pharmaceutical companies for a better management of healthcare costs and wellness [3].

R. M. Aileni (✉) · G. Suciu · V. Suciu · S. Pasca · R. Strungaru
Telecommunication and Information Technology, Politehnica University of Bucharest, 060042, Iuliu Maniu 1-3, Bucharest 060042, Romania

© Springer International Publishing AG, part of Springer Nature 2019
M. H. Rehmani, R. Dhaou (eds.), *Cognitive Radio, Mobile Communications and Wireless Networks*, EAI/Springer Innovations in Communication and Computing, https://doi.org/10.1007/978-3-319-91002-4_6

Monitoring your health through such devices helps you save time and improve care. Also, using wearable devices, you can collect data for a long time, which is much more beneficial than a single medical test for your overall health.

However, to monitor the health issues, multiple measurements are needed. These measurements are based on a set of statistics taken from different devices and mobile applications that helps monitoring the health continuously.

RFID (radio frequency identification) is a wireless technology that provides a unique identifier for an object using electromagnetic fields, and it must be scanned to retrieve the identifying information.

The implementation of radio-frequency identification technology is steadily increasing, specifically in the healthcare industry. In hospitals, RFID is used in different areas such as radiology, infection control, reducing supply overstock and track and trace prescription drugs. Also, one of the key areas where RFID plays a critical role is during a surgical procedure, because of the importance of having everything required for the specific procedure ready [4].

In this industry, specifically in hospitals, an RFID infrastructure can bring a lot of advantages, enabling the management of equipment and of document and data files, as well as tracking and management of patients. This way, the records of each patient will be maintained better, and it will reduce the human errors as well as reduce the costs and save time.

Although RFID technology is beneficial to the healthcare industry, there are also some risks which include technical, economic and legal challenges. Technical limitations are represented by RFID tag readability, interoperability with other HIT (health information technology) or interference in the electromagnetic field by medical equipment or other devices. As a solution, cognitive radio can be used to solve the interference problems with other devices [5].

A cognitive radio (CR) is a promising technology that let the possibility to not exist harmful interference by maintaining low outage probability [6]. It is a radio that can be programmed and configured dynamically to use the best wireless channels in its vicinity avoiding user interference. Such a radio automatically detects available channels in wireless spectrum, then accordingly changes its transmission or reception parameters to allow more concurrent wireless communications in a given spectrum band at one location.

The instruments to distinguish wellbeing conditions are presently achieving reasonable and their market infiltration shows up quickly developing. Wearable sensors are joined with encompassing sensors when subjects are checked in the home condition [7–9]. Specifically, the blend of wearable and encompassing sensors is a hot topic for a few applications in the field of rehabilitation [10].

Health monitoring involves the connection of electronic devices utilized for electronic medical monitoring. Scientific literature indicates that the cellular phone, which emits electromagnetic energy, can generate malfunctionality of the medical equipment used in hospitals such as ventilators, ECG monitors, cardiac monitors and defibrillators. We should take into consideration the risk of electromagnetic interference between medical devices and aggregators (4G/5G cellular phone). Starting from these assumptions, we can anticipate that the devices used for collecting data (cellular phones, tablets and notebooks) will interfere with biomedical sensors for wearable devices and as a direct result will harm the medical devices and records. However, the used solution for patient's vital signs (ECG, pulse rate, blood pressure, temperature) monitoring – WLAN (wireless local area networks) and RFID readers – can generate malfunctions for medical devices. In order to avoid the malfunctions of EMI (electromagnetic interference) in the medical devices,a solution is to set a distance between these EMI sources and medical devices.

6.2 IoT and Medical Wearable Technologies: Perspective, Requirements and Limitations

The Internet of Things (IoT) is a new concept, providing the possibility of healthcare monitoring using wearable devices. The IoT is defined as the network of physical objects which are supported by embedded technology for data communication and sensors to interact with both internal and external objects states and the environment. In the last decade, wearable devices have attracted much attention from the academic community and industry and have recently become very popular. The most relevant definition of wearable electronics is the following: "devices that can be worn or mated with human skin to continuously and closely monitor an individual's activities, without interrupting or limiting the user's motions" [11].

In our days, the range of wearable systems, including micro-sensors seamlessly integrated into computerized watches and belt-worn personal computers (PCs) with a head-mounted display, which are worn on various parts of the body, are designed for broadband operation. The field of wearable health monitoring systems is moving towards minimizing the size of wearable devices, measuring more vital signs and sending secure and reliable data through smartphone technology. Although there has been an interest in observing comprehensive biomedical data for the full monitoring of environmental, fitness and medical data, one obvious application of wearable systems is the monitoring of physiological parameters in the mobile environment. Most commercially available wearable devices are one-lead applications to monitor vital signs.

6.2.1 Wearable Devices in Health Monitoring

The measurement of human movement has several useful applications in sports and medical. Such applications include fall risk assessment, quantifying sports exercise, studying people habits, and monitoring the elderly. Wearable trackers are becoming popular. They can motivate the user during the daily workout to perform more exercise, while providing activity measurement information through a smartphone without manual calculation [12]. Also, they enable the wearer to become aware of the daily distance walked, which is very useful to ensure that the user maintains sufficient activity in the daily routine to maintain a healthy life. To accurately observe motion of the human body, gyroscopes sensors obtain data, each for a specific purpose. These sensors can be used for human activity recognition in the ubiquitous computing domain as well. Gyroscopes and magnetometers are auxiliary sensors that can be separately be combined with accelerometers to compensate the lack of accuracy in obtained data for motion tracking.

Human motion detection has a wide range of application, from sports and recreation to biomedical. In recent years, consumer electronics have employed many semiconductor-based tracking systems to allow users to access various kinds of interface control that use body motions and gestures. An important application of motion tracking is healthcare. When the trackers became more reliable using an integrated gyroscope, the primary focus for clinical applications using inertial motion tracking was gait analysis. To apply this device for clinical purposes, and gait motion, gravity-sensitive accelerometers are used to estimate the tilt angles between the gravity vector and the sensor's axes [13].

Many wearable devices have been implemented to measure critical elements in healthcare monitoring. Most of these devices are in one lead such as electrocardiogram (ECG) and electroencephalogram (EEG) measurement. There have been recent efforts in wearable devices to provide multitask vital signs measurement.

6.2.2   Challenges and Bottlenecks for Medical IoT

Leading wearable devices based on IoT platforms must provide simple, powerful application access to IoT devices. Many platforms and structures have been proposed by the scientific community, and commercial devices are already available for medical parameter measurement. However, there are serious challenges in this way. The following are four key capabilities that leading platforms must enable:

- **Simple and secure connectivity:** A good IoT platform is expected to provide easy connection of devices and perform device management functions in three levels of data collection, data transmission to a hub and permanent storage and observation in a medical station. These steps must be secured; therefore, data encryption is necessary.
- **Power consumption:** To provide the wearer with easy device management and long-term monitoring without interruption, power loss is becoming more important. This is strictly correlated to the number of parameters that is observed, efficient code programming, as well as good data packing, encryption and compression.
- **Wearability**: Wearable devices have been designed for various types of biomedical monitoring to assist users in living long, healthy lives. This point is more significant when these devices are intended to be worn by elderly users.

Therefore, such devices must be easy to wear, easy to carry and comfortable. A wearable device is expected to be small and lightweight and should be able to be used for a long time.

- **Reduced risk in data loss**: When data is collected by a microcontroller and transmitted to smartphone or cloud storage, there is a possibility of disconnection and consequently data loss. This must be reduced as much as possible to provide safe health monitoring. It may be possible through temporary data saving in the microcontroller providing a large memory.

6.2.3 Remote Health Monitoring

Remote health monitoring (Fig. 6.1) could be used to monitor noncritical patients at home rather than in the hospital, reducing strain on hospital resources such as doctors and beds. It could be used to provide better access to healthcare for those living in rural areas or to enable elderly people to live independently at home for longer. Essentially, it can improve access to healthcare resources while reducing strain on healthcare systems and can give people better control over their own health.

Regarding wearable healthcare systems, by excluding implantable sensors, there are five fundamental sensors that should be included in a wearable healthcare system: three for monitoring the vital signs (pulse, respiratory rate and body temperature) and another two for monitoring blood pressure and blood oxygen.

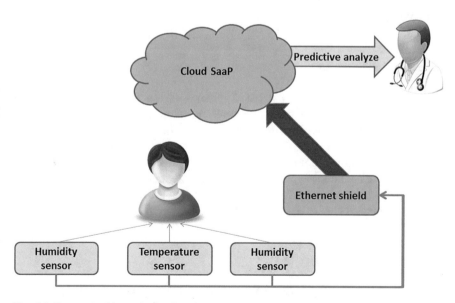

Fig. 6.1 Remote health monitoring based on sensors and cloud service

6.2.3.1 Pulse Sensors

Pulse is the most commonly read vital sign, and it can be used to detect a wide range of emergency conditions such as pulmonary embolisms and cardiac arrest. Several fitness tracking chest straps and wristwatches are equipped with pulse measurement functionality. However, these devices are not suitable for detecting health conditions, and they cannot be directly implemented into a critical health monitoring system.

Sensor types developed and analysed in recent works include photoplethysmographic (PPG), pressure, ultrasonic and RF sensors. The scientific literature indicates that PPG sensors are most suitable for pulse sensing.

6.2.3.2 Respiratory Rate Sensors

Monitoring respiratory rate could be used to detect several health conditions such as asthma attacks, hyperventilation, tuberculosis, obstruction of the airway, etc. Based on recent work, stretch sensors are strongly recommended for implementation into future systems.

6.2.3.3 Body Temperature Sensors

Monitoring body temperature can be used to detect hypothermia, heat stroke, fevers and more. Based on recent work, it is recommended that textiles should be used in order to hold temperature sensors until electronics printed on flexible polymer can be more easily manufactured.

6.2.3.4 Blood Pressure

Although it is not a vital sign itself, blood pressure is frequently measured in hospital environment along with the three parameters mentioned above. Monitoring blood pressure can prevent hypertension which is one of the most common chronic illnesses and also a known risk factor for cardiovascular disease.

Based on recent work, no developed system is suitable for accurately measuring blood pressure continuously using a comfortably wearable device.

A wearable sensor design for continuously and non-invasively monitoring blood pressure remains a challenge in the field of healthcare IoT.

6.2.3.5 Blood Oxygen

Like blood pressure, blood oxygen level is not a vital sign itself, but it is frequently measured in hospitals as an indicator of respiratory function.

Pulse oximeter sensors measure blood oxygen by obtaining PPG signals. Two LEDs are used, one red and one infrared, and they usually come in contact with the finger.

The best option for a wearable version would be a wrist sensor. Regarding pulse oximeter sensors, future research does not focus on a new sensor technology but on its wearable design.

Besides the fundamental sensors mentioned above, there are sensors that can be incorporated into systems that monitor a particular condition.

For example, the ECG electrods can be used to evaluate heart health, the EEG electrods can be used to measure brain activity and aldo can be used to monitor sleep disorders and seizures in patients who have suffered head injuries (also, such EEG systems have other specific applications such as to detect driver fatigue or stress management). Other sensors that are applicable to many common diseases or conditions and require more research are fall detection and gain detection for elderly people and non-invasive blood glucose monitor sensor.

6.2.4 Communications Standards

Communications related to IoT healthcare can be classified into two main categories: short-range communications and long-range communications.

6.2.4.1 Short-Range Communications

Regarding wearable healthcare systems, typically, short-range communications are used between sensor nodes and the central node where data processing occurs [14].

The most common short-range communications used in IoT are BLE (Bluetooth low energy) and ZigBee.

BLE was developed to provide an energy-efficient standard that could be used to coin cell battery-operated devices, including wearables.

For healthcare applications, BLE is used in star topology where the centre of the star topology is the central node which has sensors linked to it. BLE specifications are as follows: low latency of 3ms, range of 150m in open field, high data rate of 1Mbps, operates in the 2.4 GHz band and extremely low power consumption.

ZigBee was designed to provide low-cost and low-power networks for M2M communications. ZigBee has different modules (Table 6.1) which provide different characteristics in terms of range, data rate and power consumption.

In the healthcare environment, big data rate and a small range for body area network are necesary, so XBee version is preferable.

In addition, if BLE is implemented for wearable devices, ZigBee is not commonly implemented in wearable devices (e. q. smartphones) because of compatibility issues, while BLE is. So, in conclusion ZigBee is suggested that ZigBee is rather suitable for fixed location than wearable healthcare systems.

Table 6.1 Characteristics of different modules of ZigBee

Module	Range	Power for transmission	Data rate
XBee	30 m	1 mW	250 kbps
XBee-PRO	90 m	63 mW	250 kbps
XBee-PRO 900 XCS	610 m	250 mW	10 kbps

6.2.4.2 Long-Range Communications

There are also a few disadvantages of remote health monitoring. The most significant disadvantages include the security risk that comes with having large amounts of sensitive data stored in a single database, the potential need to regularly have an individual's sensors recalibrated to ensure that they're monitoring accurately and possible disconnections from healthcare services if the patient was out of cellular range or their devices ran out of battery.

6.3 Electromagnetic Interference for Medical Devices Connected Through IoT

In recent years, the mobile revolution and the explosion of wirelessly connected devices have created the desire of conveniently charging devices without the need of plugging them into a wall outlet. Wireless charging is among the fastest-growing segments of technology today, especially when it comes to portable devices such as mobile phones, tablets and laptop computers. Also, larger items such as electric vehicles will soon be available with wireless charging capabilities.

Wireless devices of all types are constantly creating signals across a relatively narrow frequency band, and one result is a challenge in recognizing and mitigating the electromagnetic interference problem. Wireless interference is invisible to the human eye and requires specialized equipment and training to accurately understand its effects on specialized medical equipment.

In research, the active and passive radio-frequency identification (RFID) tags were used and placed in proximity to medical devices ranging from infusion pumps to external pacemakers [15]. There were a total of 41 devices in 17 categories from 22 manufacturers. RFID readers were used to read tags, and out of 123 tests, 34 incidents of EMI occurred, of which 22 turned the medical equipment off or caused an inaccurate reading.

The magnetic induction from each of these devices is described by a fundamental law of electromagnetism, Faraday's law, which states that if a time-varying magnetic field passes through the surface of any conducting loop, a voltage will be induced in that loop. The voltage induced in this loop is proportional to the time-varying change in the magnetic flux.

Licensed frequency bands mean that individual companies pay a licensing fee for the exclusive right to transmit on assigned channels within that band in a given geographic area ensuring that operators do not interfere with each other's transmissions. Unlicensed frequency bands, however, do not require any permission to use, the only requirement being to meet some rules associated with the particular frequency band. IoT devices are planned to be used for considerably higher frequencies than standard testing is performed for today. The most far-reaching visions involve radio coverage in outdoor environments for frequencies up to about 30 GHz and indoors up to about 90 GHz.

The challenges of electromagnetic interference extend to the home as well. Increased efforts of creating the "medical home" must take into account the multitude of wireless devices and potential impact to medical devices placed in patients' homes. The expected result from wireless devices sharing the same frequency range of 2.4–5 GHz is unknown, and the devices can include TVs, microwaves, home wireless routers, baby monitors and so on. The electromagnetic congestion at home is in many ways like the growing electronic congestion in hospital environment as they acquire more and more electronic sensors that operate within a few feet of each other. The Internet of Things is making medicine participatory, personalized, predictive and preventive. Medical device and high-tech companies are joining forces to satisfy an exponentially growing demand that might reach to 20 billion medical connected devices by 2020.

Healthcare is the industry where the Internet of Things may provide significant advantages to patients, caregivers and medical institutions, but it is likely the most complicated and life-critical area of the IoT. As the number of radio frequency emitters increases, public health and safety must also be considered. Traditional IoT items that can be found in any company or organization (e.g. personnel tracking, electronic surveillance systems) are also present in the medical environment and may live for years as implantable such as pacemakers, which have lifetimes of a decade or more. The more interesting applications revolve around those things which are both active and interactive. These devices may be implanted or wearable.

It is a known issue that RFID and NFC readers can be a problem if operated in the near vicinity of persons with pacemakers and implantable cardioverter defibrillators. Although NFC devices typically require a shorter read range, they operate at a frequency that usually causes more problems with implanted cardiac rhythm management devices (CRMDs).

As wireless connectivity has gone from a critical care hospital to the home of a recently discharged patient, the primary connectivity device has gone from the nurses' station to a personal computer. Multiple server concepts are being developed today, including commercial body area networks that facilitate this transition.

A significant number of studies have been conducted investigating electromagnetic interference on cardiac rhythmic device systems, mostly exploring the interference caused by cellular telephones. According to these studies, sophisticated filtering techniques have not been able to eliminate the risk of electromagnetic interference due to mobile phones in modern CRMDs. Other studies have implicated various medical procedures such as electrocautery and magnetic resonance imaging as being possible sources of such interference. These procedures should be avoided by the patients. Also, many studies have also indicated RFID as a potential source of clinically significant interference.

The most important medical devices in need of high electromagnetic immunity are pacemakers and implanted cardioverter defibrillators. The need to operate correctly, without a disruption to their proper function in the presence of external magnetic field sources, is of keen interest to medical device manufacturers. Pacemakers are electronic devices surgically implanted in patients to monitor and control irregularities in a patient's natural heart activity. The primary functions of a pacemaker are sensing and controlling the heart rhythm. While in sensing modes the pacemaker monitors the patient's heart activity through lead wires and if only natural heart activity is present, it will not enter the pacing mode; only if the patient's heart rhythm is too slow or gets interrupted, the pacemaker sends an electrical impulse to the heart to regulate its heartbeat. These devices are particularly susceptible to electromagnetic interference from induction because the lead wires sense the minimal levels of electrical activity within the heart, and therefore small induced voltages can interfere with its proper functioning. The patient has little or no control over the frequency or the modulation of an electromagnetic source, so medical device manufacturers build safeguards into the design of pacemakers. These safeguards are designed to revert to a conservative mode of operation when the pacemaker no longer receives the heart muscle stimulus, providing pacing activity at a predetermined fixed rate. This procedure is known as noise reversion or safety mode.

Many emerging technologies are converging at an ever-increasing rate to realize systems such as the IoT industry, promising to transform and improve the lives of billions worldwide. However, as we approach the full realization of the IoT, we must be aware of the potential harm that such systems may cause to us.

6.4 Personal Health Monitoring in IoT: Requirements and Configurations

In the last period, various sensors have been developed for a variety of medical applications, including for WSN-based healthcare services. Such sensors are evolving enough to deliver the same services through the IoT. Also, wearable devices can come with a set of desirable features appropriate for the IoT architecture. This has led to the integration of this type of sensors into wearable products. For wearable it is used a dedicated service called wearable device access (WDA). There are many systems that can be used in a wide range of healthcare applications through various mobile devices such as smartphones and smart watches.

Data acquisition can be performed by multiple wearable sensors. These sensors measure physiological biomarkers, such as EMG (muscle activity), ECG (electrocardiogram), respiratory rate, skin temperature, blood pressure and so on. The sensors connect to the network through an intermediate device (concentrator), for example, a smartphone. Usually, to transfer the data to the device, a sensory acquisition platform is used. This platform is equipped with a short-range radio such as low-power Bluetooth or Zigbee. Further, data may be transmitted to a healthcare organization for long-term storage using Internet connectivity on the device, typically via a smartphone's WiFi or a cellular data connection. Sensors in the data acquisition part form an IoT-based architecture, because each individual sensor's data can be accessed through the Internet via a smartphone.

To augment processing or storage capability whenever the local mobile resources do not fulfil the application's requirements, a cloudlet is used. A cloudlet is a processing or storage device near a mobile client. The cloudlet can be a local processing unit which is directly accessible by the concentrator through WiFi network. In case of limitations on the device, the cloudlet can be used to transmit the data to the cloud. Cloud processing has three distinct components: analytics, storage and visualization. This system is designed for long-term storage of biomedical information but also for assisting health professionals with diagnostic information. Analytics that use sensor data are becoming more common, and these analyses can help with diagnoses and prognoses for many diseases and health conditions. Additionally, a key requirement for any such system is visualization. Data by the concentrator needs to be transferred to the cloud for long-term storage. Offloading data storage to the cloud offers benefits of accessibility and scalability on demand, both by the user and clinical institutions [16].

Due to the advent of the IoT and the latest improvements in e-health, health data sources can range from body sensor networks to medical analysis and diagnosis. Other requirements that may exist:

- Privacy, integrity and authentication are mandatory when sensitive data are exchanged across the network.
- Bounded latency and reliability need to be granted when dealing with emergency situations for the intervention to be effective.
- Interoperability is needed to enable different things to cooperate as to provide the desired service [17].

A good example of a personal health monitoring in IoT can be in the case of ECG. The ECG monitoring includes the measurement of the heart rate as well as the diagnostics of multifaceted arrhythmias through the use of electrical activity of the heart recorded by an electrocardiography. The innovation presented in [18] introduces an IoT-based electrocardiogram monitoring system composed of a wireless receiver processor and a portable wireless acquisition transmitter. The system integrates a search automation method to detect abnormal data such that cardiac function can be identified on a real-time basis.

In recent years, electronic devices have been recorded with a smartphone-controller sensor, which highlights the increase in the number of smartphones as an IoT engine. Various software and hardware products have been designed to make smartphones a versatile healthcare device. The following list contains some examples of smartphone applications:

- Pedometer: records the number of steps made by the user
- ECG self-monitoring: registering ECG data based on the built-in "ECG self-check" software

- Google Fit: tracks the user's walking, running and cycling activities
- Blood pressure watch: tracks, collects, analyses and shares blood pressure data [19]

6.5 Cognitive Radio Modelling for IoT

New technologies are shifting towards Internet of Things (IoT) and cognitive radio networks (CRNs). The relation between these three technologies is quite strong because new semantics-oriented IoT are not as effective if they are not equipped with cognitive radio capability.

As machine-to-machine has developed a lot in the recent years, mobile connectivity and Internet of Mobile Things (IoMT) have also risen. IoT refers to the interconnection of different objects through the Internet using different communication technologies. Sensors and communication modules are added to these objects.

Although wireless technologies are becoming more and more popular because of their flexibility, they still have some major problems and concerns, like bandwidth support and availability of spectrum.

Some new trends are cognitive radio networks, and they add new solutions to the problem mentioned. Some main functions of the CRNs are spectrum sensing, spectrum decision, spectrum management and spectrum mobility. Cognitive radio networks (CRNs) have the advantage of adjusting the transmission parameters according to the needs of IoT systems [20].

There are a few reasons why IoT requires CRN. CRN works by searching for available spectrum bands by dynamically changing transmitter parameters based on the interaction with the environment. Each node acts like a fast switch with simultaneous transmissions. Using dynamic spectrum access (DSA), a CR user is enabled to adapt to different network conditions. A primary user (PU) is allowed to use the spectrum with guaranteed protection. In addition, cognitive radio (CR) devices can share licensed bands by several models: commons, shared-use and exclusive-use [21].

The cognitive radio solves the problems of the spectrum scarcity and underutilization, through dynamic spectrum access, by selecting the appropriate channel available in location and time [22].

Objects used in the IoT field need to be equipped with cognition to think, learn, make decisions and understand the social and physical world as well. CR needs additional requirements such as intelligent decision-making, perception action cycles, data analytics, on-demand service provisioning, semantic derivation or knowledge discovery.

A few reasons for the development of CR-based IoT devices are listed below, with little explanation for the ulterior motive.

Bandwidth allocation: in the next few years, the number of IoT devices will grow exponentially, and it will be difficult to allocate spectrum bands to these objects. Also, it is expected for the number of PUs to increase.

In the present, technologies like Bluetooth and Zigbee have limited range. In this particular area IEEE 802.22 for a wireless regional access network (WRAN), a CRN standard has a long range and can be used for short to long range applications.

Interference-free channels can be searched through dynamic spectrum access capability.

For future IoT structures, mobility is an important part; with cognitive capability added, IoT devices can obtain easy connectivity.

Cloud servers are starting to be used often as a solution to increasing big data generated by the IoT devices. CR-based objects have the purpose to search for available storage places in cloud servers and send data through spectrum sensing [22].

Radio spectrum is the most important resource for wireless communications. For now, the radio spectrum is underutilized most of the time and can represent an appropriate solution for one of the IoT problems.

Cognitive radio (CR) is a technology which acts as a promising solution to the problem of spectrum shortage and inefficiency of utilization in wireless networks.

Cognitive radio nodes are smart wireless devices that can sense the environment, understand the network changes and make smart decisions when it comes to transmitting the data. SEAT cycles are usually used. SEAT cycle (Fig. 6.2) represents a process of scanning the spectrum (S), exchanging control information (E), agreeing upon white space (A) and finally transmitting data (T) on the existing network.

The unlicensed users are called CR users or secondary users (SUs). The cognitive radio can be used in many IoT applications [23].

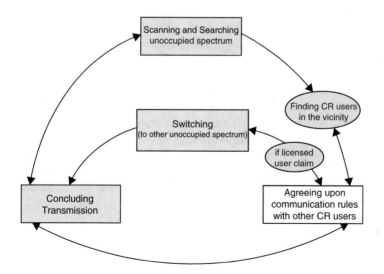

Fig. 6.2 Cognitive radio communication model

In the half-duplex communication (HDC), the SUs can sense but cannot transmit during the spectrum sensing. In the full-duplex communication (FDC) in CRN, in comparison with the HDC in CRN, the SUs can sense spectrum and transmit on a given channel [24].

- **Networking applications based on geolocation**
  Geolocation techniques can offer information about where a CR is and what is happening around it. Most or all smartphones can provide information about the location of the devices. The reports via email or SMS could be sent to a CR user and can be processed to notice some patterns or the behaviour of the user.
- **Commercial applications**
  A considerate amount of TV band has been obtained by switching TV channels from analog to digital [25].
- **Free mobile calls with improved link quality**
  CR functionality can be enabled to detect white spaces, and the free channels are detected/sensed by the CR mobile device. Also switching from GSM networks to CR networks can improve call quality and achieve a good or better coverage, e.g. inside a lift or building where the reception of GSM signals is quite weak.
- **Public safety and disaster management**
  A problem with lack of interoperability between first responder radio systems is that frequencies used for public safety use are interleaved to frequencies designed for industrial, transportation and non-military users. A cognitive radio could be a successful solution for the congestion in radio bands.

CIoT (Cognitive Internet of Things) can connect the physical world (with objects and assets) and the social world (with the human request and social conduct) by programmed arranged operation and clever administration provisioning.

CIoT describes the connections among five central subjective undertakings: discernment activity cycle, enormous information examination, semantic deduction and learning revelation, wise necessary leadership and on-request benefit provisioning.

Cognitive Internet of Things (CIoT) is another system worldview, where (physical/virtual) things or articles are interconnected and carry on as operators, with least human mediation; the elements cooperate with each other after a setting of mindful discernment activity cycle, utilize the strategy of comprehension by working to gain from both the physical condition and interpersonal organizations, store the scholarly semantics and additional information in sorts of databases and adjust to changes or vulnerabilities by means of asset effective primary leadership instruments, because of two essential targets:

1. Crossing over the physical world (with objects, assets and so forth) and the social world (with the human request, social conduct and so on), together with themselves to frame an intelligent physical-cyber-social (iPCS) framework
2. Allocating smart resources, automated networks and operating service provision

6.6   Algorithms for Cognitive Radio Used in IoT for Medical Monitoring

IoT supports many input-output devices and sensors like camera, microphone, keyboard, speaker, displays, near-field communications, Bluetooth and accelerometer. The main component of the IoT is the RFID system. RFID can automatically identify the still or moving entities. The main aim of IoT is to monitor and control objects via Internet. The idea behind it consists of interconnecting objects by sensors and monitoring via the Internet.

For the Future Internet and IoT, it is very much essential to keep track and control the immensely growing number of networked nodes so that it will be possible to network them with everyday objects in homes, offices, buildings, industries, transportation systems, etc. in a cost-effective and valuable way.

The healthcare sector is an ideal example of how cognitive networking and cognitive radio techniques can be employed to enhance the robustness, scalability and utility of medical equipment and systems using wireless communications. At present, the interest for wireless communication technologies is increased for medical applications, which can significantly enhance the patients' mobility, a key factor for speedy recovery after surgical procedures and interventions.

Examples of these applications include electrocardiograms, pulse oximeters, dosimeters and movement alarms. Additionally, the use of wearable biomedical sensors allows the remote monitoring of patients suffering from chronic diseases and the elderly at home by using telemedicine systems (Fig. 6.3).

The wireless body area network (WBAN) has produced the first draft of a document specifying the physical and medium access control layer characteristics of the radio interfaces for WBAN applications. A medical BAN (MBAN, i.e. a WBAN for medical applications) comprises multiple sensor nodes, each capable of

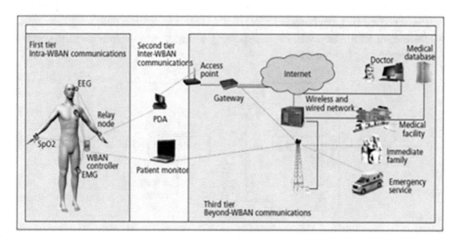

Fig. 6.3 Three-tier architecture of a telemedicine system

sampling, processing and communicating one or more vital signals. This biomedical information is transmitted to a body network controller. The MBAN constitutes the first tier of a telemedicine system, which is referred to as intra-WBAN communications. MBANs offer many benefits for healthcare; the surge in adoption rates across the healthcare sector will certainly create new interference scenarios with other collocated electronic systems.

6.6.1 Fuzzy Logic

Fuzzy logic (FL) is a concept proposed by [26] and was presented as a method of processing data. FL is inspired by human reasoning. The principle of its operation is like the way people make decisions: we have the option of choosing yes or no. In the same way, fuzzy logic, based on known information and some rules, chooses a decision between true and false. In Boolean logic, true and false values are associated with integer numbers 1 and 0 (fuzzy logic, Stanford Encyclopaedia [27]). In fuzzy logic (FL) and numbers between 0 and 1 are used, the fuzzy process taking decisions between "completely true" and "completely false". Thus, this method is efficient in making compromise-based decisions, being useful in solving multidimensional problems. LF has proved to be an extremely useful technique in cases where problems are difficult to solve by traditional mathematical methods or when human logic cannot understand the mathematical process [28].

Fuzzy logic and the FL decision process take place in three phases, as presented in Fig. 6.4:

- **Fuzzification**: is the transformation of input variables into a set of fuzzy numbers (numbers between 0 and 1) using predefined membership functions (MBF).
- **Interference**: The interference determines the fuzzy output set according to input values, result based on rules in the predefined database. The decisions at this stage are based on if-then clauses.
- **Defuzzification**: consists in transposing a fuzzy subset into a single value corresponding to an output.

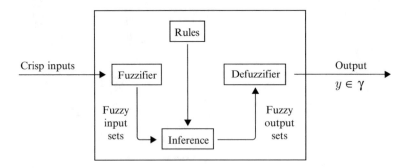

Fig. 6.4 The structure of a fuzzy logic system

Considering the continuous development of cognitive radio, fuzzy logic is a suitable algorithm to apply in this domain. This algorithm can be used to solve the many applications of cognitive radio used in IoT for medical monitoring having the capacity to respond in a flexible way to changing conditions.

Fuzzy logic is very common in medicine in the development of knowledge-based systems. This method is very well suited for interpreting data collected from patients in correct diagnosis, in real-time monitoring of patients [29].

With the help of fuzzy logic, a pulse oximeter alarm system has been created that can differentiate false alerts caused by different reasons from true alarms. This system works by defining numerical input variables such as different levels of oxygen saturation and oxygen saturation exchange rate [30].

6.6.2   Neural Networks

Neural network is a branch of the science of artificial intelligence. Artificial neural networks (ANN) characterize sets of information processing elements (analogous to human neurons), strongly interconnected and operating in parallel, which aim to interact with the environment in a manner like the human brain to solve various problems. The specific property of these networks is the ability to learn based on examples, using previous experiences to improve its results, like the way people learn, by forming synaptic connections between neurons.

There is no generally accepted definition of ANN for these types of systems, but most researchers agree with the definition of artificial neural networks as networks of simple elements strongly interconnected by means of links called interconnections through which numerical information is propagated [31]. Neural network architecture refers to the number of artificial neurons and the linkages between them in different layers. The concept of weight in this case refers to the link between neurons. This one changes during the learning process. Weight influences signal power at a connection. The signal produced at a neuron connection is a real number, so the input signals are multiplied by the weight of the corresponding link to which a bias is added, then multiplying the results to get the input into the next neuron that is subjected to a nonlinear function [32].

With the help of neural networks, numerous IoT medical monitoring devices have been deployed such as continuous glucose monitoring (CGM) devices which are useful in tracking real-time blood glucose levels. It has developed an application based on artificial neural networks that detects epileptiform discharges (EDs) when trained on EEG records marked by an electroencephalograph (EEG). These neural networks have also been used in ECG systems, bringing the possibility to select optimal parameters for each patient. The ability of neural networks to learn from previous experiences has led to the development of a respiratory rate determination system.

6.6.3   Genetic Algorithms

Genetic algorithms (GA) are based on the principles of genetics and natural selection. This algorithm is distinguished by the inspirational trait of animal life, those who know how to adapt best, resist the passage of time and evolve, while less adapted animals fail to survive with time disappearing as a result of natural selection. The probability that a species survives is directly proportional to the adaptive capability; analogy can define the concept of optimization, so a better solution we say is approaching the optimum. Therefore, we can say that a genetic algorithm is an informational model that emulates the biological evolutionary model to solve optimization or search problems [33].

Several years ago, an algorithm has been developed based on a genetic algorithm along with other methods that help extract data. It has the ability to sense the most significant cancer genes. This algorithm facilitates the proper selection of treatment and drug development for different types of cancer [34]. Another branch in which the genetic algorithm has been used to develop an IoT product is dermatology. The genetic algorithm had the best result in the new method of detecting skin tumours [35].

6.7   Cognitive Radio: Future Challenges of Personal Health Monitoring and IoT

The term cognitive radio was first used and introduced by Dr. Joseph Mitola in 1999–2000 to describe smart radios that can only detect the available wireless channels around and that can adjust their operating parameters such as frequency, transmission power and modulation strategy to ensure secure and efficient radio frequency communication. This concept has the following cognitive features: self-awareness, intelligence, learning, adaptability, reliability and efficiency [36].

With the promise to save time and reduce costs, the IoT concept has become increasingly popular among people and researchers, with the number of Internet-connected objects growing more and more every day.

In 2013, Cisco launched the IoE (Internet of Everything) connection meter. Cisco suggests that 80 smart things per second were connected to the Internet, forecasting 50 billion things will be connected to the Internet in 2020 (Cisco connection counter [37]).

IoT is growing in ever-changing areas. Today, this concept can be found in medicine, agriculture, transport, smart city management, security and emergency situations, smart grid and smart metering, etc.

A particular interest is the use of real-time healthcare and the monitoring of vital signs of patients such as temperature, blood pressure and oxygen and electrocardiogram. The device that can continuously monitor patients to transmit data to medical applications is called medical BAN or MBAN [38]. The benefits of implementing such a system include the analysis of vital long-term parameters, rapid alert of healthcare professionals to intervene in an emergency, reduced hospital stay and increased patient comfort throughout their stay in the hospital. At the same time, MBAN intervenes in the early detection of diseases.

With the adoption of this innovative solution, new challenges arise from interfering with nearby medical devices. This can lead to degradation of performance and may also become critical to the patient's health. It is therefore necessary to have a clean and little crowded spectrum.

In this regard, the IEEE research community created a working group called IEEE 802.22 WRAN and adopted radio cognitive to enhance the use of spectrum in ISM bands and unused TV bands, while companies like Motorola, Philips and Qualcomm have also shown interest in this innovative solution and are investing in technology development (Internet of Things, IERC [39]).

At the same time, the IERC (European Research Cluster on the Internet of Things) has launched a standardization project on IoT technologies (RERUM [40]). This refers to cognitive radio adaptation to IoT devices to minimize wireless interfaces and to ensure continuous connection [41].

Radio cognitive (CR) features have advanced capabilities. This technology can automatically detect wireless spectrum, adapt to real-time transmission parameters and adjust to maintain effective communication of wireless spectrum in space, time, frequency and modulation mode. CR networks are considered to be the future services for the use of IoT technologies to improve and empower them in order to efficiently use them (Cisco connection counter [37]).

IoT offers many advantages but at the same time brings many technical challenges in areas such as home networks, intelligent cities, logistics, medicine, etc.

- **Heterogeneity issues**: The development and design of IoT applications involve costs, resources, energy, infrastructure, coverage, connectivity, life expectancy, etc. A single hardware and software platform are not enough to develop them, so the best solution is the use of heterogeneous systems.
- **Flexible, dynamic and efficient networking and communication**: Most IoT applications will require a higher QoS with adaptability and reconfiguration features. The heterogeneous devices will be difficult to manage, and for the end user, plug and play objects with self-discovery, self-configuration and automatic software deployments will be important.
- **Self-organization, reconfigurability and autonomicity**: Reconfiguration is a feature required for transparent insertion of smart objects without any external action. They should have this capacity to detect neighbouring objects and reconfigure them
- **Energy efficiency**: The whole range of IoT objects, sensors, smartphones and applications in medicine and beyond must be energy-efficient, have a lifetime as much as possible and have sufficient processing power for effective management and optimization.

- **Cooperative and ambient intelligence**: Intelligent tools capable of environmental intelligence and cooperation are agents that perform management functions. Collaboration and communication between neighbouring intelligent objects improves their knowledge of the local environment and makes decision-making easier in the area of the network. The system must have the ability to feel the real world, to communicate and process unfinished information and to turn them into meaningful information, acting on them.

In IoT, the wireless sensor networks (WSNs) should use energy-efficient routing protocols by network structure (hierarchical), communication model (negotiation-based), topology-based (location-based or mobile agent-based) and reliable routing (QoS-based or multipath-based) [42]. For efficient energy use, in hospital WBAN applications for patient monitoring, the use of the hybrid rapid response routing (HRRR) protocol was proposed for delay-sensitive traffic [43, 44, 47].

A solution for energy-efficient sensor network, in sensing application, is the use of the tiny operating system (TinyOS) [23], optimized for the memory limits in WSN.

The WSNs for health application involve quality and using heterogeneously clustered routing (QHCR) protocol for efficient energy consumption based on four energy levels with clustering and multipath support [45].

The energy available can affect the performance, functionality and lifetime of WSN, and for data transmissions in IoT, health applications are required techniques for generating or harvesting energy at the sensor nodes [46].

Wireless security is an important aspect for CRNs to ensure a reliable spectrum sensing and optimal resource allocation and management. A proposed approach is to use the classification method based on clustering method in order to group the sensing nodes in clusters having a head cluster responsible with collecting sensing reports from different sensing nodes in the same cluster [43, 44, 47].

The cognitive radio networks (CRNs) are vulnerable to many security attacks (radio frequency blocked or jammed by sending a signal with the same frequency with enough power).

The most used attacks in spectrum sensing phase are the primary user emulation attack and spectrum sensing data falsification attack [43, 47, 48].

6.8   Conclusions

The reliance electronics – internet (IoT) it is a good opportunity for think at the future electronics' internet through cognitive smart radios that detect the available wireless channels because it is a challenge to connect a huge volume of electronics and to preserve the accuracy of the data and to avoid the electromagnetic interference (EMI). It is important to consider the energy efficiency aspects because the IoT applications based on sensor network will influence the digitalization in numerous domains, such as home, military, health [48, 49] and space.

The cognitive radio (CR) is a promising solution to the problem of spectrum shortage and for efficient wireless networks.

Cognitive radio nodes are smart wireless devices (smart sensors) that can sense the environment (e.g. human body physical parameters), understand the network changes and make smart decisions when it comes to transmitting the data.

In the era of cooperative Internet solutions, the cognitive radio networks are future services for the use of IoT technologies to improve the management and smart control of the wearable devices and sensor networks. These aspects can create the premises of the adaptive and responsive Future Internet smart decision system support for prioritizing critical events (e.g. patients with chronic diseases at home that can use a smart decision system for adapting the ambient parameters (humidity, temperature, ventilation) in strong relation with the parameters received in real time from smart wearable).

Acknowledgement This paper was partially supported by ESTABLISH, WINS@HI and EmoSpaces projects.

References

1. Amato A, Coronato A (2017) An IoT-aware architecture for smart healthcare coaching systems. IEEE 31st International Conference on Advanced Information Networking and Applications (AINA), pp. 1027–1034. Taipei, 2017
2. Asthana S et al (2017) A Recommendation system for proactive health monitoring using IoT and wearable technologies. IEEE International Conference on AI & Mobile Services (AIMS), pp. 14–21, Honolulu, 2017
3. Cao H et al (2009) Enabling technologies for wireless body area networks: a survey and outlook. IEEE Commun Mag 47(12):84
4. Amendola S et al (2014) RFID technology for IoT – based personal healthcare in smart spaces. IEEE Internet Things J 1(2):144–152
5. Coustasse A et al (2015) Benefits and barriers of implementation and utilization of radio-frequency identification (RFID) systems in transfusion medicine. Perspect Heath Inf Manag 1d:12
6. Chávez-Santiago R et al. (2014) Dual-band cognitive radio for wearable sensors in hospitals. 8th International Symposium on Medical Information and Communication Technology (ISMICT), pp. 1–52, Firenze, 2014
7. Arase Y et al (2011) Mobile search assistance from HCI aspect. IJSSC 1(1):18–29. https://doi.org/10.1504/IJSSC.2011.039104
8. Sabzevar AP, Sousa JP (2011) Authentication, authorisation and auditing for ubiquitous computing: a survey and vision. IJSSC 1(1):59–67. https://doi.org/10.1504/IJSSC.2011.039107
9. Suciu G et al (2015) Big data, internet of things and cloud convergence – an architecture for secure E-health applications. J Med Syst 39(11):1–8
10. Gao W, Emaminejad S (2016) Fully integrated wearable sensor arrays for multiplexed in situ perspiration analysis. Nature 529(7587):509–514
11. Anliker U et al (2004) A wearable multiparameter medical monitoring and alert system. IEEE Trans Inf Technol Biomed 8(4):415–427
12. Mayagoitia RE et al (2002) Accelerometer and rate gyroscope measurement of kinematics: an inexpensive alternative to optical motion analysis systems. J Biomech 35(4):537–542

13. Baker SB et al (2017) Internet of things for smart healthcare: technologies, challenges, and opportunities. IEEE Access 5:26521–26544
14. van der Togt R et al (2008) Electromagnetic interference from radio frequency identification inducing potentially hazardous incidents in critical care medical equipment. JAMA 299 (24):2884–2890
15. Hassanalieragh M. et al. (2015) Health monitoring and management using internet-of-things (IoT) sensing with cloud-based processing: opportunities and challenges. IEEE International Conference on Services Computing, pp. 285–292, New York, 2015
16. Bui N, Zorzi M (2011) Health care applications: a solution based on the internet of things. Proceedings of the 4th International Symposium on Applied Sciences in Biomedical and Communication Technologies, p. 131
17. Liu ML et al (2012) Internet of things-based electrocardiogram monitoring system. Chinese Patent 102(764):118
18. Islam SMR et al (2015) The internet of things for health care: a comprehensive survey. IEEE Access 3:678–708
19. Amjad M et al (2018) Wireless multimedia cognitive radio networks: a comprehensive survey. IEEE Commun Surv Tutorials 1(1):99
20. Hassan MR et al (2017) Exclusive use Spectrum access trading models in cognitive radio networks: a survey. IEEE Commun Surv Tutorials 19(4):2192–2231
21. Elderini T et al (2017) Channel quality estimation metrics in cognitive radio networks: a survey. IET Commun 11(8):1173–1179
22. Khan AA et al (2017) Cognitive-radio-based internet of things: applications, architectures, Spectrum related functionalities, and future research directions. IEEE Wirel Commun 24 (3):17–25
23. Amjad M et al (2017) Full-duplex communication in cognitive radio networks: a survey. IEEE Commun Surv Tutorials 19(4):2158–2191
24. Shah MA et al. (2013) Cognitive radio networks for internet of things: applications, challenges and future. 19th International Conference on Automation and Computing, pp. 1–6, London
25. Setiawan D et al (2009) Interference analysis of guard band and geographical separation between DVB-T and EUTRA in digital dividend UHF band. International Conference on Instrumentation, Communication, Information Technology, and Biomedical Engineering, pp. 1–6
26. Novák V et al (1999) Mathematical principles of fuzzy logic. Springer Science & Business Media, 0-7923-8595-0
27. "Fuzzy Logic". Stanford Encyclopedia of Philosophy. Bryant University. 2006-07-23. Retrieved 2008-09-30
28. Lee CC (1990) Fuzzy logic control systems: Fuzzy logic controller. IEEE Trans Syst Man Cybern 20(2):404–435
29. Phuong NH, Kreinovich V (2001) Fuzzy logic and its applications in medicine. Int J Med Inform 62(2–3):165–173
30. Kaestle S (2001), Quality indicator for measurement signals, in particular, for medical measurement signals such as those used in measuring oxygen saturation, EP1196862A1, US6725074, WO2000077659A1
31. Ciocoiu IB (2001) Reţele neurale artificiale. Cantes Publishing, Iasi, Romania
32. Burse K et al (2011) Convergence analysis of complex valued multiplicative neural network for various activation functions. IEEE International Conference on Computational Intelligence and Communication System (CICN 2011), pp. 279–282
33. Poli R et al (2008) A field guide to genetic programming. Lulu Enterprises, UK
34. Shah S, Kusiak A (2007) Cancer gene search with data-mining and genetic algorithms. Comput Biol Med 37(2):251–261
35. Handels H et al (1999) Feature selection for optimized skin tumor recognition using genetic algorithms. Artif Intell Med 16(3):283–297

36. Rawata P (2016) Cognitive radio for M2M and internet of things: a survey. Comput Commun 94:1–29
37. Cisco connections counter. https://blogs.cisco.com/news/cisco-connections-counter
38. Nekovee M (2009) A survey of cognitive radio access to TV white spaces. Ultra Modern Telecommunications & Workshops, ICUMT'09. International Conference IEEE, pp. 1–8
39. Internet of Things (2015) IERC Position Paper on Standardization for IoT technologies published by IoT European Research Cluster (IERC)
40. RERUM, Reliable, Resilient and secure IoT for smart city applications. Available at: https://ict-rerum.eu
41. Pantazis NA (2013) Energy-efficient routing protocols in wireless sensor networks: a survey. IEEE Commun Surv Tutorials 15(2):551–591
42. Umer T et al. (2016) Hybrid rapid response routing approach for delay-sensitive data in hospital body area sensor network. Proceedings of the 7th International Conference on Computing Communication and Networking Technologies, p. 3
43. Amjad M et al (2016a) TinyOS-new trends, comparative views, and supported sensing applications: a review. IEEE Sensors J 16(9):2865–2889
44. Amjad M (2017) QoS-aware and heterogeneously clustered routing protocol for wireless sensor networks. IEEE Access 5:10250–10262
45. Akhtar F, Rehmani MH (2015) Energy replenishment using renewable and traditional energy resources for sustainable wireless sensor networks: a review. Renew Sust Energ Rev 45:769–784
46. Khasawneh M, Agarwal A (2016) A collaborative approach towards securing spectrum sensing in cognitive radio networks. Proced Comput Sci 94:302–309
47. Amjad MF et al (2016b) Towards trustworthy collaboration in spectrum sensing for ad hoc cognitive radio networks. Wirel Netw 22(3):781–797
48. Akyildiz IF et al (2002) A survey on sensor networks. IEEE Commun Mag 40(8):102–114
49. Rashid B, Rehmani MH (2016) Applications of wireless sensor networks for urban areas: a survey. J Netw Comput Appl 60:192–219

Chapter 7
Millimeter Waves: Technological Component for Next-Generation Mobile Networks

Jolly Parikh and Anuradha Basu

7.1 Introduction

Envisioning a world where all connectable things will be seamlessly connected has led to the evolution of mobile broadband wireless communication to fifth generation (5G). Improved throughputs, increased capacity demands, and need for defining additional use cases for wireless access are the driving forces beyond this evolution. In order to build up a connected society, 5G services will require accessing a variety of frequency bands, in the unoccupied spectrum, so as to provide better quality of service and wider channels and hence higher data rates as compared to the legacy networks. The technical requirements of 5G networks as recommended in [39] are peak data rates greater than 10 gigabits per second (Gbps), cell-edge data rate of 100 Mbps, and end-to-end latency of 1ms. The various 5G use case scenarios, identified by the International Telecommunication Union-Radio (ITU-R), which will impact the increase in spectrum demand can be enlisted as enhanced mobile broadband to deliver applications such as high-definition videos to high-density (e.g., stadium) areas with ubiquitous coverage, higher trustworthy communication for industry/transport automation, low latency communications applications, and high/medium data rate services for massive machine to machine (M2M) or machine-type communication (MTC) for applications like e-health, vehicle to vehicle (V2V), augmented reality, and tactile Internet. Figure 7.1 shows the usage scenario of the International Mobile Telecommunication (IMT) 2020 [38] and beyond requiring access to different bands in the frequency spectrum.

mmWave technology with operations in frequency bands of range 30–300 GHz will be used to meet the demands of the wide bandwidth requiring applications of the

J. Parikh (✉) · A. Basu
Bharati Vidyapeeth's College of Engineering, New Delhi, India
e-mail: jolly.parikh@bharatividyapeeth.edu

© Springer International Publishing AG, part of Springer Nature 2019   167
M. H. Rehmani, R. Dhaou (eds.), *Cognitive Radio, Mobile Communications and Wireless Networks*, EAI/Springer Innovations in Communication and Computing, https://doi.org/10.1007/978-3-319-91002-4_7

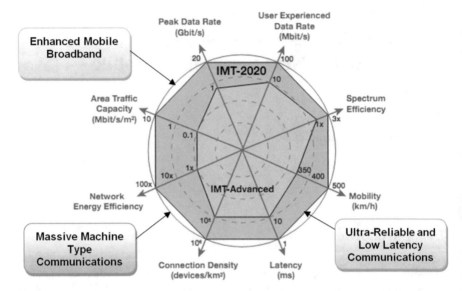

Fig. 7.1 5G usage scenarios and key capabilities of IMT 2020 as compared to IMT-Advanced [38]

future multimedia world. It is considered as an important technological component of 5G networks.

In this chapter, we have surveyed state-of-the-art works on mmWave networks for NGMNs. In summary, we make following contributions:

- We survey the various applications that are supported by networks at mmWave frequencies.
- Various frequency bands proposed by organizations around the world for mmWave operations and spectrum pooling for efficient spectrum utilization have been discussed here.
- We survey the challenges posed by the characteristics of mmWave and the solutions proposed.
- We highlight the contributions made by various organizations for standardization of channel models for 5G networks at mmWave frequencies.
- We provide insight to the state-of-the-art work done in the area of antenna technology for 5G systems at mmWave frequency.
- We discuss the research carried out for achieving energy efficiency (EE) in mmWave networks.
- We survey the literature regarding the solutions offered for optimization of mmWave networks.
- We focus upon the work carried out by researchers in the area of cognitive radio networks (CRNs) at mmWave frequencies for improving the efficiency of future generation wireless networks.

7.2   Applications of Millimeter Waves

Wide bandwidth, short wavelengths, and their performance under various atmospheric conditions may prove mmWaves to be advantageous or disadvantageous for particular application. mmWave's licensed spectra, e.g., 71–76, 81–86, and 92–95 GHz, proved to be useful for high bandwidth, point-to-point communication links, whereas the 60 GHz unlicensed spectrum is preferred for short-range data links as in Wireless Gigabit (WiGig). mmWave frequencies find their applications in the fields of entertainment, astronomy, medicine, meteorology, and various wireless communication techniques like satellite communication, cellular communication, radio-wave communication, etc. In the world of entertainment, for high-definition video transmission, mmWaves are preferred over conventional techniques of compression and then transmission of video signals. The quality of video signals is better with mmWave technology. Ultra-high-definition (UHD) video transmission from digital setup boxes, HD game stations, and other HD video sources can be carried out by integration of tiny modules designed for operation at mmWave frequencies. In the world of virtual reality (VR), mmWaves are used to provide wireless connectivity to VR headsets. This will enhance the data rates, thereby enabling transmission of high-quality video and audio from multimedia devices. This technology aims to improve the user experience. Millimeter-wave radars in the range of 76GHz–81GHz are ideal for use in vehicle control as they enable very precise detection of objects. Freescale semiconductor has come up with automotive radar chip PRDTX11101 transmitter and a multichannel receiver plus one 32-bit MCU for automotive applications at 77GHz frequencies. This chip can be used in automotive applications requiring detection ranges from 20 meters to 200 meters. With companies manufacturing chips at mmWave frequencies, vehicle-to-vehicle (V2V) communication, which is one of the important technological components of NGMNs, will gain benefits by design of autonomous vehicles. Device-to-device (D2D) communication technology and mmWave communication together can do wonders in improving the capacity of the NGMNs. There are many challenges and open issues in this area, to be worked upon. These waves being of low power have proven to be nonhazardous in security equipment like human body scanners. Hence they have found usage in security systems at public places like airports, railway stations, etc. Millimeter-wave human body scanners operating at 70–80GHz and with transmission power of 1 mW have been developed by Rhode and Schwartz for airport security systems. In the medical field, mmWaves of frequency ranges 40–70GHz are used for treatments providing relief in pain, e.g., mmTherapy or extra-high-frequency (EHF) therapy. Satellite communication which requires low latency communication links can use mmWaves in 60 GHz band for cross-linking (communications between satellites). This is because for operation at 60 GHz frequencies, at higher altitude of geosynchronous orbits, there is no oxygen in space. This results in negligible energy absorption losses at these frequencies and hence proves to be efficient links for satellite communication. High-security applications like surveillance can also benefit from this technology. Equipment designed at millimeter-wave frequencies experience attenuation even due to cables, connectors, printed circuit board (PCB) links, etc. Hence mmWaves can be used as efficient wireless links for PCB-to-PCB or chip-to-chip linking. Yet another important application of mmWaves is to provide flexible and cost-effective backhaul systems for 5G networks. An article by [28] gives an insight to this future backhaul technology and discusses the related open problems and potential solutions. Thus applications requiring high data rates, low latency, and good quality of user experience would demand for wider bandwidth far beyond the 100 MHz of bandwidth required by 4G systems.

7.3 Millimeter-Wave Frequency Spectrum

As lower frequency bands are already occupied by other applications and services, higher frequency mmWave bands, with potentially large spectrum availability, are being focused upon. The mmWave technology for 5G and beyond networks is a promising candidate for addressing the ongoing spectrum shortfall and network congestion experienced in spectrum below 6 GHz. Massive amount of raw bandwidth exists in the mmWave spectrum. Recently, 28 GHz, 38 GHz, and 70–80 GHz frequency bands are found to be more suitable for outdoor mmWave mobile communications. This is because of low attenuation loss due to atmospheric absorption (much less than 0.1 dB/200 m) as compared to that at 60 GHz (approx. 4 dB/ 200 m). Advantages of exploiting the 28 GHz bands for wireless backhauling cellular communications in range of 500 meters [74] are that it can create multipath environments, it can be used for non-line-of-sight (NLOS) communications, and most importantly it can reuse the 3rd Generation Partnership Project (3GPP) Long-Term Evolution (LTE) functionalities. Table 7.1 lists the various mmWave bands and their frequency ranges.

Different organizations like the World Radiocommunication Conference 2015 (WRC-15) in ITU-R, the European Conference of Postal and Telecommunications Administrations (CEPT) in Europe, the Federal Communications Commission

Table 7.1 mmWave frequency bands and their corresponding frequency ranges

mmWave frequency band	Frequency range(GHz)
Q band	30–50
U band	40–60
V band	50–75
E band	60–90
W band	75–110
F band	90–140
D band	110–170
G band	110–300

Table 7.2 Frequency candidates for 5G and beyond selected in the four different organizations [11]

WRC-15	CEPT	FCC	5GMF
24.25 – 27.5	24.25 – 27.5		
		27.5 – 28.35	24.75 – 31.0
31.8 – 33.4	31.8 – 33.4	31.8 – 33.4	
37.0 – 40.5		37.0 – 38.6	31.5 – 42.5
		38.6 – 40.0	
40.5 – 42.5	40.5 – 43.5	40.5 – 42.5	
42.5 – 43.5			
45.5 – 47.0			45.3 – 47.0
47.0 – 47.2	45.5 – 48.9		
47.2 – 50.2		47.2 – 50.2	47.0 – 50.2
50.4 – 52.6			50.4 – 52.6
66.0 – 76.0	66.0 – 71.0	64.0 – 71.0	66.0 – 76.0
	71.0 – 76.0	71.0 – 76.0	
81.0 – 86.0	81.0 – 86.0	81.0 – 86.0	81.0 – 86.0

(FCC) in the USA, and the Fifth Generation Mobile Communications Promotion Forum (5 GMF) in Japan [26, 27, 39], (5GMF 2016) have proposed different frequency bands for 5G and beyond communication networks. Table 7.2 summarizes these proposed frequency bands. The 5G Americas White Paper on 5G Spectrum Recommendations [4] shows the status of public proposals for 5G spectrum bands worldwide as of July 2015.

Even at mmWave frequencies, it becomes essential to implement methods for efficient utilization of the spectrum. [11] have proposed the concept of spectrum pooling at mmWaves. They have discussed the type of supporting architecture, the spectrum access techniques, and the impact of beam forming at both base stations and user terminals, when implementing spectrum pooling in mmWave networks. Simulations showing the performance of the spectrum pooling over different carrier frequencies have been discussed, emphasizing the importance of pooling at mmWave, for efficient spectrum utilization.

7.4   Characteristics of mmWaves

Propagation characteristics of mmWaves are different than those of UHF waves. The higher the frequency, the larger the absorption by obstacles, higher penetration losses, higher path loss, and stronger shadowing. Attenuation due to rain and absorption at molecular and atmospheric levels limit the range of mmWave communications [24, 36, 88]. The oxygen, water absorption, and rain losses at mmWave frequencies are shown in [87]. Figure 7.2 depicts the atmospheric attenuation of mmWaves, and Fig. 7.3 shows the rain attenuation. It has been seen that oxygen absorption comes out to be 15–30 dB/Km at 60 GHz [22].

As material penetration is low, outdoor to indoor propagation is difficult at mmWave frequencies, but interference reduces significantly as we have increased isolation between indoor and outdoor. Figure 7.4 shows the material penetration losses [67, 89] at higher frequencies.

Singh et al. [78] had demonstrated that the free space propagation loss which is proportional to the square of carrier frequency comes out to be 28 dB more at 60 GHz as compared to that at 2.4 GHz. The mmWave propagation will not exhibit larger free space loss if same physical area is used by multiple antennas as in massive multiple-input multiple-output (MIMO) and with beamforming implementation. Longer transmission paths can be achieved by implementing these techniques. Dynamic beamforming can be implemented both at base stations and at mobile stations, so as to concentrate the signals to form focused beams which will mitigate the high path loss at mmWave frequencies. The benefits of hybrid beamforming architecture for 5G networks as compared to the architecture of 4G networks have been highlighted by the work carried out in [46]. The simulation results carried out

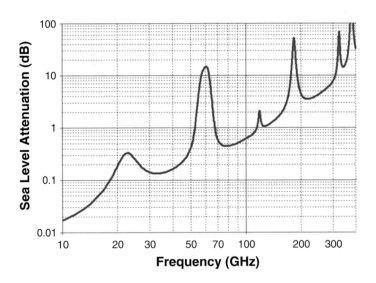

Fig. 7.2 Atmospheric and molecular absorption at mmWave frequencies [24]

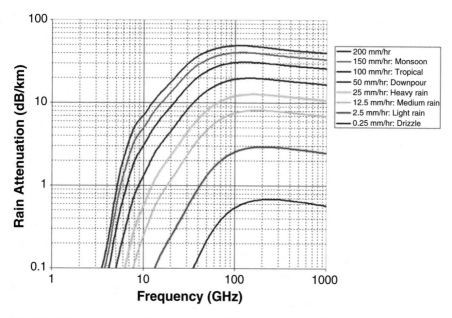

Fig. 7.3 Rain attenuation at microwave and mmWave frequencies [24]

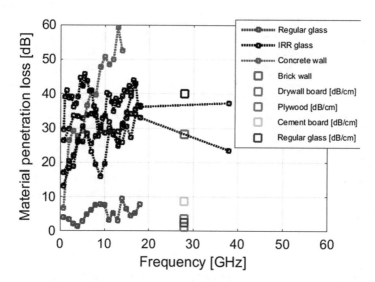

Fig. 7.4 Measured material penetration losses. (Sources: [89] and measurements by Samsung and Nokia)

supports the theory that beamforming technology will prove to be a potential technology for 5G cellular networks in terms of improvement in energy efficiency. With implementation of massive MIMO techniques, the number of radio-frequency modules will increase, resulting in increase in power consumption. The challenge of power optimization will have to be dealt with.

With increase in frequency, radio-frequency propagation behaves like optical propagation, thereby encountering blockage. This in turn causes additional losses on the path. It had been observed in [77] that human blockage affects the link budget by about 20–30 dB. Hence at mmWave communication, delay-sensitive applications, like high-definition television (HDTV), will pose a big challenge. NLOS coverage will be more challenging, and line-of-sight (LOS) coverage will have difficulties in reception due to obstructions. The NLOS channel exhibits higher attenuation as compared to a LOS channel [31]. Diffraction effects are practically negligible at mmWave frequencies. Diffuse reflection (scattering of signal energy when encountering an object) scatters the radio waves at wider range angles, thereby extending the coverage range and reaching the NLOS areas.

As per Qualcomm engineers, alternative NLOS paths – signals bouncing off buildings or other structures – have higher energy than a LOS path. These NLOS paths can be used for outdoor wireless communication. mmWave frequencies are preferred for short-range communications in small cells, wherein there are fewer users as compared to conventional microcells and macrocells. Problem of pilot contamination is overcome by mmWaves. This is because signals transmitted by mobile stations during uplink tend to fade rapidly at these frequencies, with the distance, thereby creating fewer problems for surrounding base stations which are exchanging channel information from targeted mobile stations [53]. Higher propagation losses resulting in shorter transmission paths enable spectrum reuse. Millimeter-wave technology supports indoor coverage as well as about 200–300m in outdoor coverage, based on channel conditions and height of the access point and eNodeB, above the ground.

The survey article by [56] discusses the challenges posed by the characteristics of the mmWave communications and their potential solutions. Guidelines for architectures and protocols for mmWave communications have been proposed along with few open research issues. Table 7.3 lists the differences between systems operating at different 5G mmWave frequencies, with respect to certain key parameters.

7.5 Contributions for Standardization of Channel Models for 5G Networks at mmWave Frequencies

Varied characteristics of mmWaves make NGMN channel modeling a challenging issue. The 3rd Generation Partnership Project (3GPP) initiated a working group in September 2015 to develop channel models for spectrum above 6 GHz and released TR 38.900 release 14 for channel models above 6 GHz [3]. Other groups working on

Table 7.3 Differences between systems operating at different 5G mmWave frequencies

Features	5G sub-6GHz	5G cm wave	5G mm wave
Frequency spectrum	0.5–10 GHz	3–30 GHz	30–100 GHz
Bandwidth	Approx. 5–40 MHz	Approx. 40–200 MHz	Approx. 400–2 GHz
Duplex topology	FDD/TDD	Dynamic TDD	Dynamic TDD
Transmit power (DL/UL)	>40 dBm/ 23 dBm	<approx. 30 dBm/ 23 dBm	<approx. 30 dBm/ 23 dBm
Modulated waveform (DL/UL)	OFDMA/SC-FDMA	OFDMA/OFDMA	SC-TDMA/SC-TDMA
Multiple access type	Time and frequency	Time and frequency	Time
Multi antenna technique	SU/MU beamforming and medium rank	SU/MU beamforming and high rank	SU/MU beamforming and low rank
TTI (transmission time interval)	Flexible	Approx. 0.25 ms	Approx. 0.1 ms

Table 7.4 Existing and ongoing campaign efforts worldwide for 5G channel measurements and modeling

Organization	Contribution
METIS2020 [48]	Studies on channel modeling (up to 86 GHz)
	3D modeling, spherical wave modeling, high spatial resolution
METIS (Leszek et al. 2015)	Map-based models, stochastic model, hybrid model for flexibility and scalability
COST2100 [90]	Geometry-based stochastic channel model for reproducing the stochastic properties of MIMO channels over time, frequency, and space
NIST 5G mmWave alliance [55]	Provides guidelines for measurement calibration and parameterization in various environments
NYU WIRELESS [60–72]	Provided urban propagation measurements at 28, 38, 60, and 73 GHz for indoor and outdoor large-scale and small-scale channel models. Introduced the concepts of time cluster spatial lobes (TCSL)

the same path are Mobile and wireless communications Enablers for the Twenty-twenty (2020) Information Society (METIS 2020) [48], Millimeter-Wave Evolution for Backhaul and Access (MiWEBA) [50], Millimeter-Wave Based Mobile Radio Access Network for Fifth Generation Integrated Communications (mmMagic) [51], European Telecommunications Standards Institute (ETSI) mmWave [25], IEEE 802.11ad, International Conference on Cooperative Radio communications for Green Smart Environments (IC1004) [37], National Institute of Standards and Technology (NIST) 5G mmWave Channel Model Alliance [55], and New York University (NYU) WIRELESS [45, 61, 62, 73]. Table 7.4 lists few of the existing and ongoing campaign efforts carried out worldwide for 5G channel modeling.

Urban micro (UMi), urban macro (UMa), and in-house (InH) scenarios were studied in [], and the rural macro (RMa) scenario at 73 GHz was proposed by [42]. The various statistical models derived from 28 GHz and 73 GHz mmWave

RF measurements have been proposed by M. K. Samimi in his doctoral work [69]. These can be used as benchmark to evaluate the performance of wireless broadband systems, both in link and system-level simulations, for LOS as well as NLOS environments. In [8], it was seen that with multiple clusters supporting spatial multiplexing, strong signals can be detected up to 100–200m, from potential cell sites, even in highly NLOS environments. The authors had predicted that mmWave systems would offer higher capacities than the current 4G cellular networks even with same density of small cell networks, in urban environments as of today. Other channel models based on the concept of multipath clusters have also been proposed by few researchers []. These models take into consideration the signal blockages occurring in presence of macroscopic obstacles lying in between transmitter and receiver. In [15], the key differences between massive MIMO channel models at microwave and mmWaves have been discussed along with their effects on the configuration of the future communication transceivers. The close-in (CI) free space reference distance path loss model and the close-in free space reference distance with height-dependent path loss exponent model (CIH) which can be used from 500 MHz to 100 GHz for rural mmWave coverage and interference analysis have been proposed in [43]. CIH model predicts rural macrocell path loss taking into consideration the effective path loss exponent as a function of base-station antenna height. The 3GPP technical specification 38.900 [3] included scenarios of UMi, UMa, and InH for outdoor and indoor 5G environments [] as well as RMa scenario using frequencies up to 100 GHz. It can be seen that a lot of work has been done toward standardization of channel models for NGMNs under varied scenarios, and it is still an ongoing process. Based on the existing channel models, efforts are being made toward practical implementation of wireless networks in the future. Techniques for achieving energy efficiency and optimization have been worked upon over the past few years. The following sections give an insight to the development done in these fields.

7.6   Energy Efficiency in Networks Operating at mmWave Frequencies

Network energy efficiency (EE) is being considered as one of the important system-level parameters and is attracting researchers to work upon the open issues in this field. Numerous techniques such as MIMO, beamforming, cognitive radio, cooperative heterogeneous communications, etc. have been proposed to improve the EE of networks. Figure 7.5 shows how EE can be obtained in heterogeneous networks.

Use of relay stations has also been considered as an effective way to achieve EE in cellular networks. EE in relay-assisted mmWave cellular networks is studied in [84, 85], and open issues related to EE in mmWave ultra-dense small cell networks have been identified in these research articles. Through his research work discussed in [9], he has proposed solutions for the improvement of energy-efficient properties

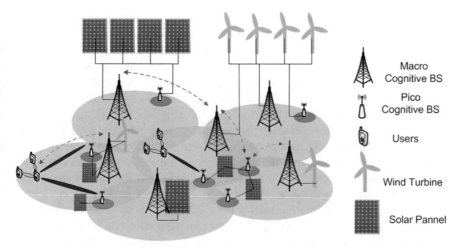

Fig. 7.5 Energy efficiency in heterogeneous networks

of mobile devices in next-generation wireless networks. The future scopes identified by this work have been researched upon over a period of time paving path for improvement of upcoming technologies like device-to-device (D2D) communications. In [18], authors have discussed the various EE techniques for 5G networks and the research challenges associated with this area which can be worked upon to achieve high EE in NGMNs. In [58] authors have carried out simulations to show that mmWave EE proves to be superior to 2 GHz system EE in environments facilitating LOS links. Networks deployed in environments with high signal to interference and noise ratio (SINR) at receiver are more energy efficient when operating at 28 GHz and 60 GHz frequencies than those operating at 2 GHz frequency but tolerating lower SINRs at receiver. In [29] the system-level dynamics of mmWave-based wearable networks have been assessed which will be constituting a large chunk of the next-generation personal networks.

7.7 Antenna Technology for 5G Systems at mmWave Frequencies

To achieve the 5G requirements of higher peak rates (>10Gbps) and cell-edge rates (100 Mbps), advanced semiconductor and antenna technology will be required. Beamforming and spatial multiplexing techniques are being considered to improve the system capacity and coverage. Large-scale phased arrays are used to implement these techniques in 5G systems. The extra path loss of 20–30 dB which occurs at 3–30 GHz/mmWave (30–1100 GHz) bands can be compensated by using large-scale phased arrays. The high gain antenna arrays, which are also highly directional, can boost the effective radiative power (ERP), thereby mitigating the losses and interference resulting in larger transmission ranges. In [1381] concepts of use of massive MIMO at mmWave frequency bands have been discussed. Marzetta had originally launched the idea of a large-scale antenna array at base stations in [47]. Theoretically, at 30 GHz carrier frequency, if more than 180 antennas can be mounted in an area, then at 80 GHz carrier frequency, the same space can be occupied by about 1300 m antennas arrayed. This concept of double massive MIMO both at transmitter and receiver had been proposed in [13]. The feasibility of such user terminal is practically constrained by certain parameters which include large power consumption, power amplifiers with low efficiency, more complex hardware, and so on. Major design considerations and challenges of mmWave 5G antennas for cellular handsets have been discussed in [35]. A compact RF transmitter and receiver for 5G communication systems, which can be applied in Q-band mmWave massive MIMO systems of 16 channels, have been proposed in [21]. Many industries have developed modules for supporting mmWave technology communications. Qualcomm Atheros developed ARR9642 802.11n transceiver which includes Wilocity 60 GHz transmitter chip to form three bands (2.4, 5, and 60 GHz bands) that suit hot spots, routers, and other WLAN products. Hittite Microwave Corporation has come up with HMC6000LP711E transmitter and HMC6001LP711E receiver integrated chips (ICs), both operating in 57–64 GHz industrial and scientific band (ISM band). These are used for designing WiGig devices and data communication equipments for backhaul and HD video transmissions.

7.8   Cognitive Radios and mmWave Technology

Cognitive radios (CRs) work on the technique of continually sensing the spectrum and detecting and utilizing unoccupied channels [34, 49] and provide solutions to the present spectrum congestions. As compared to software-defined radios (SDRs), CRs face more challenges [66]. Energy detection and feature detection are some of the techniques which enable spectrum sensing [19]. The comprehensive survey carried out by authors of [52] is an in-depth study of communication over CR-based wireless networks such as cognitive radio sensor networks (CRSNs), mesh networks, CR cellular networks, and CR ad hoc networks. Wireless mobile and cognitive radio networks (WMCRNs) have been studied in aspect to their multimedia applications, routing protocols, MAC protocols, quality of experience (QoE), spectrum sensing approaches, and cross layer designs. It is an exhaustive work focusing upon the open issues and challenges in the world of CRNs. Many future research directions in this area have been highlighted here. Authors of [20] have proposed the use of mmWave and Wi-Fi CogCell hybrid architecture for 5G. It would enable smooth network management, fast channel access, and device discovery over 24GHz channel and data transmission over 60 GHz channel. Control plane functionality will be by Wi-Fi, while the data plane functionality will be supported by 60 GHz. In [83], the authors discuss two underlay cognitive transceiver designs that enable mmWave spectrum access along with interference combating. The architecture proposed in this work is a hybrid analog digital precoding architecture. [23] have provided a comprehensive survey and tutorial on spectrum inference in CRNs. Few critical research challenges and open issues in the area of spectrum inference in future wireless networks have been discussed here to enable future researchers to work on them. Work is ongoing in the field of CRs and mmWave communication. Few of the research areas identified in this field are full-duplex communication CR, energy-efficient CR, CR in wireless sensor networks, and CR in smart grids [41].

7.9   Optimization in mmWave Networks

Optimization in mmWave networks is yet another challenge faced while designing a network at mmWave frequencies. Till date a lot of work has been done in optimizing the mmWave networks in terms of energy efficiency, sum rate, spectral efficiency, interference encountered, and circuits and components used in hardware equipments. Researchers are still striving to design the optimal 5G mmWave networks. To obtain higher energy efficiency in mmWave massive MIMO systems, the concept of beam selection is adopted. Taking into consideration the interferences from multiple users, the authors in [30] have proposed two interference-aware (IA) beam selection techniques to achieve near-optimal sum-rate performances in massive MIMO-implemented mmWave systems. As compared to the conventional millimeter beam selection techniques proposed in [75], these systems offer higher sum rate and EE, the two important network performance metrics in cellular networks. In [76], simulations have been carried out to illustrate the performance of hybrid precoding algorithms proposed by the authors. It is observed that near-optimal spectral efficiency can be achieved by using these methods under perfect channel state information. Yet another article by [7, 65] discusses the power optimization techniques for next-generation wireless networks. The book by [16] provides a comprehensive coverage of all aspects to achieve optimization in mmWave networks. It covers views of experts from industry as well as academia. Optimization for cognitive radio networks (CRNs), considering wideband spectrum sensing, with noise variance uncertainty has been carried out by authors of [12]. Solutions for throughput maximization of the CRN have been discussed here.

7.10   Projects Carried Out by Different Groups

Many organizations across the world have been working to standardize and meet the requirements of 5G networks. The mmMagic project has been one of the largest 5G PPP (public-private partnership) phase I projects, focusing on key technology components and architecture of 5G mobile communications systems operating between 6 and 100 GHz. It has made substantial contributions to 5G global

Table 7.5 Projects of mmWave group

Group	Project period	Area of work
5G Berlin	Jan 2018–Dec 2028	Building of 5G test area and opening of 5G center
5G Champion	Jun 2016–Jun 2018	Development of key enabling technologies for the proof of concept environment at 2018 Winter Olympics. New architectural approach for efficient end-to-end performance for 5G radio access, core network, and satellite technologies
AMMCOA	Apr 2017–Mar 2020	To develop mobile infrastructure-less wireless 5G networking solutions for tactile communication
LZ-5G Testbed	Jul 2016–Jun 2018	Efforts for further development of 5G mobile communications into early trial-based industry partners
5G Cross Haul	Jul 2015–Jan 2018	Developing a 5G integrated backhaul and fronthaul transport network enabling a flexible and software-defined reconfiguration of networking elements
mmMagic	Jul 2015–Jun 2017	Evaluation of development of new mobile radio access technologies for wireless communications in frequency range of 6–100 GHz

standardization (3GPP) and WRC19 preparatory work (ITU-R). As per the research work carried out at [57], new frame structures will be required for supporting new 5G services like (ultra-reliable and low-latency communication) URLLC and mmWave band operations. 3GPP TR38.802 [2] describes the technical specifications of new radio access technologies from physical layer aspects. The tutorial paper [86] summarizes the recent technical development in the field of mmWave communications for mobile networks. Many challenges of this area have been identified, and till date solutions provided by different organizations and industries, contributing toward standardization of deployment of mobile networks at mmWave, have been highlighted in this in-depth study work. Research conducted by Qualcomm technologies have shown that about 81% of outdoor downlink coverage is possible by using 5G new radio mmWave with existing 4G Long-Term Evolution macro- and small cell sites [59]. The prototype system developed by Qualcomm will contribute to the rapid commercialization of 5G new radio technologies. Table 7.5 gives a brief view of the projects carried out by the mmWave group toward standardization and meeting the 5G network requirements.

7.11 Future Areas of Research for 5G Networks at mmWave Frequencies

Few of the research areas which have been identified during the research work being carried out till date are listed as follows:

Architecture and air interface of networks operating at mmWave:

- Designing network architecture to support mmWave (as non-standalone and standalone) systems.
- Techniques for optimizing the air interfaces in 5G networks and beyond (waveforms, channel coding, frame structure, etc.).
- Pilot contamination at mmWave frequencies.
- Channel estimation for mmWave frequencies.
- Designing channel models to support various scenarios in mmWave environment.
- In order to further validate the RMa, CI, and CIH models proposed in [43], by the NYU WIRELESS NYUSIM software [80], additional measurements across new frequencies, transmitter heights, and distances are needed.

Mobility and coverage at mmWaves:

- Studies need to be carried out to quantify the performance of systems with indoor mobile users as well as users falling in out of coverage areas. The solutions can be found based on the concept of either multihop relaying or fall back to conventional microwave cells.
- Effective solutions for application of mmWaves in high-mobility environments taking into consideration the different propagation losses.
- Energy-efficient techniques for supporting long transmission distances at mmWave frequencies need to be identified.

Improvements in directivity and data rates:

- Double massive MIMO systems at mmWaves
- Developing of multi-antenna beamforming architectures and providing beam tracking/aligning solutions

Hardware modeling:

- Modeling of hardware impairments in the transmitter RF chains and in antenna designs. At receiver, hybrid architecture using fewer RF chains needs to be used. Low-power analog processing circuits need to be designed in mmWave hardware. For future generation of networks, a cost-efficient, fully digital RF implementation should be worked upon.

7.12 Conclusion

This chapter provides insight to the efforts made by researchers toward overcoming the challenges in the usage of mmWave technology to meet the 5G requirements. The challenges faced by the use of mmWave frequencies for bandwidth-hungry applications of future-generation wireless networks, the solutions offered, and the future research areas of mmWave technology have been surveyed herein. To summarize, we have tried to provide till date progress done for defining mmWave technology as one of the chief technological components for NGMNs. Hope this will pave the path for designing of efficient, heterogeneous networks of future generations and will contribute to achieving a smart environment where things will be connected and communicating with each other efficiently.

References

1. 3GMF White Paper (2016) 5G Mobile Communications systems for 2020 and beyond
2. 3GPP (2017) Study on new radio access technology Physical layer aspects. 3GPP V14.1.0, TR 38.802, June 2017
3. 3GPP Technical specification group radio access network (2016) Channel model for frequency spectrum above 6 GHz (Release 14). 3rd Generation Partnership Project (3GPP), TR 38.900 V14.2.0. http://www.3gpp.org/DynaReport/38900.htm. Accessed 1 Oct 2017
4. 5G Americas (2017) White Paper on 4G Spectrum Recommendations
5. Aalto University, AT&T, BUPT, CMCC, Ericsson, Huawei et al (2016) 5G channel model for bands up to 100 GHz. http://www.5gworkshops.com/5GCM.html. Accessed 2 Oct 2017
6. Aalto University, BUPT, CMCC, Nokia et al (2015) 5G Channel model for bands up to 100 GHz. Technical report. report, 6 December 2015
7. Abrol A, Jha RK (2016) Power optimization in 5G networks: a step towards GrEEn communication. IEEE Access 4:1355–1374
8. Akdeniz MR, Liu Y, Samimi M, Sun S, Rangan S et al (2014) Millimeter wave channel modeling and cellular capacity evaluation. IEEE J Sel Areas Commun 32(6):1164–1179
9. Andreev S (2012) Energy efficient and cooperative solutions for next-generation wireless networks, vol 1060. Tampere University of Technology Publication, Tampere
10. Ayach O, Rajagopal S, Abu-Surra S et al (2014) Spatially sparse precoding in millimeter wave MIMO systems. IEEE Trans Wirel Commun 13(3):1499–1513
11. Boccardi F et al (2016) Spectrum pooling in MmWave networks: opportunities challenges and enablers. https://arxiv.org/abs/1603.01080
12. Bogale TE, Vandendorpe L, Le LB (2015) Wideband sensing and optimization for cognitive radio networks with noise variance uncertainty. IEEE Trans Commun 63(4):1091–1105
13. Buzzi S, D'Andrea C (2016a) Doubly massive mmWave MIMO systems: using very large antenna arrays at both transmitter and receiver. In: 2016 I.E. Global Telecommunications Conference (GLOBECOM 2016), Washington DC, USA, 5–7 December 2016
14. Buzzi S, D'Andrea C (2016b) Are mmWave low-complexity beamforming structures energy-efficient? Analysis of the downlink MU-MIMO. In: Proceedings of the International Workshop on Emerging Technologies for 5G Wireless Cellular Networks, in conjunction with 2016 I.E. GLOBECOM, Washington DC, USA, 4 December 2016
15. Buzzi S, D'Andrea C (2017) Massive MIMO 5G cellular networks: mmwave vs. micro wave frequencies. https://doi.org/10.3969/j
16. Buzzi M, Samanta K (2016) Advanced technologies for mmWave circuits and system integration. Academic Press, ISBN:0124166539 9780124166530
17. Buzzi S, D'Andrea C, Foggi T, Ugolini A, Colavolpe G (2016a) Spectral efficiency of MIMO millimeter-wave links with single-carrier modulation for 5G networks. In: Proceedings of 20th International ITG Workshop on Smart Antennas (WSA 2016), Munich, Germany, March 2016
18. Buzzi S et al (2016b) A survey of energy-efficient techniques for 5G networks and challenges ahead. IEEE J Sel Areas Commun 34(4):697–709
19. Cabric D, Mishra SM, Brodersen RW (2004) Implementation issues in spectrum sensing for cognitive radios. In: Conference Record of 38th Asilomar Conference on Signals, Systems, and Computers, November 2004

20. Chandra K, Prasad R, Quang B, Niemegeers IGMM (2015) CogCell: cognitive interplay between 60GHz Picocells and 2.4/5GHz hotspots in the 5G era. IEEE Commun Mag 53 (7):118–125
21. Chen J, Lin W, Yan P, Xu J, Hou D, Hong W (2017) Design of mm-Wave transmitter and receiver for 5G. In: 10th Global Symposium on Millimeter-Waves, China, 24–26 May 2017, https://doi.org/10.1109/GSMM.2017.7970330
22. Daniels RC, Heath RW (2007) 60 GHz wireless communications: emerging requirements and design recommendations. IEEE Veh Tech Mag 2(3):41–50
23. Ding G et al (2017) Spectrum inference in cognitive radio networks: algorithms and applications. IEEE Commun Surv Tutorials 99
24. E-band technology (2017) E-band communications. http://www.e-band.com/index.php?id=86. Accessed 10 Sep 2017
25. ETSI (2015) New ETSI group on millimetre wave transmission starts work. http://www.etsi.org/news-events/news/866-2015-01-press-new-etsigroup-on-millimetre-wave-transmission-starts-work. Accessed 25 Sep 2017
26. European Commission (2016) 5G for Europe: an action plan
27. Federal Communications Commission (2016) Report and order and further notice of proposed rulemaking. FCC, Washington, DC, pp 16–89
28. Feng W et al (2016) Millimetre-wave backhaul for 5g networks: challenges and solutions. Sensors 16(6):892
29. Galinina O, Pyattaev A, Johnsson K, Turlikov A, Andreev S, Koucheryavy Y (2016) Assessing system-level energy efficiency of Mmwave-based wearable networks. IEEE J Sel Area Commun 34(4):923–937
30. Gao X, Dai L, Chen Z, Wang Z, Zhang Z (2016) Near-optimal beam selection for beamspace mm wave massive MIMO systems. IEEE Commun Lett 20(5):1054–1057
31. Geng SY, Kivinen J, Zhao XW, Vainikainen P (2009) Millimeter wave propagation channel characterization for short-range wireless communications. IEEE Trans Veh Technol 58(1):3–13
32. Haneda K et al (2016a) Indoor 5G 3GPP-like channel models for office and shopping mall environments. In: IEEE International Conference on Communications Workshops (ICCW), Kuala Lumpur, Malaysia, May 2016
33. Haneda K et al (2016b) 5G 3GPP-like channel models for outdoor urban microcellular and macrocellular environments. In: IEEE 83rd Vehicular Technology Conference (VTC2016-Spring), Nanjing, China, May 2016, pp 1–7
34. Haykin S (2005) Cognitive radio: brain-empowered wireless communications. IEEE J Sel Areas Commun 23(2):201–220
35. Hong W, Baek K-h, Ko S (2017) Millimeter-Wave 5G antennas for smartphones: overview and experimental demonstration. IEEE Trans Antennas Propag 65:6250. https://doi.org/10.1109/TAP.2017.2740963
36. Humpleman RJ, Watson PA (1978) Investigation of attenuation by rainfall at 60 GHz. Proc Inst Electr Eng 125(2):85–91 February 1978
37. IC1004 (2013) http://www.ic1004.org/
38. ITU-R M.2083-0 (2015) IMT vision-framework and overall objectives of the future development of IMT for 2020 and beyond
39. ITU-R Provisional Final Acts (2015) World Radio Communication Conference (WRC-15), p 426
40. Lee J, Gil GT, Lee YH (2014) Exploiting spatial sparsity for estimating channels of hybrid MIMO systems in millimeter wave communications. In: 2014 I.E. Global Communications Conference, Austin, TX, 8–12 December 2014, pp 3326–3331
41. Lodro M (2016) Cognitive radio and millimeter waves: research directions. https://wwwresearchgatenet/publication/299410411, March 21, 2016
42. MacCartney G, Jr, Sun S, Rappaport T, Xing Y, Yan H, Wang R, Yu D (2016) Millimeter wave wireless communications: new results for rural connectivity. In: Proceedings of the 5th

workshop on all things cellular: operations, applications and challenges, pp 31–36, ACM, October 03–07, 2016
43. MacCartney G Jr. and Rappaport T (2017) Study on 3GPP Rural Macrocell Path Loss Models for Millimeter Wave Wireless Communications. In: 2017 I.E. International Conference on Communications (ICC), Paris, pp. 1–7, May 2017
44. MacCartney GR Jr et al (2015) Indoor office wideband millimeter-wave propagation measurements and models at 28 GHz and 73 GHz for ultra-dense 5G wireless networks. IEEE Access 3:2388–2424
45. MacCartney GR Jr, Rappaport T, Sun S, Deng S (2015) Indoor office wideband millimeter-wave propagation measurements and channel models at 28 and 73 GHz for ultra-dense 5G wireless networks. IEEE Access 3:2388–2424
46. Malatala M, Deruyck M, Tanghe E, Martens E, Joseph W (2017) Performance evaluation of 5G millimeter-wave cellular access networks using a capacity-based network deployment tool. Hindawi Mobile Information Systems, Article ID 3406074, https://doi.org/10.1155/2017/3406074, Volume 2017
47. Marzetta T (2015) Massive MIMO: an introduction. Bell Labs Tech J 20:1538–7305/14, ALCATEL-LUCENT
48. METIS2020 (2015) METIS channel model. Tech Rep METIS2020, Deliverable D1.4 v3. https://www.metis2020.com/wpcontent/uploads/METIS. Accessed 25 Sep 2017
49. Mitola J, Maquire GQ (1999) Cognitive radio: making software radios more personal. IEEE Pers Commun 6(4):13–18
50. MiWEBA (2014) Channel modeling and characterization. Tech. Rep. MiWEBA, Deliverable D5.1.http://www.miweba.eu/wp-content/uploads/2014/07/MiWEBA. Accessed 25 Sep 2017
51. mmMagic (2014) https://5g-ppp.eu/mmmagic/. Accessed 25 Sep 2017
52. Muhammad A, Rehmani MH and Shiwen Mao (2018) Wireless Multimedia Cognitive Radio Networks: A Comprehensive Survey. IEEE Communications Surveys & Tutorials, Issue 99
53. Naqvi SAR, Hassan SA, ul Mulk Z (2016) Pilot reuse and sum rate analysis of mmWave and UHF based massive MIMO systems. In: IEEE 83rd Vehicular Technology Conference (VTC Spring), Nanjing, May 2016, pp 1–5
54. Nguyen HC et al (2016) An empirical study of urban macro propagation at 10, 18 and 28 GHz. In: IEEE 83rd Vehicular Technology Conference (VTC2016-Spring), Nanjing, China, May 2016
55. NIST (2015) https://www.nist.gov/ctl/5g-mmwave-channel-model-alliance. Accessed 25 Sep 2017
56. Niu Y, Li Y, Jin D, Li Su, Vasilakos A (2015) A survey of millimeter wave (mmWave) communications for 5G: opportunities and challenges. Springer US, https://doi.org/10.1007/s11276-015-0942-z
57. Nomor Research (2017) 5G Frame Structure. White paper, August 2017
58. Omar M, Imran A, Refai H (2016) mmWave based vs 2 GHz networks: what is more energy efficient. In: Wireless Communications and Mobile Computing Conference (IWCMC), Cyprus, 5–9 September 2016
59. Qualcomm (2017) Mobilizing 5G NR mmwave network coverage simulation. White paper, https://www.qualcomm.com/media/documents/files/white-paper-5g-nr-millimeter-wave-network-coverage-simulation.pdf. Accessed 2 Nov 2017
60. Rappaport T, Sun S, Mayzus R et al (2013) Millimeter wave mobile communications for 5G cellular: it will work! IEEE Access 1:335–349
61. Rappaport T, MacCartney G Jr, Samimi M, Sun S (2015a) Wideband millimeter-wave propagation measurements and channel models for future wireless communication system design. IEEE Trans Commun 63(9):3029–3056
62. Rappaport T, Heath RW Jr, Daniels RC, Murdock JN (2015b) Millimeter wave wireless communications. Pearson/Prentice Hall, Upper Saddle River

63. Rappaport T et al (2015c) Wideband millimeter-wave propagation measurements and channel models for future wireless communication system design. IEEE Trans Commun 63 (9):3029–3056

64. Raschkowski L et al (2015) Report number: ICT-317669-METIS/D1.4 ver 3, Affiliation: ICT-317669: Mobile and wireless communications Enablers for the Twenty-twenty Information Society

65. Ratheesh R, Vetrivelan P (2016) Power optimization of wireless network. IJET 8(1):247–256

66. Razavi B (2009) Challenges in the design of cognitive radios. In: IEEE 2009 Custom Integrated Circuits Conference (CICC), Italy, https://doi.org/10.1109/CICC.2009.5280806, 13–16 September 2009

67. Rodriguez I et al (2014) Radio propagation into modern buildings: Attenuation measurements in the range from 800 MHz to 18 GHz. In: IEEE 80th Vehicular Technology Conference (VTC Fall), September 2014

68. Sakaguchi K et al (2017) Where, when and how mmWave is used in 5G and beyond. arXiv:1704.08131

69. Samimi M (2016) Statistical channel models for 5G. Dissertation, New York University

70. Samimi M, Rappaport T (2015) 3-D Statistical channel model for millimeter-wave outdoor communications. In: IEEE International Conference on Communications (ICC), London, June 2015

71. Samimi M, Rappaport T (2016a) Local multipath model parameters for generating 5g millimeter-wave 3gpp-like channel impulse response. In: 10th European Conference on Antennas and Propagation (EuCAP 2016), Davos, Switzerland, April 2016

72. Samimi M, Rappaport T (2016b) 3D millimeter-wave statistical channel model for 5G wireless system design. IEEE Trans Microw Theory Tech 64(7):2207–2225

73. Samimi M, Rappaport T, MacCartney GR Jr (2015) Probabilistic omnidirectional path loss models for millimeter-wave outdoor communications. IEEE Wirel Commun Lett 4(4):357–360

74. Samsung Electronics Co., Ltd. (2015) 5G vision white paper. http://www.samsung.com/global/businessimages/insights/2015/Samsung-5G-Vision-0.pdf. Accessed 11 Oct 2017

75. Sayeed A, Brady J (2013) Beamspace MIMO for high-dimensional multiuser communication at millimeter-wave frequencies. In: Proceedings of the IEEE GLOBECOM, December 2013, pp 3679–3684

76. Shi Q, Hong M (2018) Spectral efficiency optimization for millimeter wave multi-user MIMO systems. arXiv:1801.07560 [cs.IT]

77. Singh S, Ziliotto F, Madhow U, Belding EM, Rodwell M (2009) Blockage and directivity in 60 GHz wireless personal area networks: from cross-layer model to multi hop MAC design. IEEE J Sel Areas Commun 7(8):1400–1413

78. Singh S, Mudumbai R, Madhow U (2011) Interference analysis for highly directional 60-GHz mesh networks: the case for rethinking medium access control. IEEE/ACM Trans Netw (TON) 19(5):1513–1527

79. Sun et al (2016) Investigation of prediction accuracy, sensitivity, and parameter stability of large-scale propagation path loss models for 5G wireless communications (Invited Paper). IEEE Trans Veh Technol 65(5):2843–2860

80. Sun S, MacCartney GR, Jr., Rappaport T (2017) A novel millimeter wave channel simulator and applications for 5G wireless communications. In: 2017 I.E. International Conference on Communications (ICC), Paris, pp 1–7, May 2017

81. Swindlehurst L, Ayanoglu E, Heydari P, Capolino F (2014) Millimeter-wave massive MIMO: the next wireless revolution? IEEE Commun Mag 52(9):56–62

82. Thomas T et al (2016) A prediction study of path loss models from 2–73.5 GHz in an urban-macro environment. In: IEEE 83rd Vehicular Technology Conference (VTC2016-Spring), Nanjing, China, May 2016

83. Tsinos C, Maleki S, Chatzinotas S, and Ottersten B (2016) Hybrid analog-digital transceiver designs for cognitive large-scale antenna Array systems. arXiv:1612.02957v1 [cs.IT] 9 December 2016
84. Turgut E, Gursoy MC (2015) Energy efficiency in relay-assisted mmwave cellular networks. CoRR, vol. abs/1510.05961
85. Wei Z et al (2016) Energy-efficiency of millimeter-wave full-duplex relaying systems: challenges and solutions. IEEE Access 4:4848–4860
86. Xiao M, Mumtaz S, Huang Y, Dai L, Li Y, et al (2017) Millimeter wave communications for future mobile networks. arXiv:1705.06072v1 [cs.IT], 17 May 2017
87. Zhang YP, Wang P, Goldsmith A (2015) Rainfall effect on the performance of millimeter-wave MIMO systems. IEEE Trans Wireless Commun 14(9):4857–4866
88. Zhao Q, Li J (2006) Rain attenuation in millimeter wave ranges. In: Proceedings of IEEE International Symposium Antennas, Propagation EM Theory, October 2006, pp 1–4
89. Zhao H et al (2013) 28 GHz millimeter wave cellular communication measurements for reflection and penetration loss in and around buildings in New York City. In: IEEE International Conference on Communications (ICC), June 2013, pp 5163–5167
90. Zhu M, Tufvesson F, Eriksson G (2013) The COST 2100 channel model: parameterization and validation based on outdoor MIMO measurements at 300 MHz. IEEE Trans Wirel Commun 12 (2):888–897

Chapter 8
Spectrum Sensing in Cognitive Radio Networks Under Security Threats and Hybrid Spectrum Access

Kuldeep Yadav, Abhijit Bhowmick, Sanjay Dhar Roy, and Sumit Kundu

8.1 Introduction

Cognitive radio (CR) has been proposed to resolve the spectrum scarcity problem faced by the wireless networks for last few years [1–3]. In cognitive radio network (CRN), SUs are allowed to utilize the PU spectrum band without hampering the PU quality of service (QoS) [4, 5]. To find the presence of PU, sensing of spectrum using conventional energy detection (CED) is a suitable choice while the signal from PU is unknown to SU [6]. The performance of improved energy detector (IED) is studied in [7, 8]. It is shown in [7, 8] that the use of IED instead of CED improves detection probability and decreases the total error probability (sum of false alarm probability and missed detection probability). In IED, the signal power raise factor is replaced by an arbitrary positive power $'x'$. When $x = 2$, the IED becomes CED. A SU periodically senses the presence of PU and dynamically selects the best available spectrum for its transmission. A SU can transmit either in underlay mode or in overlay mode. In the case of hybrid transmission [9, 10], a SU works in underlay mode if PU is sensed to be active or in the overlay mode if PU is sensed to be idle. In detection cycle, if SU spends more time on spectrum sensing, the detection performance improves at the cost of lower throughput since the time for data transmission reduces and vice versa. The trade-off between sensing time and throughput is addressed in [11], where the authors studied the optimal sensing time which maximizes the throughput of SU.

K. Yadav · S. D. Roy · S. Kundu
Department of ECE, NIT Durgapur, Durgapur, West Bengal, India
e-mail: kuldeep.yadav@ieee.org

A. Bhowmick (✉)
Department of Communications, SENSE, VIT University, Vellore, Tamil Nadu, India

© Springer International Publishing AG, part of Springer Nature 2019
M. H. Rehmani, R. Dhaou (eds.), *Cognitive Radio, Mobile Communications and Wireless Networks*, EAI/Springer Innovations in Communication and Computing,
https://doi.org/10.1007/978-3-319-91002-4_8

CRNs are inclined to genuine security threats due to malicious attacks [12, 13]. Since SUs are self-subjective and reconfigurable, some malicious users can manipulate their transmitter parameters and mimic incumbent signal to pretend as an authorized PU. The reason is one of a kind normal for CRNs architecture that permits different obscure wireless devices to entrepreneurially access to the PU spectrum bands. In this way, alternate SUs are prevented from accessing the PU spectrum band. Such attack is called as primary user emulation (PUE) attack [14, 15]. The motive of PUE attacker is to persuade the SUs that the channel is occupied by some selfish users when it is really accessible to utilize the whole spectrum band for their own purpose. Till date, several research literatures have been led to develop a viable way to deal with PUE attacker. To recognize PU and PUE signals, the authors in [15] proposed a transmitter verification technique. The proposed strategy takes the advantage of the geological area of a PU by exploiting two methods: distance ratio test and distance difference test. In [16], the authors proposed a location-based defense scheme that uses the physical location of PU transmitter and received signal strength characteristics. In [17], the authors proposed an analytical approach based on Fenton approximation and Markov inequality and obtained a lower bound on the probability of a successful PUE attack on a SU by a set of cooperating PUE attacks. In [18], the authors proposed a Neyman-Pearson composite hypothesis test and a Wald's sequential probability ratio test to detect PUE attack. To mitigate the destructive effect of PUE attack, most of the research works have been conducted based on the assumption that the physical location or unique properties of the PU transmitter is known to SU. But, an appropriate strategy capable of accurate PU detection, without any prior information about location and properties of PU signal, is more important. In [14], the authors proposed an attack-aware threshold selection scheme that requires no prior information about physical location or specific properties of PU signal. The PUE attacker compels the SUs to falsely detect the PU signal and cease their transmissions. This malicious activity results in decreased throughput of the SU. In [19], the authors proposed a simplified counter approach to maximize the throughput of the SUs facing PUE attacks. In [19], a weight factor scheme is used to combine the cooperating SUs data at the fusion center and maximize the SU throughput while maintaining an interference constraint at the PU. In [20], the authors proposed a cooperative multiband spectrum sensing scheme for a CRN operating in the presence of PUE attack to maximize the aggregate achievable throughput. In [19, 20], authors have investigated the throughput performance under PUE attack with the motive of maximizing the throughput of a SU in overlay mode. Further in [9, 20] authors investigated the performance of SU in terms of throughput with hybrid spectrum access scheme without considering the impact of PUE attack.

In this chapter, the network performance is analyzed in terms of throughput incorporating the effects of PUE attack under hybrid spectrum access scheme for CED as well as IED-based spectrum sensing. In hybrid spectrum access scheme, the SU switches between overlay or underlay modes based on spectrum sensing decision available at the SU. A novel analytical expression for throughput of a SU under the considered scenario, i.e., joint impact of the PUE and hybrid spectrum access scheme, is developed. To defend PUE attacks, an optimal threshold selection approach is proposed in this chapter, which reduces the false alarm probability and enhances the throughput of SU. Impact of several parameters such as sensing time, attacker's strength, attacker's presence probabilities, maximum allowable SU transmit power, and tolerable interference limit at PU receiver on the throughput of SU is also investigated. The effectiveness of our proposed scheme based on optimal threshold selection for compensating the degradation caused by PUE is also shown. Further, the network performances under the presence of PUE attack is also compared with the network performance without PUE attack. Major contributions of this chapter are highlighted below:

- Network throughput performance is investigated under PUE attack for hybrid spectrum access.
- A new analytical expression of SU throughput is developed under PUE attack and hybrid spectrum access. Joint impact of the PUE and hybrid spectrum access scheme on throughput is shown.
- Optimal threshold is estimated to improve the performance of a SU under PUE attack.
- Impact of several parameters such as sensing time, attacker's strength, attacker's presence probabilities, etc. on the throughput of SU is captured.
- A comparative analysis of throughput is carried out for CED- and IED-based spectrum sensing.

The remaining portion of the chapter is organized as follows: Section 8.2 presents the system model and related discussions. In Sect. 8.3, proposed optimal threshold selection approach is described. In Sect. 8.4, analytical expression for the throughput of a SU under the considered scenario is derived. In Sect. 8.5, results and discussions are presented. Future research direction is indicated in Sect. 8.6. At last, the conclusion is made in Sect. 8.7.

8.2 System Model

The detection cycle of a SU is shown in Fig. 8.1. A SU first senses the spectrum status for τ unit of time, and during the remaining frame duration, i.e., $(T - \tau)$, it transmits if PU is sensed to be absent. The network model of our interest consists of a SU transmitter (SU-Tx), a SU receiver (SU-Rx), a PU transmitter (PU-Tx), a PU receiver (PU-Rx), and a PUE transmitter (PUE-Tx) as shown in Fig. 8.2. The SU-Tx performs spectrum sensing to detect the presence of PU-Tx. If PU-Tx is detected to be present, the SU-Tx works in an underlay mode to transmit information to SU-Rx. It needs to adjust its transmit power to maintain an interference constraint at PU-Rx. However, if PU-Tx is detected to be absent, the SU-Tx works in an overlay mode. PUE-Tx transmits an emulated primary signal during a spectrum sensing interval to SU to prevent it from accessing the PU spectrum bands. It is assumed that the information regarding presence or absence of PU is available to PUE-Tx.

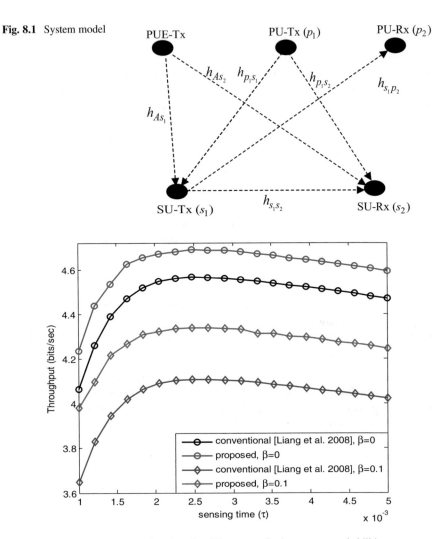

Fig. 8.1 System model

Fig. 8.2 Throughput versus sensing time for different attacker's presence probabilities

It is assumed that all the links in the network are independent and identically distributed (i.i.d.) Rayleigh fading channel with zero mean and unity variance. Let the channel coefficients of the channels between SU-Tx and SU-Rx, SU-Tx and PU-Rx, PU-Tx and SU-Rx, PU-Tx and SU-Tx, PUE-Tx and SU-Rx, and PUE-Tx and SU-Tx be represented as $h_{s_1 s_2}, h_{s_1 p_2}, h_{p_1 s_2}, h_{p_1 s_1}, h_{As_2}$ and h_{As_1}, respectively. The respective channel power gains are exponentially distributed, which are given as $g_{s_1 s_2} = |h_{s_1 s_2}|^2, g_{s_1 p_2} = |h_{s_1 p_2}|^2, g_{p_1 s_2} = |h_{p_1 s_2}|^2, g_{p_1 s_1} = |h_{p_1 s_1}|^2, g_{As_2} = |h_{As_2}|^2$ and $g_{As_1} = |h_{As_1}|^2$, respectively.

Let the received signal at SU-Tx be sampled at f_s, and K be the number of samples, $K = \tau f_s$. Depending on the presence or absence of the PU and PUE signal, the received signal at SU-Tx can be expressed as following [14]:

$$y(n) = \begin{cases} w(n) & H_{s_0} : \text{only noise} \\ h_{p_1 s_1} s_1(n) + w(n) & H_{s_1} : \text{PU} + \text{noise} \\ h_{As_1} s_2(n) + w(n) & H_{s_2} : \text{PUE} + \text{noise} \\ h_{p_1 s_1} s_1(n) + h_{As_1} s_2(n) + + w(n) & H_{s_3} : \text{PU} + \text{PUE} + \text{noise} \end{cases} \quad (8.1)$$

where $y(n)$ be the received signal and $n = 1, 2, \ldots K$. The additive noise $w(n)$, is assumed to be i.i.d. random process with mean 0 and variance σ_u^2. Here, $s_1(n)$ is the PU signal, and it is considered to be an i.i.d. random process with mean 0 and variance $\sigma_{s_1}^2$, ; similarly, the PUE signal $s_2(n)$ is also assumed to be an i.i.d. random process with mean 0 and variance $\sigma_{s_2}^2$. It is considered that $s_1(n)$, $s_2(n)$ and $w(n)$ are circularly symmetric complex Gaussian.

The test statistics $'R'$ of the SU-Tx at the output of energy detector is given as [11]

$$R = \frac{1}{K} \sum_{n=1}^{K} |y(n)|^x \quad (8.2)$$

where $'x'$ is an arbitrary positive power and the energy detector is known as IED, and it becomes CED when $x = 2$.

The test statistics obtained in (8.2) can be approximated as a Gaussian distributed random variable for large number of samples [11]. With Gaussian approximation, the mean and variance of the probability density function (PDF) of R for all the four conditions are given as follows: $\mu_0 = \frac{2^{x/2} \sigma_u^x}{\sqrt{\pi}} \Gamma\left(\frac{x+1}{2}\right)$ and

$\sigma_0^2 = \frac{2^x \sigma_u^{2x}}{K \sqrt{\pi}} \left[\Gamma\left(\frac{2x+1}{2}\right) - \frac{1}{\sqrt{\pi}} \Gamma^2\left(\frac{x+1}{2}\right) \right]$ under H_{s_0}, $\mu_1 = \mu_0 \left(\gamma_1 |h_{p_1 s_1}|^2 + 1\right)^{x/2}$,

$\sigma_1^2 = \sigma_0^2 \left(\gamma_1 |h_{p_1 s_1}|^2 + 1\right)^x$ under H_{s_1}, $\mu_2 = \mu_0 \left(\gamma_2 |h_{As_1}|^2 + 1\right)^{x/2}$,

$\sigma_2^2 = \sigma_0^2 \left(\gamma_2 |h_{As_1}|^2 + 1\right)^x$ under H_{s_2}, $\mu_3 = \mu_0 \left(\gamma_1 |h_{p_1 s_1}|^2 + \gamma_2 |h_{As_1}|^2 + 1\right)^{x/2}$, and $\sigma_3^2 = \sigma_0^2 \left(\gamma_1 |h_{p_1 s_1}|^2 + \gamma_2 |h_{As_1}|^2 + 1\right)^x$ under H_{s_3}. Here $\gamma_1 = \sigma_{s_1}^2 / \sigma_u^2$ and $\gamma_2 = \sigma_{s_2}^2 / \sigma_u^2$ are signal-to-noise ratio for PU and PUE signals, respectively. The PDF of R under all possible four conditions are given below:

$$f_{R|H_{S_0}}(y) = \frac{1}{\sqrt{2\pi \sigma_0^2}} \exp\left(-\frac{(x - \mu_0)^2}{2\sigma_0^2}\right) \quad (8.3)$$

$$f_{R|H_{S_1}}(y) = \frac{1}{\sqrt{2\pi \sigma_1^2}} \exp\left(-\frac{(x - \mu_1)^2}{2\sigma_1^2}\right) \quad (8.4)$$

$$f_{R|H_{S_2}}(y) = \frac{1}{\sqrt{2\pi\sigma_2^2}} \exp\left(-\frac{(x-\mu_2)^2}{2\sigma_2^2}\right) \tag{8.5}$$

$$f_{R|H_{S_3}}(y) = \frac{1}{\sqrt{2\pi\sigma_3^2}} \exp\left(-\frac{(x-\mu_3)^2}{2\sigma_3^2}\right) \tag{8.6}$$

To find the status of PU, the test statistic of SU-Tx is compared with a predefined sensing threshold (λ). The performance metrics, i.e., false alarm probability $\left(P_{f_1}(\tau) \text{ or } P_{f_2}(\tau)\right)$ and detection probability $\left(P_{d_1}(\tau) \text{ or } P_{d_2}(\tau)\right)$ under different hypothesis of the SU can be expressed as follows:

$$P_{f_1}(\tau) = P(D^{\mathrm{on}}|H_{s_0}) = Q\left(\frac{(\lambda-\mu_0)}{\sigma_0}\right) \tag{8.7}$$

$$P_{d_1}(\tau) = P(D^{\mathrm{on}}|H_{s_1}) = Q\left(\frac{(\lambda-\mu_1)}{\sigma_1}\right) \tag{8.8}$$

$$P_{f_2}(\tau) = P(D^{\mathrm{on}}|H_{s_2}) = Q\left(\frac{(\lambda-\mu_2)}{\sigma_2}\right) \tag{8.9}$$

$$P_{d_2}(\tau) = P(D^{\mathrm{on}}|H_{s_3}) = Q\left(\frac{(\lambda-\mu_3)}{\sigma_3}\right) \tag{8.10}$$

where $Q(.)$ is the Q-function and $P_{f_1}(\tau)$ and $P_{f_2}(\tau)$ indicate the probability of false alarm under hypotheses H_{s_0} and H_{s_2}, respectively. Similarly, the detection probability under hypotheses H_{s_1} and H_{s_3} is indicated by $P_{d_1}(\tau)$ and $P_{d_2}(\tau)$, respectively.

The presence and absence of PU signal in the given channel are indicated by two hypotheses H_1 and H_0, respectively. Correspondingly, the presence and absence of the PUE signal are indicated by the events A^{on} and A^{off}, respectively. Let us consider two conditional probabilities, α and β, to represent the presence of the PUE signals under hypothesis H_1 and H_0, i.e., $\alpha = P(A^{\mathrm{on}}|H_1)$ and $\beta = P(A^{\mathrm{on}}|H_0)$, respectively. Thus, the absence of the PUE signals under hypothesis H_1 can be expressed as $P(A^{\mathrm{off}}|H_1) = 1 - P(A^{\mathrm{on}}|H_1) = 1 - \alpha$ and under hypothesis H_0 as $P(A^{\mathrm{off}}|H_0) = 1 - P(A^{\mathrm{on}}|H_0) = 1 - \beta$.

Thus, the total false alarm probability (P_f) and total detection probability (P_d) under the PUE attack can be expressed as following [14]:

$$\begin{aligned}
P_f(\tau) = P(D^{\mathrm{on}}|H_0) &= P(D^{\mathrm{on}}|H_0, A^{\mathrm{off}})P(A^{\mathrm{off}}|H_0) + P(D^{\mathrm{on}}|H_0, A^{\mathrm{on}})P(A^{\mathrm{on}}|H_0) \\
&= P(D^{\mathrm{on}}|H_{s_0})(1-\beta) + P(D^{\mathrm{on}}|H_{s_2})\beta \\
&= (1-\beta)P_{f_1}(\tau) + \beta P_{f_2}(\tau)
\end{aligned}$$

$$\tag{8.11}$$

$$\begin{aligned}
P_d(\tau) = P(D^{\mathrm{on}}|H_1) &= P(D^{\mathrm{on}}|H_1, A^{\mathrm{off}})P(A^{\mathrm{off}}|H_1) + P(D^{\mathrm{on}}|H_1, A^{\mathrm{on}})P(A^{\mathrm{on}}|H_1) \\
&= P(D^{\mathrm{on}}|H_{s_1})(1-\alpha) + P(D^{\mathrm{on}}|H_{s_3})\alpha \\
&= (1-\alpha)P_{d_1}(\tau) + \alpha P_{d_2}(\tau)
\end{aligned}$$

$$\tag{8.12}$$

8.3 Optimal Threshold Selection Approach

The optimal threshold is the threshold of the energy detector which minimizes the total error. An error occurs when the SU declares the presence of PU, while PU signal is not actually present or declares the absence of the PU, while PU is actually transmitting signals. So, the total error probability can be written as

$$
\begin{aligned}
P_e(\tau) &= P(D^{\text{on}}|H_0)P(H_0) + P(D^{\text{off}}|H_1)P(H_1) \\
&= P_f(\tau)P(H_0) + (1 - P_d(\tau))P(H_1)
\end{aligned}
\tag{8.13}
$$

In the conventional method, PUE signal is not considered, i.e., α and β are equal to zero. Hence, total error probability is defined as

$$
P_e(\tau) = P_{f_1}(\tau)P(H_0) + (1 - P_{d_1}(\tau))P(H_1)
\tag{8.14}
$$

We set the first derivative of the total error probability to zero for finding conventional optimal threshold. Thus, we get from (8.14) as

$$
\begin{aligned}
&\frac{\partial P_e(\tau)}{\partial \lambda} = 0 \\
&P(H_0)\frac{\partial P_{f_1}(\tau)}{\partial \lambda} - P(H_1)\frac{\partial P_{d_1}(\tau)}{\partial \lambda} = 0 \\
&P(H_0)\frac{\partial P_{f_1}(\tau)}{\partial \lambda} = P(H_1)\frac{\partial P_{d_1}(\tau)}{\partial \lambda} \\
&P(H_0)\frac{\partial Q((\lambda - \mu_0)/\sigma_0)}{\partial \lambda} = P(H_1)\frac{\partial Q((\lambda - \mu_1)/\sigma_1)}{\partial \lambda}
\end{aligned}
\tag{8.15}
$$

We know $Q(x) = \frac{1}{2}\text{erfc}(x/\sqrt{2})$, so (8.15) can be rewritten as

$$
P(H_0)\frac{\partial \left[\frac{1}{2}\text{erfc}((\lambda - \mu_0)/\sqrt{2}\sigma_0)\right]}{\partial \lambda} = P(H_1)\frac{\partial \left[\frac{1}{2}\text{erfc}((\lambda - \mu_1)/\sqrt{2}\sigma_1)\right]}{\partial \lambda}
\tag{8.16}
$$

We also know $\frac{\partial \text{erfc}(z)}{\partial z} = -\frac{2}{\sqrt{\pi}}e^{-z^2}$, so (8.16) can be rewritten as

$$
P(H_0)\left[-\frac{2}{\sqrt{\pi}}e^{-\frac{(\lambda - \mu_0)^2}{2\sigma_0^2}}\left(\frac{1}{\sqrt{2}\sigma_0}\right)\right] = P(H_1)\left[-\frac{2}{\sqrt{\pi}}e^{-\frac{(\lambda - \mu_1)^2}{2\sigma_1^2}}\left(\frac{1}{\sqrt{2}\sigma_1}\right)\right]
\tag{8.17}
$$

After simplifying (8.17), conventional optimal threshold is found as

$$
\lambda = \frac{(\mu_0\sigma_1^2 - \mu_1\sigma_0^2) + \sqrt{M}}{(\sigma_1^2 - \sigma_0^2)}
\tag{8.18}
$$

where $M = (\mu_0\sigma_1^2 - \mu_1\sigma_0^2)^2 + (\sigma_1^2 - \sigma_0^2)\left(\mu_1^2\sigma_0^2 - \mu_0^2\sigma_1^2 + 2\sigma_0^2\sigma_1^2 \ln\left(\frac{\sigma_1 P(H_0)}{\sigma_0 P(H_1)}\right)\right)$.

In the presence of PUE signal, the total error probability can be written as

$$P_e(\tau) = \left[(1-\beta)P_{f_1}(\tau) + \beta P_{f_2}(\tau)\right]P(H_0)$$
$$+ \left[1 - (1-\alpha)P_{d_1}(\tau) - \alpha P_{d_2}(\tau)\right]P(H_1) \qquad (8.19)$$

The first derivative of the total error probability is set to zero for finding proposed optimal threshold. After differentiating $P_e(\tau)$ with respect to τ, we obtain

$$\frac{P(H_0)}{P(H_1)} = \frac{(1-\alpha)\sigma_3 e^{-\frac{(\lambda-\mu_1)^2}{2\sigma_1^2}} + \alpha\sigma_1 e^{-\frac{(\lambda-\mu_3)^2}{2\sigma_3^2}}}{(1-\beta)\sigma_2 e^{-\frac{(\lambda-\mu_0)^2}{2\sigma_0^2}} + \beta\sigma_0 e^{-\frac{(\lambda-\mu_2)^2}{2\sigma_2^2}} \left(\frac{\sigma_0\sigma_2}{\sigma_1\sigma_3}\right)} \qquad (8.20)$$

It is seen that obtaining a closed form expression of optimal threshold λ from (8.20) is difficult. So we proceed to obtain optimal threshold λ from (8.20) by numerical methods. Equation (8.20) is our novel development for finding an optimal threshold of energy detection under the presence of PUE attacker. Next we evaluate the throughput of CRN with PUE attackers. For our proposed case, optimal threshold as obtained by numerical evaluation of (8.20) is plugged in the false alarm probability and detection probability expressions as given vide Eqs. 8.7, 8.8, 8.9, and 8.10.

8.4 Throughput of Secondary User Under PUE Attack

It is considered that the SU transmits in hybrid spectrum access scheme, i.e., overlay mode (SU transmits with maximum power as PU is absent) and underlay mode (SU transmits with controlled power as PU is present). In overlay mode, the SU transmits with maximum power, i.e., $p_s = P_{max}$; on the other hand, it transmits with a controlled transmit power, i.e., $p_s = I_{th}/g_{s_1p_2}$, in underlay mode. Here, I_{th} is the tolerable interference limit, imposed by PU on SU to maintain its (PU) quality of service. Based on the sensing decisions, the switching mode between underlay and overlay mode can be found from Table 8.1, i.e., Case1, Case2, Case3, and Case4 fall under overlay mode, while Case5, Case6, Case7, and Case8 fall under underlay

Table 8.1 Combination of sensing decision, true channel state with PUE attack

Possible case	Sensing decision	True channel state	PUE signal	Transmission mode
1.	Idle	Idle	Absent	Overlay
2.	Idle	Idle	Present	Overlay
3.	Idle	Busy	Absent	Overlay
4.	Idle	Busy	Present	Overlay
5.	Busy	Idle	Absent	Underlay
6.	Busy	Idle	Present	Underlay
7.	Busy	Busy	Absent	Underlay
8.	Busy	Busy	Present	Underlay

Table 8.2 Notations

Symbol	Description
p_p	PU transmit power
p_s	SU transmit power
ρ	Relative strength of PUE attacker w.r.t PU signal
p_a	PUE attacker transmit power
I_{th}	PU interference limit
γ_j	Signal-to-interference-noise ratio for j-th case, $j = 1,2,...,8$
N_0	Power spectral density of the AWGN noise
B	Bandwidth
C_j	Mean capacity of SU for j-th case, $j = 1,2,....8$
F_{γ_j}	CDF of γ_j
R_j	Throughput of SU for j-th case, $j = 1,2,....8$
E_i	Exponential integral

mode. It may be noted that the transmission for Case3 and Case4 is useless since PU is active. As SU transmission is based on the sensing decision, collision may occur in the Case3 and Case4 which results in no useful transmission. Thus, in overlay mode, only the Case1 and Case2 are considered in the present paper. Notations for different parameters are given in Table 8.2.

To obtain the throughput of a SU, it is required to know the mean capacity of the SU. The mean capacity of the SU can be expressed as [21]

$$C_j = \frac{B}{\log_e 2} \int_0^\infty \frac{1 - F_{\gamma_j}(z)}{1+z} dz \qquad (8.21)$$

We now present the expressions for capacity under six Cases, i.e., for Case1 and Case2 of overlay mode and Case5 to Case8 of underlay mode as indicated in Table1.

For Case1, γ_1 can be written as

$$\gamma_1 = \frac{p_s g_{s_1 s_2}}{N_0} \qquad (8.22)$$

The CDF of γ_1 can be expressed as

$$\begin{aligned} F_{\gamma_1}(z) &= P_r\left(\frac{p_s g_{s_1 s_2}}{N_0} < z\right) \\ &= P_r\left(g_{s_1 s_2} < zN_0/p_s\right) \\ &= 1 - \exp(-zN_0/p_s) \end{aligned} \qquad (8.23)$$

Therefore, the mean capacity (C_1) can be written by using (8.21) as

$$C_1 = \frac{B}{\log_e 2} \int_0^\infty \frac{1 - (1 - \exp(-zN_0/p_s))}{1+z} dz \qquad (8.24)$$

Using ([22], 3.352.4) and after some algebra, (8.24) can be simplified as

$$C_1 = -\frac{B}{\log_e 2} \exp(N_0/p_s) E_i(-N_0/p_s) \qquad (8.25)$$

Thus, the throughput of SU under Case1 can be written as

$$R_1 = \left(\frac{T-\tau}{T}\right) C_1 (1 - Q_{f_1}) P(H_0)(1 - \beta) \qquad (8.26)$$

For Case2, γ_2 can be written as

$$\gamma_2 = \frac{p_s g_{s_1 s_2}}{N_0 + p_a g_{As_2}} \qquad (8.27)$$

The CDF of γ_2 can be expressed as

$$
\begin{aligned}
F_{\gamma_2}(z) &= P_r\left(\frac{p_s g_{s_1 s_2}}{N_0 + p_a g_{As_2}} < z\right) \\
&= \int_0^\infty P_r\left(g_{s_1 s_2} < z\left(\frac{N_0 + p_a u}{p_s}\right)\right) f_{g_{As_2}}(u) du \\
&= \int_0^\infty \left(1 - \exp\left(-z\left(\frac{N_0 + p_a u}{p_s}\right)\right)\right) \exp(-u) du \\
&= 1 - \left(\frac{p_s}{z p_a + p_s}\right) \exp(-z N_0/p_s)
\end{aligned}
\qquad (8.28)
$$

Therefore, the mean capacity (C_2) can be written by using (8.21) as

$$
\begin{aligned}
C_2 &= \frac{B}{\log_e 2} \int_0^\infty \frac{\left(\frac{p_s}{z p_a + p_s}\right) \exp(-z N_0/p_s)}{1+z} dz \\
&= \frac{B(p_s/p_a)}{(1 - p_s/p_a)\log_e 2} \int_0^\infty \left(\frac{\exp(-z N_0/p_s)}{z + p_s/p_a} - \frac{\exp(-z N_0/p_s)}{z+1}\right) dz
\end{aligned}
\qquad (8.29)
$$

using ([22], 3.352.4), and after some algebra, (8.29) can be written as

$$C_2 = \frac{Bp_s}{(p_a - p_s)\log_e 2} \left[\exp(N_0/p_s)E_i(-N_0/p_s) - \exp(N_0/p_a)E_i(-N_0/p_a) \right] \quad (8.30)$$

Thus, the throughput of SU under Case2 can be written as

$$R_2 = \left(\frac{T - \tau}{T} \right) C_2 \left(1 - Q_{f_2} \right) P(H_0)\beta \quad (8.31)$$

For Case5, γ_5 can be written as

$$\gamma_5 = \frac{p_s g_{s_1 s_2}}{N_0} \quad (8.32)$$

where, $p_s = I_{th}/g_{s_1 p_2}$.
The CDF of γ_5 can be expressed as

$$\begin{aligned} F_{\gamma_5}(z) &= P_r \left(\frac{p_s g_{s_1 s_2}}{N_0} < z \right) \\ &= P_r \left(\frac{g_{s_1 s_2}}{g_{s_1 p_2}} < zN_0/I_{th} \right) \end{aligned} \quad (8.33)$$

After some algebra, (8.33) can be written as

$$F_{\gamma_5}(z) = zN_0/(I_{th} + zN_0) \quad (8.34)$$

Therefore, the mean capacity (C_5) can be written by using (8.21) as

$$\begin{aligned} C_5 &= \frac{B}{\log_e 2} \int_0^\infty \frac{1 - zN_0/(I_{th} + zN_0)}{1 + z} dz \\ &= -\frac{B(I_{th}/N_0)\log(I_{th}/N_0)}{(1 - I_{th}/N_0)\log_e 2} \end{aligned} \quad (8.35)$$

Thus, the throughput of SU under Case5 can be written as

$$R_5 = \left(\frac{T - \tau}{T} \right) C_5 Q_{f_1} P(H_0)(1 - \beta) \quad (8.36)$$

For Case6, γ_6 can be written as

$$\gamma_6 = \frac{p_s g_{s_1 s_2}}{N_0 + p_a g_{As_2}} \quad (8.37)$$

where, $p_s = I_{th}/g_{s_1 p_2}$.
The CDF of γ_6 can be expressed as

$$F_{\gamma_6}(z) = P_r\left(\frac{p_s g_{s_1 s_2}}{N_0 + p_a g_{As_2}} < z\right)$$

$$= P_r\left(\frac{g_{s_1 s_2}}{g_{s_1 p_2}} < z(N_0 + p_a u)/I_{th}\right) f_{g_{As_2}}(u)du$$

$$= \int_0^\infty \frac{z(N_0 + p_a u)}{I_{th} + z(N_0 + p_a u)} \exp(-u)du$$

$$= \int_0^\infty \frac{N_0/p_a}{u + (I_{th} + zN_0)/zp_a} \exp(-u)du + \int_0^\infty \frac{u}{u + (I_{th} + zN_0)/xp_a} \exp(-u)du$$

$$\tag{8.38}$$

using ([22], 3.352.4) for first part and ([22], 3.352.5) for second part, and after some algebra, (8.38) can be written as

$$F_{\gamma_6}(z) = \begin{array}{l} 1 - [\exp((I_{th} + zN_0)/zp_a)E_i(-(I_{th} + zN_0)/zp_a)] \\ \times [(N_0/p_a) - (I_{th} + zN_0)/zp_a] \end{array} \tag{8.39}$$

Therefore, mean capacity (C_6)can be written by using (8.21) and after some algebra as

$$C_6 = \frac{B}{\log_e 2} \int_0^\infty \frac{\left(-\frac{I_{th}}{zp_a}\right)\exp\left(\frac{I_{th}+zN_0}{zp_a}\right)E_i\left(-\frac{I_{th}+zN_0}{zp_a}\right)}{1+z} dz \tag{8.40}$$

Thus, the throughput of SU under Case6 can be written as

$$R_6 = \left(\frac{T-\tau}{T}\right)C_6 Q_{f_2} P(H_0)\beta \tag{8.41}$$

For **Case7**, mean capacity (C_7) can be written as similar as **Case6**:

$$C_7 = \frac{B}{\log_e 2} \int_0^\infty \frac{\left(-\frac{I_{th}}{zp_p}\right)\exp\left(\frac{I_{th}+zN_0}{zp_p}\right)E_i\left(-\frac{I_{th}+zN_0}{zp_p}\right)}{1+z} dz \tag{8.42}$$

Thus, the throughput of SU under Case7 can be written as

$$R_7 = \left(\frac{T-\tau}{T}\right)C_7 Q_{d_1} P(H_1)(1-\alpha) \tag{8.43}$$

For **Case8**, γ_8 can be written as

$$\gamma_8 = \frac{P_s g_{s_1 s_2}}{N_0 + P_p g_{p_1 s_2} + P_a g_{As_2}} \tag{8.44}$$

The CDF of γ_8 can be written as

$$
\begin{aligned}
F_{\gamma_8}(z) &= P_r\left(\frac{P_s g_{s_1 s_2}}{N_0 + P_p g_{p_1 s_2} + P_a g_{As_2}} < z\right) \\
&= P_r\left(\frac{\left(I_{th}/g_{s_1 p_2}\right) g_{s_1 s_2}}{N_0 + P_p g_{p_1 s_2} + P_a g_{As_2}} < z\right) \\
&= P_r\left(\frac{g_{s_1 s_2}}{g_{s_1 p_2}} < z\left(N_0 + P_p u + P_a v\right)/I_{th}\right) f_{g_{p_1 s_2}}(u)
\end{aligned}
\tag{8.45}
$$

$$
\begin{aligned}
\times f_{g_{As_2}}(v) du dv \quad &= \int_0^\infty \int_0^\infty \frac{z\left(N_0 + P_p u + P_a v\right)}{I_{th} + z\left(N_0 + P_p u + P_a v\right)} \exp(-u)\exp(-v) du dv \\
&= 1 \\
&+ \int_0^\infty \left(I_{th}/z P_p\right)\exp\left(\frac{I_{th} + z N_0 + z P_a v - z P_p v}{z P_p}\right) E_i\left(-\frac{I_{th} + z N_0 + z P_a v}{z P_p}\right) dv
\end{aligned}
$$

Therefore, mean capacity (C_8) can be written by using (8.21) as

$$C_8 = \frac{B}{\log_e 2} \int_0^\infty \int_0^\infty \frac{\left(-\frac{I_{th}}{z P_p}\right)\exp\left(\frac{I_{th}+z N_0+z P_a v-z P_p v}{z P_p}\right) E_i\left(-\frac{I_{th}+z N_0+z P_a v}{z P_p}\right)}{1+z} dv dz \tag{8.46}$$

Thus, the throughput of SU under Case8 can be written as

$$R_8 = \left(\frac{T-\tau}{T}\right) C_8 Q_{d_2} P(H_1)\alpha \tag{8.47}$$

Therefore, the throughput of a SU for underlay mode can be expressed as

$$R_{underlay} = R_5 + R_6 + R_7 + R_8 \tag{8.48}$$

Similarly, the throughput of a SU for overlay mode can be expressed as

$$R_{overlay} = R_1 + R_2 \tag{8.49}$$

Hence, the overall throughput of a SU can be expressed as

$$R_{overall} = R_{underlay} + R_{overlay} \tag{8.50}$$

We now present numerical results based on our analytical development presented above.

8.5 Results and Discussions

In CRN, when PU is present, the PUE attack increases the detection probability by increasing the energy level sensed by SU. So, if a PUE attacker wants to decrease the performance of the CRN, it should not send any signal over the sensing time when PU is present. On the other hand, if PU is present and PUE attacker sends a signal over sensing time, it causes interference to the PU. PUE will tend to send PU like signal when actual PU is absent. We assume that the behavior of PU is known to PUE. So, in this chapter our concern is to study the effect of attacker in absence of PU. Thus, we focus on the effect of β, which indicates the probability of presence of attacker in absence of PU. In this section, MATLAB-based simulation results are carried out to validate the analytical results based on our formulation presented above. Performance is investigated under hybrid spectrum access scenario. Unless otherwise stated, the parameters used are $f_s = 1$ MHz, $T = 50$ ms, $K = 2500$ (from Figs. 8.4, 8.5, 8.6, 8.7 and 8.8), $p_p = -12$ dB, $P_{max} = 20$ dB, $I_{th} = -25$ dB, $P(H_0) = 0.6$, and $P(H_1) = 0.4$.

Throughput performance of conventional scheme [11] and proposed scheme is compared in Fig. 8.2. It is also found that for a particular sensing time for both type of schemes, throughput of a SU decreases as β increases. It is found that the proposed scheme outperforms the conventional scheme.

Figure 8.3 shows the throughput of SU as a function of sensing time for different attacker's presence probabilities, i.e., $\beta = 0.3, 0.5$ and a fixed attacker strength ($\rho = 0.2$). It is observed that the throughput of SU decreases as β increases. As β increases, PUE attacker sends PUE signal over more time in the sensing interval, so the total false alarm probability (P_f) increases which results in degradation of throughput. It is also found that throughput of SU is improved in the case of the proposed optimal threshold selection approach as compared to the conventional method.

Figure 8.4 shows the impact of attacker strength (ρ) on SU throughput. The performance is investigated for different attacker's presence probabilities, i.e., $\beta = 0.3, 0.5$. It is observed that the throughput of SU decreases as the attacker strength increases. As the attacker strength increases, the SU is likely to infer the presence of PU under H_0 with higher probability which increases the total false alarm probability and results in degradation of throughput of SU. It is also observed that throughput of SU decreases as attacker's presence probability increases. It is found that the throughput of SU is improved for the proposed optimal threshold selection approach as compared to the conventional approaches.

Figure 8.5 shows the variation of throughput of SU with IED parameter for several attacker probabilities, i.e., $\beta = 0.1, 0.3$, and a fixed attacker strength ($\rho = 0.2$). The nature of throughput is concave with respect to IED parameter (x). Thus, there is an optimal value of x for a given β for which throughput is maximized. It is also observed that the throughput of SU decreases as β increases. As β increases, the total false alarm probability increases which in turn degrades the

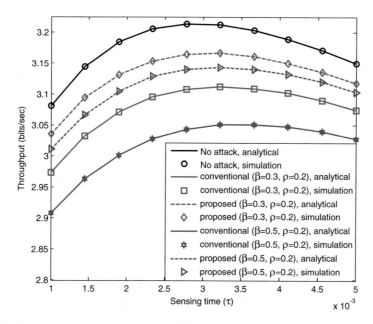

Fig. 8.3 Throughput versus sensing time for different attacker's presence probabilities

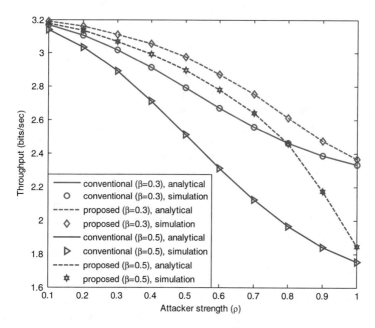

Fig. 8.4 Throughput versus attacker strength for probabilities of different attacker's presence probabilities

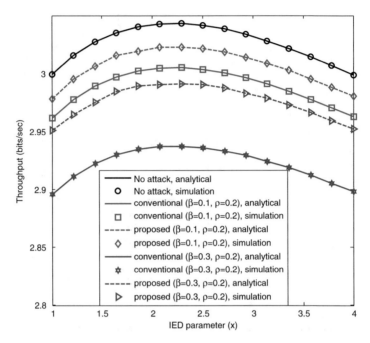

Fig. 8.5 Throughput versus IED parameter for different attacker's presence probabilities

throughput. It is found that the throughput of SU is improved for the proposed optimal threshold selection approach as compared to the conventional approaches.

Figure 8.6 shows the throughput of SU as a function of IED parameter for different attacker strengths, i.e., $\rho = 0.2, 0.3$ and a fixed attacker presence probability ($\beta = 0.2$). It is observed that the throughput of SU decreases as the attacker strength increases. As the attacker strength increases, the SU is likely to infer the presence of PU under H_0 with higher probability which increases the total false alarm probability and results in degradation of throughput of SU. The throughput of the proposed method is improved as compared to the conventional approach.

In Fig. 8.7, the throughput of SU is shown as a function of maximum transmits power of SU for a fixed attacker's presence probability ($\beta = 0.5$) and fixed attacker strength ($\rho = 0.8$). It is observed that the throughput of SU increases as the maximum transmit power of SU increases. As $P_{\max}$ increases, the received signal-to-noise interference at SU-Rx increases, which in turn increases the channel capacity of SU. As channel capacity increases, the throughput of SU increases. The throughput performance of the proposed optimal threshold selection approach outperforms the conventional approaches by 23% at $P_{\max} = 5$ dB.

Figure 8.8 shows the throughput of SU as a function of tolerable interference limit (I_{th}) of PU for a fixed attacker's presence probabilities ($\beta = 0.5$) and fixed attacker strength ($\rho = 0.8$). It is observed that the throughput of a SU increases as PU interference limit increases. An increase in tolerable interference limit allows the

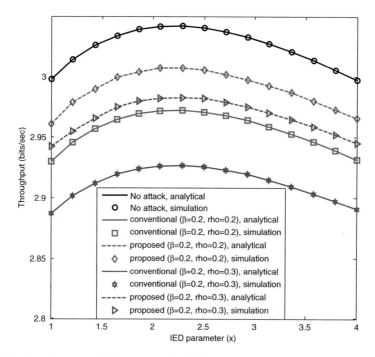

Fig. 8.6 Throughput versus IED parameter for different attacker strength

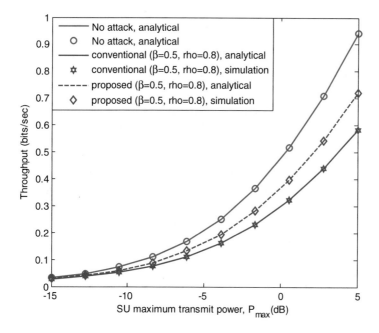

Fig. 8.7 Throughput versus SU maximum transmit power

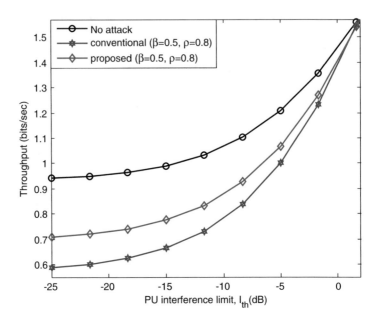

Fig. 8.8 Throughput versus PU interference limit

SU to transmit with higher power and results in increase in throughput of SU. It is found that the throughput of SU is improved for the proposed approach as compared to the conventional approach by 16% at $I_{th} = -15$ dB.

8.6 Future Research Direction

This work can be extended for a cooperative network such as wireless sensor network, vehicular network [23, 24], etc. The number of devices in a network is increasing day by day; thus, machine-to-machine (M2M) and device-to-device (D2D) communications are now become a part of future Internet of Things (IoT) framework [25]. In IoT, a network may suffer from spectrum congestion as well as security issues. Thus, the present study can help in designing the future IoT framework. Energy harvesting issue [26] can also be incorporated in the present work to study the network performance under green communications.

8.7 Conclusions

The throughput performance of a cognitive radio networks is investigated under PUE attack for conventional energy detection as well as improved energy detection where SU spectrum access is hybrid, i.e., combination of underlay and overlay. A novel analytical expression for throughput of a SU under the considered scenario is developed and evaluated. Impact of several parameters such as sensing time, IED parameter, attacker strength, attacker's presence probabilities, maximum allowable SU transmit power, and tolerable interference limit at PU on the throughput of a SU is shown. There is a noticeable impact of PUE attack on the throughput of a SU. There is an optimal sensing time for which the throughput of a SU is maximized. There is also an optimal value of IED parameter for which the throughput of a SU is maximum. Selection of an optimal threshold improves the throughput performance of a SU under PUE attack. The degradation in performance due to PUE attack can be compensated by the choice of an optimal threshold. Future direction of the present work is also indicated.

References

1. Mitola J, Maguire GQ (1999) Cognitive radio: making software radios more personal. IEEE Pers Commun 6(4):13–18. https://doi.org/10.1109/98.788210
2. Haykin S (2005) Cognitive radio: brain-empowered wireless communications. IEEE J Sel Areas Commun 23(2):201–220. https://doi.org/10.1109/JSAC.2004.839380
3. Amjad M, Rehmani MH, Mao S (2018) Wireless multimedia cognitive radio networks: a comprehensive survey. IEEE Commun Sur Tutorials:1–49. https://doi.org/10.1109/COMST.2018.2794358
4. Federal Communications Commission (2003) Facilitating opportunities for flexible, efficient, and reliable spectrum use employing cognitive radio technologies, notice of proposed rulemaking and order, FCC 03–322
5. Arslan H (2007) Cognitive radio, software defined radio, and adaptive wireless systems. Signals and communication technology. Springer, New York. https://doi.org/10.1007/978-1-4020-5542-3
6. Urkowitz H (1967) Energy detection of unknown deterministic signals. Proc IEEE 55 (4):523–531. https://doi.org/10.1109/PROC.1967.5573
7. Chen Y (2010) Improved energy detector for random signals in Gaussian noise. IEEE Trans Wirel Commun 9(2):558–563. https://doi.org/10.1109/TWC.2010.5403535
8. Singh A, Bhatnagar MR, Mallik RK (2011) Cooperative spectrum sensing with an Improved energy detector in cognitive radio network. In: Proceedings of NCC 2011, Indian Institute of Science, Bangalore, pp 1–5. https://doi.org/10.1109/NCC.2011.5734777
9. Senthuran S, Anpalagan A, Das O (2012) Throughput analysis of opportunistic access strategies in hybrid underlay overlay cognitive radio networks. IEEE Trans Wirel Commun 11 (6):2024–2035. https://doi.org/10.1109/TWC.2012.032712.101209
10. Bhowmick A, Prasad B, Roy SD, Kundu S (2016) Performance of cognitive radio network with novel hybrid Spectrum access schemes. Wirel Pers Commun 91(2):541–560. https://doi.org/10.1007/s11277-016-3476-5
11. Liang Y-C, Yonghong Z, Peh CY, Hoang AT (2008) Sensing- throughput tradeoff for cognitive radio network. IEEE Trans Wirel Commun 7(4):1326–1337. https://doi.org/10.1109/TWC.2008.060869
12. Jianwu L, Zebing F, Zhiyong F, Ping Z (2015) A survey of security issues in cognitive radio networks. China Commun 12(3):132–150. https://doi.org/10.1109/CC.2015.7084371
13. Sharma RK, Rawat DB (2015) Advances on security threats and countermeasures for cognitive radio networks: a survey. IEEE Commun Surv Tutorials 17(2):1023–1043. https://doi.org/10.1109/COMST.2014.2380998

14. Sharifi AA, Sharifi M, Niya MMJ (2016) Secure cooperative spectrum sensing under primary user emulation attack in cognitive radio networks: attack-aware threshold selection approach. AEU Int J Electron Commun 70(1):95–104. https://doi.org/10.1016/j.aeue.2015.10.010
15. Chen R, Park JM (2006) Ensuring trustworthy spectrum sensing in cognitive radio networks. 1st IEEE workshop on networking technologies for software defined radio networks. pp 110–119. https://doi.org/10.1109/SDR.2006.4286333
16. Chen R, Park JM, Reed JH (2008) Defense against primary user emulation attacks in cognitive radio networks. IEEE J Sel Areas Commun 26(1):25–37. https://doi.org/10.1109/JSAC.2008.080104
17. Anand S, Jin Z, Subbalakshmi KP (2008) An analytical model for primary user emulation attacks in cognitive radio networks. In: Proceedings of the IEEE 3rd international symposium of new frontiers in dynamic spectrum access networks. pp 1–6. https://doi.org/10.1109/DYSPAN.2008.16
18. Jin Z, Anand S, Subbalakshmi KP (2009) Mitigating primary user emulation attacks in dynamic spectrum access networks using hypothesis testing. ACM SIGMOBILE Mob Comput Commun Rev 13(2):74–85. https://doi.org/10.1145/1621076.1621084
19. Shrivastava S, Rajesh A, Bora PK (2015) A simplified counter approach to primary user emulation attacks from secondary user perspective. IEEE 26th annual international symposium on personal, indoor, and mobile radio communications. pp 2149–2154. https://doi.org/10.1109/pimrc.2015.7343653
20. Saber MJ, Sadough SMS (2016) Multiband cooperative spectrum sensing for cognitive radio in the presence of malicious users. IEEE Commun Lett 20(2):404–407. https://doi.org/10.1109/LCOMM.2015.2505299
21. Suraweera HA, Smith PJ, Shafi M (2010) Capacity limits and performance analysis of cognitive radio with imperfect channel knowledge. IEEE Trans Veh Technol 59(4):1811–1822. https://doi.org/10.1109/TVT.2010.2043454
22. Gradshteyn IS, Ryzhik IM (2007) Table of integrals, series, and products, 7th edn. Academic Press, Waltham
23. Amjad M, Afzal MK, Umer T, Kim BS (2017) QoS-aware and heterogeneously clustered routing protocol for wireless sensor networks. IEEE Access 5:10250–10262. https://doi.org/10.1109/ACCESS.2017.2712662
24. Amjad M, Sharif M, Afzal MK, Kim SW (2016) TinyOS-new trends, comparative views, and supported sensing applications: a review. IEEE Sensors J 16(9):2865–2889. https://doi.org/10.1109/JSEN.2016.2519924
25. Khan AA, Rehmani MH, Rachedi A (2017) Cognitive-radio-based internet of things: applications, architectures, Spectrum related functionalities, and future research directions. IEEE Wirel Commun 24(3):17–25. https://doi.org/10.1109/MWC.2017.1600404
26. Ren J, Hu J, Zhang D, Guo H, Zhang Y, Shen XJ (2018) RF energy harvesting and transfer in cognitive radio sensor networks: opportunities and challenges. IEEE Commun Mag 56(1):104–110. https://doi.org/10.1109/MCOM.2018.1700519

Kuldeep Yadav received his BTech degree in electronics and communication engineering in 2011 from Gautam Buddha Technical University, Uttar Pradesh, India, and MTech degree in telecommunication engineering in 2014 from NIT Durgapur, India. He is currently working as a research scholar in the Department of Electronics and Communication Engineering, NIT, Durgapur, India. His research interests include cognitive radio networks, spectrum sensing, and security issues in spectrum sensing.

Abhijit Bhowmick received his B.E. (Hons) degree in electronics and telecommunication engineering in 2002 from University of Burdwan, West Bengal, India, and MTech degree in telecommunication engineering in 2009 from NIT, Durgapur, and PhD in wireless communication engineering in 2016 from NIT, Durgapur, respectively. He worked for Cubix Control System Pvt. Ltd. from 2004 to 2006. After that he joined the Department of Electronics and Comm. Engineering, Bengal College of Engg. and Tech., Durgapur, as a lecturer in 2006. He joined VIT University, Vellore, TN, India, in the School of Electronics Engineering (SENSE) as an Associate Professor (SG) in June, 2016. As of today, he has published more than 35 research papers in various journals and conferences. He is a reviewer of several IEEE, Springer, Wiley, and Elsevier journals.

Sanjay Dhar Roy was born in Balurghat, West Bengal, India, in 1974. Sanjay Dhar Roy received his B.E. (Hons) degree in electronics and telecommunication engineering in 1997 from Jadavpur University, Kolkata, India, MTech degree in telecommunication engineering in 2008, and PhD in wireless communication engineering in 2011 from NIT, Durgapur, respectively. He worked for Koshika Telecom Ltd. from 1997 to 2000. After that he joined the Department of Electronics and Communication Engineering, National Institute of Technology, Durgapur, as a lecturer in 2000 and is currently an assistant professor there. His research interests include radio resource management, handoff, smart antenna techniques, beamforming, and cognitive radio networks. As of today, he has published 100 research papers in various journals and conferences. Mr. Dhar Roy is a member of IEEE (Communication Society) and is a reviewer of IET Communications, Electronics Letters and Journal of PIER, IJCS, and Wiley.

Sumit Kundu received his B.E. (Hons) degree in electronics and communication engineering in 1991 from NIT, Durgapur, India, and MTech degree in telecommunication systems engineering, and PhD in wireless communication engineering from IIT, Kharagpur, India, respectively. He has been a faculty in the Department of ECE, National Institute of Technology, Durgapur, since 1995 and is currently a full-time professor there. His research interests include wireless ad hoc and sensor networks, cognitive radio networks, cooperative communication, energy harvesting, and physical layer security in wireless networks. As of today, he has published more than 150 research papers in various journals and conferences. He is a senior member of IEEE (Communication Society) and is a reviewer of several IEEE and Elsevier journals.

Chapter 9
Optimum Spectrum Sensing Approaches in Cognitive Networks

Ishrath Unissa, Syed Jalal Ahmad, and P. Radha Krishna

9.1 Introduction

The demand for wireless application is increasing day by day with limited spectrum availability. Cognitive radio network (CRN) is a better solution for utilizing the spectrum efficiently. CRN consists of two main users: primary (or licensed) user and secondary (or unlicensed) user. Secondary user dynamically accesses the band allocated to the primary users without causing any interference to them. For this purpose, the CRN has to perform specific sequence called cognitive radio cycle [23] for effective use of the available band. This cycle consists of four main steps, namely, spectrum sensing, spectrum management, spectrum sharing, and spectrum mobility.

- *Spectrum Sensing*: CRN needs to sense the spectrum whether the primary user is active or inactive. It should basically detect the underutilized band for the secondary user to communicate.
- *Spectrum Management*: In this step, CRN has to decide whether the detected band is suitable for information transfer or not.
- *Spectrum Sharing*: After selection of suitable band, secondary user shares the spectrum.

I. Unissa (✉)
Mahatma Gandhi Institute of Technology, Hyderabad, India

S. J. Ahmad
MIET, Jammu, India
e-mail: jalal.ece@mietjammu.in

P. Radha Krishna
Infosys Limited, Hyderabad, India
e-mail: radhakrishna_p@infosys.com

© Springer International Publishing AG, part of Springer Nature 2019   209
M. H. Rehmani, R. Dhaou (eds.), *Cognitive Radio, Mobile Communications and Wireless Networks*, EAI/Springer Innovations in Communication and Computing,
https://doi.org/10.1007/978-3-319-91002-4_9

- *Spectrum Mobility*: Secondary users have dynamic access over the bands. To avoid the interference caused to the primary user, it has to shift to another underutilized spectrum to continue its communication process when the primary user is in an active state.

Spectrum sensing is a crucial stage in CRN. The main aim of spectrum sensing is to provide an increased number of bands to the secondary user and to avoid interference with the primary user. This can be done by continuously observing the band provided to primary users. However, there are limitations listed in spectrum sensing [18, 30], namely, *channel uncertainty, noise uncertainty, aggregate interference uncertainty*, and *sensing interference limit*.

1. *Channel Uncertainty*: Uncertainty in the strength of the received signal is due to noise interference of channel which may wrongly percept that the primary channel is active and it is not suitable for secondary user communication. Hence, CRN should sense the spectrum efficiently by keeping channel noise in view. There is a trade-off between spectrum efficiency and interference avoidance [20]. If the observation time of band is increased, then interference can be avoided but at the cost of reduced efficiency of the spectrum. On the other hand, if the transmission time is longer, spectrum efficiency is increased with more probability of interference.
2. *Noise Uncertainty*: Sensitivity can be calculated as the minimum signal-to-noise ratio (SNR) at which the primary band can be detected with accuracy by secondary users as [30]

$$\text{Sensitivity} = \text{SNR}_{\min} = T_p P_l (I_r + D_{\max})/P_n \qquad (9.1)$$

where T_p is transmitted power of the primary user, P_l is the path loss, I_r is the interference range of the secondary user, $D_{\max}$ is the maximum distance between the primary transmitter and its corresponding receiver, and P_n is the noise power. Equation 9.1 states that the noise power should be known to evaluate the sensitivity which is not possible practically, and it can be done by considering worst case noise assumption for designing more sensitive detector.

3. *Aggregate Interference Uncertainty*: Increasing number of secondary users may create uncertainty in sensing the primary band as several secondary users may be acting on it at the same time. Therefore, spectrum detection should be more accurate to avoid interference to the primary user.
4. *Sensing Interference Limit*: Interference is caused to the primary receiver when it is active and communication is established by the secondary user with it. There are two limitations associated with it: (1) the secondary user may not know the exact position of the primary receiver, and (2) secondary user is unaware when the receiver is inactive. These factors play an important role in determining the sensing interference limit.

9.1.1   Spectrum Sensing in CRNs

Spectrum sensing methods are broadly categorized into two: general and optimized (Fig. 9.1). In general cooperative sensing methods, two or more CRNs work collectively to get the information about primary user's activity and share it with each other either in centralized or in a distributed manner. Secondary user (SU) depends on itself for detecting primary user activity in the cooperative method. It is further classified into two: (a) primary user (PU)-dependent methods which require primary user information such as modulation type, pulse shape, and packet format and (b) PU-independent methods which require no prior information. PU dependent are further classified into the matched filter and cyclostationary methods [21]. Energy detection methods are an example of PU-independent methods in which only noise is known to the receiver. Interference method mainly concentrates on the receiver interference and sets one interference temperature limit. If limit exceeds, PU is active and SU cannot utilize it.

There are mainly four techniques for optimized methods, namely, cluster sensing, optimized parametric, sensing in cognitive ad hoc network, and optimized sensing disruption for CRN. Cluster sensing method groups similar cognitive radios (CR) together and works in cooperation for detecting the vacant frequency spectrum. In the optimized parametric method, different parameters are optimized to achieve spectrum efficiency with minimum noise interference. Sensing in cognitive ad hoc network provides optimum sensing scheme for CRN. Collective decision of CRN is considered for network performance with more accuracy, noise immunity, and spectrum efficiency. In optimized sensing disruption for CRN, the ability of false detection in spectrum sensing is improved to increase the unused bands for SUs.

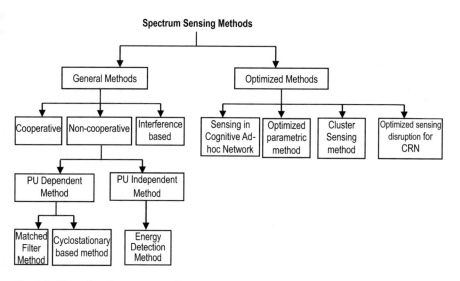

Fig. 9.1 Hierarchy of spectrum sensing methods

To enhance the spectrum sensing, the research community has proposed various optimization techniques by considering different parameters such as energy, SNR, interference ratio, and spectrum band to improve the spectrum utilization efficiency and minimize interference in single and multiple bands by considering abovementioned constraints. In this chapter, we:

a. Highlight the issues and challenges for spectrum sensing in CRNs.
b. Discuss existing optimal spectrum sensing strategies and present observations by considering different parameters including energy.
c. Present a comparative analysis of optimal sensing techniques.

The rest of the chapter is organized as follows: Sect. 9.2 presents the related work. In Sect. 9.3, we discuss different optimized techniques in spectrum sensing and provide a comparative analysis of these techniques. Section 9.4 presents future work and we conclude our chapter in Sect. 9.5.

9.2 Related Work

There are numerous techniques to detect the underutilized frequency band and utilize it for the data transmission by the secondary user. Energy detection method requires small sensing time and easy to implement with simple calculations. However, it is not valid when SNR is less. This is because of uncertainty in the channel due to noise and its inefficiency to detect wide distributed spectrum [12, 42, 43]. In cooperative spectrum sensing, a group of SUs works together to reduce channel uncertainty which may fail to detect hidden node [10, 12, 42, 43]. In coherent or matched filter method, the time to reach maximum gain is less as it correlates the predefined signal with undefined signals. This may not be appropriate as it requires predefined information of primary user signal and increases the complexity.

Cyclostationary-based detection techniques can segregate the noise signal from the modulated signal and can operate in small SNR value. It also requires a part of primary source information and more observation time. Thanayankizil and Kailas [37] discussed an interference band detection and specified a temperature interference threshold level below which the primary user is inactive and secondary user can operate on it. The main short come of this method is that CR should have knowledge of the exact distance of the primary user to calculate the interference and impoverish the spectrum. On the other hand, multi-tapper spectrum sensing method [38] is based on the received samples, which are gathered in the form of Slepian base vector. The Fourier transform of this vector has its maximum energy concentrated between $f_c - W$ and $f_c + W$, where f_c is the cutoff frequency and W is the bandwidth of the spectrum. The complexity of this method increases as the number of samples received is large.

Tian and Giannakis [38] proposed a wavelet-based spectrum detection technique in which wavelet detects the edges in power spectral density (PSD) of spread spectrum. Radio identification-based detection which is used by trust project can obtain the features such as transmission frequency and range modulation method. CRN uses these extracted features and selects the suitable band for its operation. Zeng et al. [42, 43] proposed a comparative sensing for multi-hop CRN with high resolution, and the signal is recovered with the sample rate lower than that of the Nyquist sampling rate (i.e., the sampling rate should be equal or more than twice of the maximum frequency to recover the original message signal). Cabric et al. [13] presented a collaborative detection model with improvement in gain with a suitable threshold. However, it may be difficult to implement it practically as the centralized authority may not be present for cooperation.

Min and Shin [24] designed a framework using cooperation between the sensors sensing scheduling. This approach mainly relies on received signal strength. Akyildiz et al. [3] discussed the uncertainty at the receiver, as its location is unknown or not considered. A receiver detection method is proposed in [41] which takes the position of the primary receivers into consideration by using oscillation power leakage of the primary receiver. The model proposed by Gandetto and Regazzoni [17] discussed the end-to-end cooperative sensing technique in CRN to improve its performance. The approach proposed by Zhu et al. [44] is based on Markov chain analysis of the network reserving link to improve outturn of CRN. Sun [35] presented various algorithms to sense wideband spectrum with their advantages and issues.

Ning et al. [25] proposed a technique for band sensing to improve spectrum underutilization by adopting cooperative and optimal sensing mechanism. Chien-Min et al. [14] presented a model for ad hoc networks based on Cooperative Power and Contention Control Medium Access Control (CPC-MAC) protocol and placing the sensor nodes at appropriate positions which minimize the interference. However, cooperation among the nodes may be difficult as the nodes are dynamic to enter or leave the network. Zeng et al. [42, 43] discussed various issues and their resolutions for spectrum sensing in CRNs. Song et al. [33] presented a cooperative sensing scheme using two-level Hard Information Combination (HIC) based on energy detection. However, it may fail to validate when the there is more number of sensor nodes.

Amjad et al. [7] presented a survey on wireless multimedia cognitive radio networks (WMCRNs) and discussed the parameters required for transmitting the multimedia information using CRNs. Akhtar et al. [2] discussed the white space that can be utilized by CRNs and presented categorization of white space using dimensions of a signal, status of the license, and transmission scheme. Hassan et al. [19] described spectrum access models and presented exclusive spectrum trading technique for CRNs using game theory concepts. Amjad et al. [5, 6] addressed the importance and need of full-duplex communication in CRNs and presented antenna design models, communication approach, radio parameters, and medium access control protocols for supporting full duplex in CRNs.

9.3 Optimized Spectrum Sensing Approaches

In this section, we present four optimized sensing techniques that define methodologies to sense the underutilized spectrum efficiency by considering different parameters.

9.3.1 A Multilayered Framework for Optimal Sensing in Cognitive Ad Hoc Networks

Optimal sensing in the ad hoc network is given in two levels of optimization. In the first level, optimization is done in the physical layer in which sensing of bands per SU is considered. The second level of optimization is carried out in the data link layer. In this level, fusing of the data from each CR is done. As CR present in the network follows a common sensing mechanism, a generalized limit is set by gathering the information of band availability from all the SU present in the network. Cooperative sensing technique is used to improve the trust when there is an uncertainty in the channel and also avoid the issue related to the hidden node which is caused by SU shadowing.

The local decisions of PU's band availability may not be trustworthy when SNR is less. However, combining these local decisions and making an overall decision is much faithful. Gathering of such information, known as data fusion [32], can be done in two ways: one is making hard decisions (i.e., output 0 or 1), and another is by making a soft decision which gives output 0 or 1 and also gives the exactness of the decision made. The accuracy of the decision made increases as the number of SUs increase in the network. If the count of SUs increases beyond a certain limit, the information obtained is inadequate.

The parameters used at Level 1 for band sensing are threshold limit, SNR, sensing scheme, and channel models. These parameters are used to check the output of local band sensing in terms of detection probability (D_p) and false alarm probability (F_p). Level 2 parameters are number of inputs, information fusion rule, and sensing schedule. To achieve the target accuracy, the minimum number of SUs and its sensing mechanism are monitored. The information obtained from these SUs is fused to provide an optimized decision on band availability. The information thus obtained after every interval of time improves the resource utility and efficiency and reduces Bayesian risk. The framework for this approach is given in Fig. 9.2. Assume that there is M number of SUs trying to operate in the available band of PU. These SUs sense "s" band of frequencies say $\{b_{f1}, b_{f2}, ...b_{fs}\}$. The distance among the SUs is very less when compared to their distance from PU such that the calculations made are correlated. Each sensor is independent to get the information about spectrum availability. This information is collected and communicated from each sensor/SU through a single control link to set the global threshold.

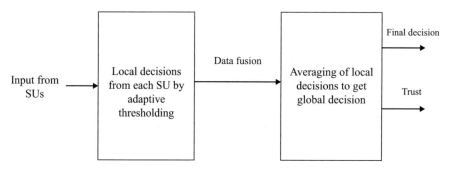

Fig. 9.2 High-level view of multilayered framework

9.3.1.1 Optimization at Level 1

Level 1 consists of two functions: (1) sensing the spectrum locally at each SU and (2) adaptive threshold.

Sensing the Spectrum Locally

Energy detector is used for sensing the band where each SU make a hard decision to get the knowledge whether the band is occupied or not by PU in frequency range b_{fs}. It requires prior knowledge of noise power spectrum to set the threshold value, say γ. Consider a case where kth SU makes the spectrum decision of lth frequency band with threshold value γ_k with output z_k:

$$d_{kl} = \begin{cases} 0, \text{Decide } H_0 \text{ if } z_k < \gamma_k \\ 1, \text{Decide } H_1 \text{ if } z_k > \gamma_k \end{cases} \qquad (9.2)$$

where H_0 and H_1 are the hypothesis for absence and presence of PU in the spectrum, respectively.

The data sensed by kth SU can be represented as row matrix-vector $[d_{k1}, d_{k2}, \ldots \ldots d_{ks}, D_{pk}, F_{pk}]$, where the first s elements gives the information about PUs over s frequency bands. If the element is 0, then there is no PU activity, otherwise the PU is active. Here, D_{pk} is detection probability and F_{pk} is false detection probability of kth SU. These probabilities remain the same for s frequency spectrum and give reliability of decision made by kth SU. The row matrix thus obtained is shared to other SUs.

Adaptive Threshold

There may be an error in the noise power evaluation which is used to set the threshold value of the energy detector. To avoid this problem, a common decision

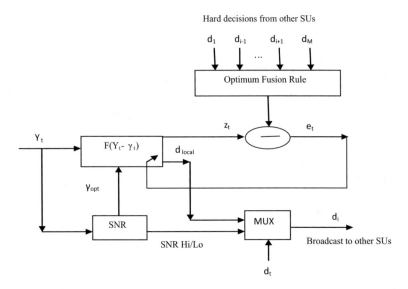

Fig. 9.3 Adaptive threshold based on group decisions [32]

is taken based on the SU group present in the network and adapted by each SU. According to the variation in the ambiance, the decision is updated continuously to avoid the errors. The adaptive threshold is set by taking group decisions as a guidance signal. This method of setting up the adaptive threshold is immune to channel variations and resistive to calculation errors.

Figure 9.3 shows the adaptive threshold based on group decision. Here, Y_t is the output of energy detector, d_t is the group decision, and z_t is the soft decision using current threshold γ_t at time instant t (i.e., sampling interval of the detector). Sigmoid function is used to get the error which is the difference between d_t and z_t. This can be represented as

$$e_t = d_t - z_t \tag{9.3}$$
$$\gamma_{t+1} = \gamma_t - 2e_t \mathcal{E} z_t (1 - z_t) \tag{9.4}$$

where e is the scaling factor. Here, γ_{t+1} gives the spectral occupancy decision for time interval $t + 1$.

We can observe that the adaptive limit is unchanged in the following two cases:

1. If $e_t \sim 0$ where the decision made locally by each CR, it is similar to the fused decision made globally.
2. If z_t is almost 1, then the global decision is the same as the local decision when SNR is high. If z_t is 0, then local decision opposes the global decision. In such case, the local decision is neglected and global decision is only applied.

The local information is updated periodically to avoid errors in the feedback path.

9.3.1.2   Optimization at Level 2

Optimization at Level 2 consists of data fusion, estimator, and sensing scheduler. Optimal schedule for sensing is followed by each SU.

Data Fusion

A spectral occupancy matrix (SOM) is formed at each secondary user by gathering the entire row matrix from other SUs. The dimensions of SOM are $(r \times (s + 2))$ where the present data from the kth SU is given by kth row as shown in the below matrix-vector:

$$
\begin{array}{c}
\text{Decision vector by } k^{\text{th}}\text{node} \longrightarrow
\end{array}
\left(
\begin{array}{l}
V_{11} \; V_{12} \; V_{13} \; \ldots \ldots V_{1s} \; D_{p1} \; F_{p1} \\[4pt]
V_{21} \; V_{22} \; V_{23} \; \ldots \ldots V_{2s} \; D_{p2} \; F_{p2} \\[4pt]
\ldots \ldots \ldots \ldots \ldots \ldots \ldots \ldots \ldots \\[4pt]
V_{r1} \; V_{r2} \; V_{r3} \; \ldots \ldots V_{rs} \; D_{pr} \; F_{pr}
\end{array}
\right)
$$

$$\uparrow$$
Decision vector of Ith frequency band

By the use of data fusion technique, the band availability information of lth frequency is adapted at each SU. The main aim is to collect the data from various sources to get the finest possible resolution P_r, which can be computed as

$$
P_r = \sum V_{kl} \log_e \left[\frac{D_{pk}(1 - F_{pk})}{F_{pk}(1 - D_{pk})} \right] \tag{9.5}
$$

Note that P_r less than λ indicates the absence of PU and P_r greater than λ indicates the presence of PU in the spectrum. Here, λ is pre-opted limit, based on which the spectral occupancy (binary) decision is made. The decision from the SU with high D_p value and low F_p value has more weight as the weighted average of local decision from r SUs is taken into account for making combined decision.

Optimal Number Estimator

The function of the estimator is to make estimations of the SU count for minimum risk when average D_p and F_p is given. We assume that PU on/off plan is available. The information gain G_i in terms of Bayesian risk [31, 40] can be computed as

$$G_i = (1 - Q_e(i))/2 \tag{9.6}$$

where Q_e is the total probability of error while sensing the data (in the fused decision). In sensing process and transmission of information, SU consumes resources of the network. Therefore, out of M number SUs, only I number of SUs will sense the spectrum at a time, and remaining SUs are at rest so that the resource consumption is reduced and accuracy of fused data is maintained.

Sensing Scheduler

If same SUs are used to sense the spectrum at each time interval, their energies may drain out quickly. To avoid this situation, an efficient sensing schedule is to be designed. In this approach, a simple scheduler is proposed where the probability of each SU is I/M to sense the data where I is estimated a minimum number of SUs and M is the size of the network.

9.3.1.3 Observations

1. Noise uncertainty in energy detector can be mitigated by an adaptive threshold.
2. Only I number of SUs sensing the spectrum out of M number of SUs is sufficient to retain the target accuracy so that the resource utilization can be reduced.
3. The value of Q_e (error probability in the fused decisions) is reduced at lower group SNR value (in dB). At higher group SNR, Q_e is almost zero.
4. If the number of SUs increases, then the value of Q_e decreases when compared to a fixed threshold measurement. This is because it does not consider the changes in the local environment which impacts local sensing and hence susceptible to more computational errors.
5. If group SNR is less, then the optimal number of SUs required for sensing the spectrum is less when compared to fixed threshold method for obtaining the desired accuracy.
6. The probability of error is less when I number of SUs sense the spectrum instead of all the SUs sensing the band.

At lower group SNR, the adaptive threshold method is more efficient with minimum error. Hence, the minimal number of SUs with suitable SNR will make other SUs in the group to take trustworthy decisions as the guidance signal makes use of the combined decision made by the SUs.

9.3.2 Optimal Sensing Disruption for a CR Adversary

This approach is presented by Peng et al. [26] which focus on the rival nodes that interfere with the spectrum sensing mechanism. These nodes put energies in the available spectrum bands. Such bands appear to be occupied the secondary user while sensing and thereby reducing the number of available bands in the view of SUs. This process is termed as sensing disruption. To mitigate this problem, the count of false detection is increased considering the adversary node's power limitations. Most of the sensing cases are exposed to adversary attack. There are two types of disruptions: (1) disruption in sensing channel and (2) disruption in cooperative sensing. In the case of disruption in sensing channel, the enemy node sends microwave signals in the band monitored by SUs. These signals may be a copy of PU signals to mislead the SUs. In the case of disruption in cooperative sensing, two decisions are made with respect to the primary user's activity: (a) local decision and (b) global decision. After sensing the spectrum locally, the SUs share the information about spectrum availability to other SUs present in the network to get the globalized decision. The adversary node may act as one of the SUs and share fake information about spectrum availability to other SUs. Hence the global decision is affected. In this approach, more focus is given to the channel disruption as it is the basic for making local and cooperative decision about PU activity.

9.3.2.1 System Model

Generally, the spectrum can be divided into two: (a) the bands which are busy (i.e., occupied by the primary user) and (b) the bands which are vacant where the primary user is absent. The adversary targets the vacant band and sends signals into it so that they appear to be busy when SUs sense them. Such frequency bands are known as spoofed bands.

Assume that, at a given time instance, there is Y number of allowable bands. Let the mean of false detection be Y_f on which adversary concentrates to reduce the number of available bands. Y_f is given as

$$Y_f = \sum_{f=1}^{Y} d_f \tag{9.7}$$

where d_f is the probability of false detection of fth allowable spectrum. And,

$$\sum_{f=1}^{Y} D_f = D \tag{9.8}$$

where D_f is the adversary power radiated on fth band. Here, $D_f \geq 0$ and $f = 1, 2, \ldots$. Y, and D represents different spoofing powers of the adversary.

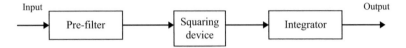

Fig. 9.4 Energy detector

9.3.2.2 Optimal Sensing

Energy Detection at the Receivers
Figure 9.4 shows the energy detector that is used for spectrum detection at SU receiver.

Prefilter is used for limiting the noise bandwidth to N_B (hertz), which is common for allowable bands. The output of the filter is given to squaring device and then integrator. The output is compared with the preset threshold to determine whether the band is vacant or busy. If the output is greater than or equal to the threshold value, then the spectrum is busy. Otherwise, it is vacant. If the adversary node is absent, then thermal noise is only present in the allowable bands. Note that thermal noise is the same for all the bands which are allowable. False alarm detection denotes only thermal noise present in the vacant bands, and probability of false detection includes both the thermal noise and spoofing effect. On the other hand, the probabilities of false alarm and false detection are same for vacant bands. Equation 9.9 represents the probability of false alarm (given by Urkowitz [39]):

$$f_p = Q\left(\frac{K}{2\sqrt{TW}\sigma_n^2} - \sqrt{TW})\right) \tag{9.9}$$

where $Q(.)$ is the Gaussian tail function, i.e.,

$$Q(x) = \frac{1}{\sqrt{2\pi}} \int_x^{+\infty} \exp\left(-\frac{t^2}{2}\right) dt \tag{9.10}$$

Here, TW is the product of integration time interval and bandwidth. The threshold value is already defined for the band f which is allowable, and we have identical noise power for all the allowable bands say K.

9.3.2.3 Probability of False Detection

Assume that the adversary node is present. The input to the squaring device $S_f(t)$ of fth band is given as the sum of the thermal noise signal and the spoofed signal:

$$S_f(t) = \beta_{pf}(t) + m_f(t) \tag{9.11}$$

where p_f is the spoofed signal (assumed to be zero-mean Gaussian distribution), m_f is the thermal noise signal, and β is the path loss between adversary node and SU.p_f and

m_f are independent of one another. The test static technique [11] is used where the sum of square of the signals received by the SU is asymptotically normal distribution with zero mean and variance $D_f + \sigma_n^2$. Hence d_f can be given as

$$d_f(D_f) = Q\left(\frac{K}{2\sqrt{TW}(\beta^2 D_f + \sigma_n^2)} - \sqrt{TW}\right) \tag{9.12}$$

9.3.2.4 Optimal Sensing Disruption (Partial Band)

Substituting Eq. 9.7 in Eq. 9.4 gives the optimum sensing disruption. By using Lagrange multipliers and Karush-Kuhn-Tucker (KKT) conditions [22, 28], minimal spoofing power is given as

$$D_k^* = \begin{cases} D/y, k \in \phi_{\lambda, Y} \\ 0, \text{otherwise} \end{cases} \tag{9.13}$$

where $\phi_{\lambda, Y} = \{k | \lambda_k^* = 0\}$ (λ_k^* is the Lagrange multiplier) and y is the number of spoofed bands.

Optimal Number of Spoofed Bands Y^*
The minimum number of bands that are spoofed by rival node by maximizing D_f is evaluated as given below:

$$Y_f = Y^* Q\left(\frac{\sqrt{TW}\frac{(K-V)}{\sigma_n^2} - 2TW\beta^2 D/(Y^*\sigma_y^2)}{2TW(1 + \beta^2 D/(Y^*\sigma_y^2)}\right)$$
$$+ (Y - Y^*)Q\left(\sqrt{TW}\left(\frac{K}{2TW\sigma_y^2} - 1\right)\right) \tag{9.14}$$

where $V = (TW + 2\sqrt{TW} + 8TW)\sigma_y^2$.

Average Number of Additional False Detections ΔY_f
In the absence of adversary node, only thermal noise is present. Average number of extra false detection by spoofing ΔY_f to separate noise and spoofing effect is given as

$$\Delta Y_f = Y_f - Y_{fp}$$
$$\Delta Y_f = \frac{\alpha \beta^2 D}{\sqrt{2\pi}(\beta^2 C^* + \sigma_y^2)^2} \exp\left(-\left(\frac{\frac{\alpha}{(\beta^2 C^* + \sigma_y^2) + b}\right)^2}{2}\right) \tag{9.15}$$

where C^* is a constant, calculated by using the parameters α, b, σ_y^2, β, and f_p, and Y_{fp} is the number of bands which are considered to be busy only due to thermal noise.

9.3.2.5 Observations

1. If the number of allowable bands (Y) is small, then the adversary node is capable of spoofing all the allowable bands. If the number of allowable bands is large, then optimal number of bands (almost constant(Y^*)) are only spoofed by the adversary.
2. Let Z be the ratio of spoofing power to noise power (i.e., $\left(D/\sigma_y^2\right)$). If the number of spoofed bands increases, then Z will also increases.
3. The mean of false detection increases with increment in the allowable bands.
4. When Y is large, then the linearity of ΔY_f increases with the spoofing power.
5. If the spoofing power increases, then the number of the percentage of available bands sharply decreases.

9.3.3 Parametric Optimization for Spectrum Sensing

In this approach [34], a centralized authority is assumed to be present in the network which communicates with all the SUs in its network range and decides the spectrum availability to avoid both uncertainties at the receiver and limitations of sensing capabilities. Spectrum sensing time is divided into *observation time* and *transmission time*.

The SUs sense the spectrum using energy detection method. A wide range of frequencies can be sensed by using software-defined radios (SDR) by altering the frequency using software-defined functions. These radios can sense and transmit the signals. The SUs have prior information about the PUs like frequency bands on which PUs work, minimum SNR, interferences, and PU activity. PU activity is calculated based on Poisson distribution periods which are spread exponentially. The rate at which the primary user is active (birth rate, say δ) and the rate at which primary user is idle (death rate, say γ) are exponentially distributed. Figure 9.5 represents the block diagram of the system model. There are three levels: Level 1 is

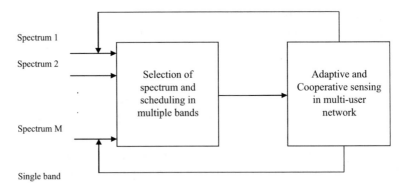

Fig. 9.5 Block diagram of system model

parametric optimization of sensing in single spectrum band, Level 2 is spectrum selection and scheduling in multiple spectrum bands, and Level 3 is cooperative and adaptive sensing in multiuser network.

9.3.3.1 Level 1

The efficiency for sensing the spectrum is given by

$$\eta = \frac{T_t}{T_t + T_o} \tag{9.16}$$

where T_t is the transmission time and T_o is the observation time. In this approach, a MAP (maximum a posteriori [27]), an optimal energy detection, is used for sensing the spectrum in accordance with PU activity. The main aim here is to focus on spectrum efficiency maximization and accuracy of spectrum sensing.

Deriving the posteriori probability [15] by assuming B_w as bandwidth of the spectrum is given as

$$P_a = \frac{\delta}{\delta + \gamma}$$
$$P_i = \frac{\gamma}{\delta + \gamma} \tag{9.17}$$

where P_a is the probability of time period when PU is active and P_i is the probability of time period when PU is inactive.

The false alarm probability (F_p) and spectrum detection probability (D_p) are represented in Eqs. 9.18 and 9.19, respectively:

$$D_p(\alpha) = D_p.P_a \tag{9.18}$$
$$F_p(\alpha) = F_p.P_i \tag{9.19}$$

Here, α is the threshold of MAP energy detector.

In energy detection, the output of the integrator is termed as chi-square distribution [16]. We can imprecise the chi-square distribution as Gaussian distribution [36] by the use of central limit theorem for a huge number of samples:

$$Z \sim \begin{cases} \mathcal{N}(n\sigma_n^2, 2\sigma_n^4) & H_0 \\ \mathcal{N}(n\sigma_n^2 + \sigma_s^2), 2n(\sigma_n^2 + \sigma_s^2)^2) & H_1 \end{cases} \tag{9.20}$$

where n is the number of samples, σ_n^2 is the noise variance, and σ_s^2 is the received signal ($r(t)$) variance. Hence, from Eqs. 9.18, 9.19, and 9.20, equations for D_p and F_p can be rewritten as

$$F_p(B_w, T_o, \gamma, \delta) = \frac{\gamma}{\delta + \gamma} Q\left(\frac{\alpha - 2T_o B_w \sigma_n^2}{\sqrt{4T_o B_w \sigma_n^4}}\right) \tag{9.21}$$

$$D_p(B_w, T_o, \gamma, \delta) = \frac{\delta}{\delta + \gamma} Q\left(\frac{\alpha - 2T_o B_w (\sigma_n^2 + \sigma_s^2)}{\sqrt{4T_o B_w (\sigma_n^2 + \sigma_s^2)^2}}\right). \tag{9.22}$$

Interference Model

Interference constraint is also considered for optimizing the parameters for sensing the spectrum. Since sensing is periodic, interference occurs in two cases: (a) busy state interference (B_i) in which the spectrum is being used by the PU, whereas SU fails to sense the presence of PU and starts transmitting the data, and (b) ideal state interference (I_i) in which the spectrum is not being utilized by the PU and SU is transmitting the data. There is a possibility that the PU becomes active when SU is transmitting:

$$E(B_i) = P_i F_p\left(\frac{\gamma}{\delta + \gamma} Te^{-\gamma T} + \frac{\delta}{\delta + \gamma} T\right) \tag{9.23}$$

$$E(I_i) = P_i(1 - F_p)(1 - e^{-\delta T})\frac{\delta}{\delta + \gamma} T \tag{9.24}$$

The ratio of interference R_i can be generated by combining Eqs. 9.23 and 9.24 as given below:

$$R_i = \frac{\gamma}{\delta}\left[e^{-\phi T} F_p + (1 - e^{-\phi T})\frac{\delta}{\delta + \gamma}\right] \tag{9.25}$$

where ϕ is max(γ, δ). The range of R_i is $P_i \geq R_i \geq \frac{P_i}{P_a}$.

The ratio of lost spectrum R_l is represented as Eq. 9.26 and varies over the range of $P_a \geq R_l \geq \frac{P_a}{P_i}$:

$$R_l = \frac{\delta}{\gamma}\left[e^{-\phi T} F_p + (1 - e^{-\phi T})\frac{\gamma}{\delta + \gamma}\right] \tag{9.26}$$

where R_i and R_l represent duality property based on the value of δ and γ.

Optimization of Sensing Parameters

The parametric optimization can be obtained by the generation of operating region by considering the parameters such as observation time and operating region of transmission time. Observation time and operating region for transmission time are given in Eqs. 9.27 and 9.28, respectively:

$$T_o = \frac{1}{B_w \Delta^2} \left[Q^{-1}(\bar{F}_p) + (\Delta + 1) \left(\frac{P_i \bar{F}_p}{P_a} \right) \right]^2 \tag{9.27}$$

where Δ is the SNR equal to $\frac{\sigma_s^2}{\sigma_n^2}$:

$$\bar{F}_p(T) = P_a - P_a \left(1 - \frac{T_p}{P_i} \right) e^{\phi T} \tag{9.28}$$

9.3.3.2 Level 2

Two techniques are used to hold the multiple available spectrums: (a) wideband sensing and (b) sequential band sensing. In wideband sensing, the transceivers sense a wide range of spectrums at a particular interval of time with fixed observation and transmission time. However, this method may lead to interference constraint violations. In sequential band sensing, only a single band is sensed by the transceiver at a specified time and adapts optimized parameters of the spectrum accordingly. In this approach, all the SUs follow sequential sensing, and the *CRs use time divided scheduling* allocates minimum time period for observation and transmission.

Opportunity cost is defined as the addition of the expected sensing capabilities of the spectrum which are blocked if any one of the spectrum band is selected by the SUs. Based on this opportunity cost scheduling, the SUs sense the band. In the beginning of the spectrum sensing cycle, the SU checks the present time period availability, and if it is busy, then all the opportunistic bands are moved to block period. Once the time period becomes available, the SU senses the spectral band based on the opportunity cost. Once the observation time is over, the SU goes to transmission state, and after its completion, the time period is again available for competing bands. This algorithm is known as least cost first serve (LCFS).

9.3.3.3 Level 3

In conventional cooperative sensing method, the band is assumed to be available if none of the PUs is active. Even if one PU is active, the SU cannot use the spectrum. In this approach, a different cooperative sensing method is discussed based on binary distribution. According to this method, the SUs should be located in the region near to PU. In the case where there are multiple PU activities, the base station must calculate the sensing decision at different regions separately, and the cooperation gain is obtained.

The sensing parameters should be re-optimized as the probabilities of false alarm and detection of spectrum varies. Generally, the spectrum sensing information varies due to SU mobility and transmission. Hence, the base station updates the optimal parameters at every change to maintain constant interference level.

9.3.3.4 Observations

1. The interference model validates for both idle and busy state of the spectrum.
2. Optimized parameters that are used for sensing result in improved efficiency.
3. The LCFS algorithm shows higher capacity when compared to first come first serve (FCFS) algorithm which does not consider opportunity cost and gives the time slot to the spectrum with highest block time.
4. The optimal transmission time increases when the number of cooperative users increases.

9.3.4 Cluster-Based Spectrum Sensing

This technique [9] uses a group of SUs with a centralized authority called "cluster head." Based on the similarity in the group information, the clusters are formed. If the similarities in the group are more, the more it is distinct from other clusters. SUs send the information (band availability information) to the cluster head by using a common channel. In this approach, twofold cluster structure is used: (1) the cluster structure which is aware of the ambiance and (2) the cluster structure which is energy efficient. The following are few assumptions that are made for the efficient clustering of SUs:

1. Each SU should accurately obtain the information about the spectrum availability from its locality.
2. A minimum of one channel which is common to all the SUs present in the cluster must be available.
3. At least one cluster head should be present in the cluster. The information about the spectrum availability must be given to the cluster head first, and then the information is broadcasted to other SUs present in that cluster.
4. To increase the life span of the SUs, a well-organized structure drains the energy at a minimum for the entire communication.

Cluster classification is given below [9]:

(a) Well-Separated Cluster: The cluster groups the nodes in which the nodes are more similar to one another than any other group in another cluster. At times, a threshold is set to confirm that nodes in the cluster are very similar to each other than any other group.

(b) Prototype-Based Cluster: Groups in a cluster have a similar pattern than any other cluster pattern. The pattern of the cluster is generally mean of the entire points present in the cluster.

(c) Graph-Based Cluster: The information is represented in terms of a graph considering the group as a node and the connections among them as edges. Further, the nodes are connected to only one cluster and are not connected to any other cluster outside.

(d) Density-Based Cluster: A region of low density is present around a denser region of the cluster. These clusters do not have a common region because a link between them is disappeared as noise.

(e) Conceptual Cluster: If the cluster shares the property with another, then it is termed as conceptual (or shared) cluster.

The comparison of the four optimized techniques is shown in Table 9.1.

9.4 Challenges and Future Scope

Cognitive radio sensors are intelligent as they dynamically change the parameters for adapting the prevailing conditions. This concept can be employed in numerous applications to obtain desirable outcomes. However, the limitations in CR spectrum sensing such as computational complexities, updating the threshold regularly, and maintaining a trade-off between spectrum observation and spectrum utilization periods may introduce difficulties. Apart from spectrum sensing, spectrum management, and spectrum mobility, also becomes a challenge.

Cognitive radio sensors can be incorporated into OFDM MIMO systems for reducing bit error rate (BER). The principles of game theory in CRNs may improve its performance. There is a scope for CR sensor deployment in satellite communication systems, the Internet of Things (IoT). Quality of service (QoS) aware routing protocol for heterogeneous clustered wireless sensor networks (WSNs [5, 6]) is energy efficient. Such approaches reduce the delay and increase the throughput and stability of the network. Application of WSNs in urban areas, its optimal deployment techniques, problems, and solutions are discussed in [29]. Deployment of tiny operating systems in CR sensors increases its flexibility and area of the application when compared to the conventional OS present in the network. Features, scheduling algorithms, and structure for TinyOS are addressed in [8]. Important parameters of cognitive radios, its application fields, protocols at different layers, and dynamic architecture with its limitations are examined in [1, 4] which provides a new research direction and also wide opportunities in real-life scenarios.

Table 9.1 Comparative analysis of optimized techniques

		Technique			
S. No	Parameter	Spectrum sensing in ad hoc networks	Optimizing sensing disruption in CRN	Optimized parametric sensing	Cluster-based sensing
1	Approach/method	At physical and data link layers	Sensing disruption in the presence of adversary node	Sensing the spectrum in single-level band and multilevel bands	Optimization of spectrum by clustering
2	Detector used	Energy detector	Energy detector	Maximum a posteriori (MAP) energy detector	–
3	Sensing technique	Cooperative	–	Cooperative using centralized authority or cluster head	Cooperative (decentralized cooperative)
4	Framework	Multilevel framework, with two levels	–	Multilevel framework, with three levels	–
5	Advantages	Accurate, resource efficiency and immune to noise power measurement and reliable	Maximize the number of false detection and efficient	Maximum sensing efficiency providing opportunities in multi-spectrum/multiuser network along with interference constraint consideration	Stable and scalable, low complexity
6	Disadvantages	The adaptive threshold must be updated at regular intervals of time to avoid errors	If adversary node is not present in the network, then this method is not validated	Calculation complexity and threshold must be updated periodically	Cluster formation may be difficult

9.5 Conclusion

Spectrum sensing is a very crucial step in cognitive radio networks. As the need for efficient spectrum utilization increases, appropriate techniques must be selected to sense the spectrum. In this chapter, four different optimization techniques are studied by considering various technologies and parameters. We also discussed the comparative analysis of these techniques. Optimized parametric sensing technique has more spectrum efficiency when compared to other methods and can be extended to multiuser/spectrum networks. This approach gives optimized equations for various parameters such as transmitting time, observation time, cooperative gain, and efficiency. Spectrum sensing in ad hoc networks is reliable and efficient in sensing the spectrum. Optimizing sensing disruption in CRN is accurate in detecting the number of spoofed bands by an adversary node in the network which gives false information about spectrum availability. Cluster-based sensing gives optimization by cluster formation with groups of nodes which has similar characteristics. One can use these techniques based on the desired parameters to use spectrum efficiently.

References

1. Akan OB, Karli OB, Ergul O (2009) Cognitive radio sensor networks. IEEE Netw 23(4). https://doi.org/10.1109/MNET.2009.5191144
2. Akhtar F, Rehmani MH, Reisslein M (2016) White space: definitional perspectives and their role in exploiting spectrum opportunities. Telecommun Policy 40(4):319–331
3. Akyildiz IF, Lee W-Y, Vuran MC, Mohanty S (2006) Next generation/dynamic spectrum access/cognitive radio wireless networks: a survey. Comput Netw 50(13):2127–2159
4. Akyildiz IF, Su W, Sankarasubramaniam Y, Cayirci E (2002) A survey on sensor networks. IEEE Commun Mag 40(8):102–114
5. Amjad M, Afzal MK, Umer T, Kim BS (2017a) QoS-aware and heterogeneously clustered routing protocol for wireless sensor networks. IEEE Access 5:10250–10262
6. Amjad M, Akhtar F, Rehmani MH, Reisslein M, Umer T (2017b) Full-duplex communication in cognitive radio networks: a survey. IEEE Commun Surv Tutorials 19(4):2158–2191
7. Amjad M, Rehmani M H, Shiwen M (2018) Wireless multimedia cognitive radio networks: a comprehensive survey. IEEE Commun Surv Tutorials. https://doi.org/10.1109/COMST.2018.2794358
8. Amjad M, Sharif M, Afzal MK, Kim SW (2016) Tiny OS-new trends, comparative views, and supported sensing applications: a review. IEEE Sensors J 16(9):2865–2889
9. Babu RG, Amudha V (2016) Spectrum sensing cluster techniques in cognitive radio networks. Procedia Comput Sci 87:258–263
10. Bagwari A, Singh B (2012) Comparative performance evaluation of spectrum sensing techniques for cognitive radio networks. In: Proceedings of 4th IEEE International conference on Computational Intelligence and Communication Networks (CICN-2012), pp 98–105
11. Boyd S, Vandenberghe L (2004) Convex optimization. Cambridge University Press, Cambridge
12. Cabric D, Mishra SM, Brodersen RW (2004) Implementation issues in spectrum sensing for cognitive radios, In: 38th Asilomar conference on signals, Systems and computers, Pacific Grove, pp 772–776
13. Cabric D, Tkachenko A, Brodersen RW (2006) Spectrum sensing measurements of pilot, energy and collaborative detection. In: IEEE Military Communications conference (MILCOM 2006), Washington DC, USA
14. Chien-Min W, Hui-Kai S, Maw-Lin L, Yi-Ching L, Chih-Pin L (2014) Cooperative power and contention control MAC protocol in multichannel cognitive radio ad hoc networks. In: 8th International conference on Innovative Mobile and Internet Services in Ubiquitous Computing (IMIS 2014), pp 305–309
15. Cox DR (1962) Renewal theory. Wiley, New York
16. Digham F F, Alouini M S, Simon M K (2003) Energy detection of unknown signals over fading channels. In: IEEE International conference on communications (ICC '03), Anchorage, AK, USA, May 2003, pp 3575–3579
17. Gandetto M, Regazzoni C (2007) Spectrum sensing: a distributed approach for cognitive terminals. IEEE J Select Areas Commun 25(3):546–557

18. Ghasemi A, Sousa ES (2008) Spectrum sensing in cognitive radio networks: requirements, challenges, and design trade-offs. IEEE Commun Mag 46(4):32–39

19. Hassan MR, Karmakar GC, Kamruzzaman J, Srinivasan B (2017) Exclusive use spectrum access trading models in cognitive radio networks: a survey. IEEE Commun Surv Tutorials 19 (4):2192–2231

20. Hossain E, Niyato D, Han Z (2009) Dynamic spectrum access and management in cognitive radio networks. Cambridge University Press, Cambridge

21. Kanti J, Tomar GS (2016) Various sensing techniques in cognitive radio networks: a review. Int J Grid Distrib Comput 9(1):145–154

22. Lee W-Y, Akyildiz IF (2008) Optimal spectrum sensing framework for cognitive radio networks. IEEE Trans Wirel Commun 7(10):3845–3857

23. Marinho J, Monteiro E (2012) Cognitive radio: a survey on communication protocols, spectrum decision issues, and future research directions. Wirel Netw 18(2):147–164

24. Min A, Shin K (2009) An optimal sensing framework based on spatial RSS profile in cognitive radio networks. In: Proceedings of the 6th Annual IEEE communications society conference on Sensor, Mesh and Ad Hoc Communications and Networks (SECON'09), Rome, Italy, pp 207–215

25. Ning Z, Yu Y, Song Q, Peng Y, Zhang B (2015) Interference-aware spectrum sensing mechanisms in cognitive radio networks. Comput Electr Eng 42:193–206

26. Peng Q, Cosman PC, Milstein LB (2010) Optimal sensing disruption for a cognitive radio adversary. IEEE Trans Veh Technol 59(4):1801–1810

27. Proakis JG (2001) Digital communications, 4th edn. McGraw-Hill, New York

28. Rao SS (1983) Optimization: theory and applications, 2nd edn. Wiley, Hoboken

29. Rashid B, Rehmani MH (2016) Applications of wireless sensor networks for urban areas: a survey. J Netw Comput Appl 60:192–219

30. Rawat DB, Yan G (2009) Signal processing techniques for spectrum sensing in cognitive radio systems: challenges and perspectives. In: First Asian Himalayas International conference on Internet – The next generation of mobile, wireless and optical communications networks (AC-ICI-2009), Kathmandu, Nepal, 3–5 Nov 2009

31. Sasirekha GVK, Jyotsna B (2010) Optimal number of sensors in energy efficient distributed spectrum sensing. In: 3rd International Symposium on Applied Sciences in Biomedical and Communication Technologies (ISABEL), Roma, Italy, Nov 2010

32. Sasirekha GVK, Jyotsna B (2011) Optimal spectrum sensing in cognitive ad-hoc networks: a multi-layer framework. In: 4th International Conference on Cognitive Radio and Advance Spectrum Management (CogART '11), Article 31, Barcelona, Spain, Oct 2011

33. Song C, Alemseged Y D, Tran H N, G. Villardi, C. Sun, S. Filin, and H. Harada (2010) Adaptive two thresholds based energy detection for cooperative spectrum sensing. In: 7th IEEE Consumer Communications and Networking Conference (CCNC 2010), Las Vegas, NV, USA

34. Sriram K, Whitt W (1986) Characterizing superposition arrival processes in packet multiplexers for voice and data. IEEE J Select Areas Commun SAC 4(6):833–846

35. Sun H (2013) Wideband spectrum sensing for cognitive radio networks: a survey. IEEE Wirel Commun 20(2):74–81

36. Tang H (2005) Some physical layer issues of the wide-band cognitive radio system. In: First IEEE International symposium on new frontiers in Dynamic Spectrum Access Networks (DySPAN), pp 151–159

37. Thanayankizil L, Kailas A (2008) Spectrum sensing techniques (II): receiver detection and interference management report. http://aravind.kailas.googlepages.com/ece_8863_report.pdf

38. Tian Z, Giannakis G (2007) Comprehensive sensing for wideband cognitive radios. In: Proceedings of IEEE IntConf acoustics, speed, signal processing, Honolulu, HI, April 2007, pp 1357–1360

39. Urkowitz H (1967) Energy detection of unknown deterministic signals. Proc IEEE 55 (4):523–531

40. Varshney PK (1997) Distributed detection & data fusion. Springer, New York

41. Wild B, Ramchandran K (2005) Detecting primary receivers for cognitive radio applications. In: First IEEE International symposium on new frontiers in Dynamic Spectrum Access Networks (DySPAN 2005), pp 124–130

42. Zeng Y, Liang YC, Hoang AT, Zhang R (2010a) A review on spectrum sensing for cognitive radio: challenges and solutions. EURASIP J Adv Signal Process. https://doi.org/10.1155/2010/381465

43. Zeng F, Tian Z, Li C (2010b) Distributed compressive wideband spectrum sensing in cooperative multi-hop cognitive networks. In: 2010 I.E. International Conference on Communications (ICC), Cape Town, South Africa

44. Zhu X, Shen L, Yum TP (2007) Analysis of cognitive radio spectrum access with optimal channel reservation. IEEE Commun Lett 11(4):304–306

Chapter 10
Learning Strategies in Cognitive Radio Involving Soft Computing Techniques

Mithra Venkatesan, Anju Vijaykumar Kulkarni, and Radhika Menon

10.1   Existing Scenario in Wireless Networks

Currently in wireless networks, fixed spectrum assignment policy is widely used. In static spectrum allocation strategy, regulation of spectrum is done by government organizations, and spectrums are consigned to license owner or services for the entire duration of time. The complete radio spectrum is divided into bands of frequencies each catering to a particular type of service. This type of frequency allocation is done on international and national basis. High-level guidance is set by the international bodies, while the detailed policies are decided by the national bodies [1].

The Federal Communications Commission (FCC) constituted a spectrum policy task force comprising of interdisciplinary FCC employees. The main purpose of establishing such a team was to find improved methods to manage the spectrum. The major finding in the report of the team was that certain frequency ranges in the spectrum remained unused for a long duration. Further, some frequency ranges are sparsely used, while few frequency bands are heavily used. The usage of allotted bands of frequency varies from 15% to 85% [2].

Though fixed spectrum allocation policies were used extensively, the drastic rise in demand for spectrum leads to increased straining of the effectiveness of static spectrum allocation policies. Hence the limited availability of bands of frequencies and the presence of sparsely occupied bands were the major contributing factors leading to a radical shift in the communication paradigm of dynamic spectrum next-generation networks and Cognitive Radio systems as shown in Fig. 10.1. These next-generation networks aim toward providing high bandwidth and improved

M. Venkatesan (✉) · A. V. Kulkarni · R. Menon
Dr. D.Y. Patil Institute of Technology, Pimpri, Pune, Maharashtra, India

© Springer International Publishing AG, part of Springer Nature 2019   233
M. H. Rehmani, R. Dhaou (eds.), *Cognitive Radio, Mobile Communications and Wireless Networks*, EAI/Springer Innovations in Communication and Computing,
https://doi.org/10.1007/978-3-319-91002-4_10

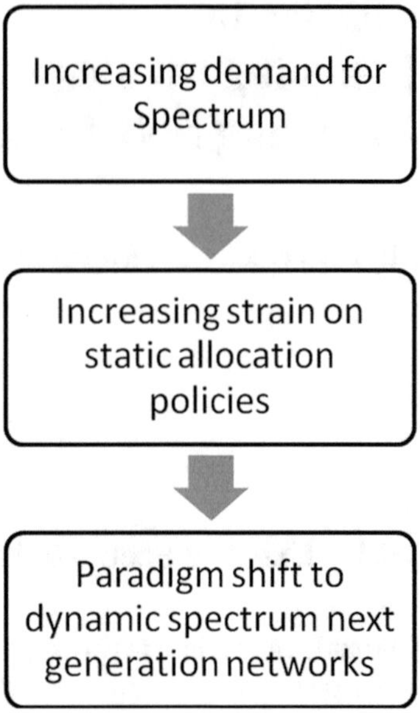

Fig. 10.1 Existing scenario
in wireless networks

Increasing demand for
Spectrum

Increasing strain on
static allocation
policies

Paradigm shift to
dynamic spectrum next
generation networks

spectrum usage. This is by providing opportunities for other users to access the licensed band of frequencies with minimum intervention to the licensed users. These arrangements work toward exploiting the underutilized electromagnetic spectrum. This sparsely used electromagnetic spectrum lead to the concept of spectrum hole or white spaces. Spectrum hole is a range of frequency allocated to a licensed user for a specific duration and position, when the spectrum is not being used by the user [1]. Spectrums can be used in an enhanced fashion, if these unused bands of spectrum can be used another user called secondary user who is not being serviced.

Cognitive Radio which is built over and above a software-defined radio is based on the technology toward exploiting these spectrum holes by the secondary users. The major task of dynamic frequency band allocation can be accomplished by a Cognitive Radio through efficient spectrum organization. Spectrum organization is the process of identifying how the spectrum is being used and by whom. The major aim of spectrum organization is to utilize the available spectrum to the maximum with as many competent users with minimum interference between different users. Cognitive Radios have a major role to play in this dynamic spectrum supervision. This dynamic approach has the potential to utilize the otherwise wasted resources, and with the availability of more spectrum for use, it reduces barriers to entry of new ventures.

The basic concept of dynamic spectrum access is utilizing the concept of spectrum sharing among the licensed and unlicensed users. The licenser client or consumer is the principal client who has access to the frequency bands on high priority, while secondary consumers access the spectrum when the primary consumer does not require it.

Implementation of dynamic spectrum access is synonymous with Cognitive Radios. The term dynamic spectrum access implies that no static allocation or assignment of frequencies is made. However, users with static and non-static frequency assignment can coexist. The radio accessing the spectrum dynamically can be licensed or unlicensed depending on the rules of the regulatory bodies. It is now possible to foresee a wireless arena where only dynamic spectrum access is in use and no static allocations are done. Further, dynamic spectrum access is an entirely fresh method of managing the spectrum which can be fully accomplished and enabled using Cognitive Radio. Hence in the background of emerging spectrum regimes, Cognitive Radio has a range of vital roles to play toward addressing the increasing traffic requirements of wireless schemes.

10.2 Motivation

- The main impetus behind the development of learning schemes capable of prediction of various key parameters is to assist in building a forecast model. These learning schemes are to be made part of a cognitive engine.
- It is also to be investigated if these forecasts done by the learning will play a role in spectrum allocation after spectrum sensing.
- Such complex learning algorithms could predict various parameters associated with the Cognitive Radio and aid in predictive modeling.
- These learning models are to be integrated in real time in a cognitive engine and are to be explored on how these predictions made by the learning algorithms contribute to spectrum allocation after spectrum sensing.
- From the literature review, majority of learning schemes are accomplished using soft computing techniques. Soft computing techniques are most relevant in highly varying radio environment. They can be used in any phase of cognition and enable advanced learning.
- It is seen that neural network-based learning schemes have been popularly developed. However, it is observed in the literature that very little work is done on unsupervised algorithms and there is huge scope for integration of learning schemes in cognitive engine.
- There is a growing need and motivation to develop efficient and powerful learning outlines and schemes which will predict operating parameters with high prediction accuracy and good design flexibility.
- These learning outlines can be an integral part of a cognition engine and when integrated can greatly contribute toward enhanced dynamic spectrum allocation.
- The increasing requirement for robust and competent learning scheme contributing to enhanced cognitive engine leading to improved dynamic spectrum allocation has been the foremost motivation for attempts behind the proposed work.

10.3   Need and Relevance

These learning schemes widely employ soft computing-based techniques. These techniques are most relevant for highly varying radio environment. Further they can be used in any phase of cognition and can enable advanced learning. These learning schemes can aid a Cognitive Radio to analyze and differentiate different radio configurations and help the radio choose the most suitable configuration to work with. These learning schemes enable predictive modeling. Hence in the current wireless scenario, where there is huge scarcity of spectrum, such learning schemes for Cognitive Radio are greatly needed to build an intelligent, smart radio which is capable of dynamic spectrum allocation at a very efficient and systematic manner. Therefore, development of robust and flexible learning schemes for Cognitive Radio is very relevant for all next-generation high-end applications.

10.4   Fundamentals of Cognitive Radio

This expression of Cognitive Radio was first formulated by Mitola in 1999 [3]. Cognitive Radio is a smart wireless communication arrangement that is conscious of its adjoining surroundings. It uses techniques to sense and learn from the surroundings and alter and adjust its states to changes in the surroundings. This process happens in real time by making alterations to certain working and functional factors like power transmitted, frequency of the carrier, method of modulation, etc. [1].

The major characteristic of a Cognitive Radio is its cognitive capability whereby the radio is able to capture the information from its radio surroundings. This ability cannot be accomplished by just simply examining the frequency bands. It can be achieved by the usage of advanced techniques to capture the variations in the radio environment. Because of this cognitive capability, the portion of spectrum unused for a specific period or position can be found. Hence the best frequency band and working factors can be chosen. Another vital attribute of Cognitive Radio is reconfigurability [3]. This quality bestows the radio with spectrum awareness wherein the radio is capable of configuring itself according to the modifications in the radio surroundings. Through reconfigurability the Cognitive Radio will be able to use variety of frequencies and different access methods which can be implemented according to its hardware structure. The different modifiable parameters that can be included into a Cognitive Radio are working frequency, modulation technique, transmission power, and communication methodology.

10.5   Cognitive Cycle

Toward realizing the cognitive capability and reconfigurability in Cognitive Radio, certain tasks are to be performed for adaptive functioning in open frequency bands. These tasks performed constitute the cognitive cycle. The three major processes in the cognitive cycle are sensing the frequency bands, analysis of the parameters sensed and taking actions and decisions accordingly as illustrated in Fig. 10.2.

Monitoring the existing spectrum bands and capturing their information resulting in finding of white spaces are the main functions of spectrum sensing. Subsequently, the features of the detected white spaces are found. This constitutes the step of analysis of spectrum. Subsequently, the Cognitive Radio determines the various key parameters like rate at which data is transmitted, mode of transmission, and bandwidth of the transmission. Finally, suitable band of frequencies is chosen according to spectrum features and user needs. This task is performed in the phase of spectrum allocation.

Essentially, there are three online cognitive tasks which are taken up by the Cognitive Radio [4]. Firstly, radio scene analysis is performed in spectrum sensing phase which includes finding the spectrum holes as well as finding the inference temperature of the radio surroundings. This is shown in Fig. 10.3. Following this, the

Fig. 10.2 Main functions in cognitive cycle

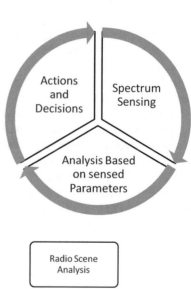

Fig. 10.3 Process in cognitive cycle

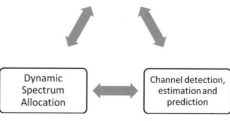

second phase involves channel detection which covers evaluation of channel-state information and forecast of channel capability which can be used by the transmitter. These two tasks are performed by the receiver. The control of transmitted power and management of frequency bands is performed in the transmitter side.

It is evident that cognitive unit is present in both transmitting and receiving end and both must work in synchronization. This synchrony is established through a feedback mechanism between the receiver and transmitter. The receiver sends the needed information on the quality of the forward link to the transmitter with the help of feedback mechanism. Hence Cognitive Radio involves feedback communication [5].

A Cognitive Radio provides room for different levels of cognition. The most basic level of cognition will involve user picking a white space and the processes in the cognitive sequence is built around the white space. At the higher level, user may use different technologies to build its cognitive sequence over a wideband spectrum hole or set of narrowband spectrum hole toward best performance in terms of proper organization of spectrum and control of transmit power in a safe fashion [6].

10.6   Artificial Intelligence and Soft Computing Techniques

The ability to gain knowledge and comprehend, to unravel problems and to make judgment is intelligence. The science which empowers machine to acquire the intelligence and accomplish some of the tasks of humans is artificial intelligence (AI) [7]. In some cognitive chores, if the machine can attain human-level performance, a machine is thought to be intelligent. In AI, machines solve difficult problems in a fashion similar to humans.

This method usually involves making use of uniqueness from human brainpower and alters them as algorithms as applicable to computers [7]. Machine learning (ML) is a subdiscipline of AI which works on transforming a large amount of data to a model. Artificial neural networks, support vector machines, classification, and some more similar methods are some of the constituents of machine learning. There are two sub-parts of hard computing as well as soft computing to artificial intelligence. The hard computing part of Artificial Intelligence is made of constituents such as Expert Systems, Formal Logic etc. which tries to provide technically right answers and donot focus on solving the problem reasonably. The hard computing methods need a clearly mentioned analytical prototype and take a huge computational time. Only in an idealistic environment these analytical prototypes are applicable. However, in real world, situations exist in non-ideal setting. The subdiscipline of AI focusing on heuristics, imperfect solutions to complex problems, is soft computing which has tolerance to imperfections, ambiguity, half-done truth, and approximations to attain robust, low-cost solutions. Soft computing works on building a replica of the human mind. Soft computing is a blend of techniques that are used to model and facilitate answers to real-life problems. Soft computing is a grouping of techniques that works in union and provides, in one form or another, information processing capacity involving real-life vague scenarios. The two salient features of soft computing are its ability to learn from experimental data and its power of generalization. The ability to forecast outputs with new inputs based on learning accomplished through previous inputs is generalization.

The principle elements or tools of soft computing are artificial neural networks, fuzzy logic, evolutionary computation, support vector machines, machine learning, probabilistic reasoning, swarm intelligence, and chaos theory [8].

10.7   Role of Soft Computing Techniques in Cognitive Engine

The major task in any cognitive engine involves sensing the surrounding, scrutinizing the parameters that have been sensed, making decisions based on the parameters sensed, and empowering cognitive, readaptable, and learning ability. These various tasks can be accomplished by different machine learning and soft computing techniques. Mostly soft computing plays a salient role in spectrum sensing, cognitive engine as well as in dynamic spectrum allocation. The following section gives detailing on the methods existing and adopted.

10.7.1   Spectrum Sensing

Different soft computing-based approaches are being used toward spectrum sensing such as Bayesian networks, fuzzy logic-based networks, support vector machine (SVM)-based networks, and artificial neural networks. Fuzzy logic and fuzzy logic-based networks are used in different forms and for various purposes to accomplish spectrum sensing.

The availability of bandwidth toward spectrum sensing is incorporated through fuzzy logic [9]. Different inputs are given to the fuzzy logic model such as signal-to-noise ratio, probability of detection, time taken to perform detection, and previous information. The outputs from the model are the various methods of spectrum sensing like energy detection, feature recognition, match finding, etc. The input factors are initially fuzzified and then subsequently if-then rules are applied according to which the method to be performed for spectrum sensing is being chosen.

To eliminate the problem of noise power uncertainty, a joint spectrum sensing method involving fuzzy set is formulated [10]. In this method, a fuzzy energy finder is being used with limit based on noise power uncertainty bound. This method gives superior performance compared to existing cooperative spectrum sensing-based methods. Toward selecting channels from available backup and candidate channels, methodology on the basis of fuzzy logic is used. In this method, precedence is given to channels resulting in improved network performance [11]. The ranking of the channel is done based on characteristics of the primary users and the strength of the signal received. Such networks have better performance than networks based on random channel selection. To reduce problems due to fading and shadowing, distributed spectrum sensing based on energy detection using fuzzy-based fusion rule is explored [12]. The fuzzy-based fusion rule takes energy received and signals to noise ratio as its inputs and makes its decision toward detection. Though the time taken for detection is large, the method makes good detection.

Exploring the problems of fading, cooperative spectrum sensing methods are modified by introducing a fuzzy logic-based hopping series to the detector [13]. Two types of hopping techniques of random sequencing and sequential sequencing approach are used, where an improvement in detection probability is achieved over standard cooperative spectrum sensing techniques. Artificial neural network-based algorithms and schemes have also been widely popular in solving various issues related to spectrum sensing.

Artificial neural networks have been established at all cognitive user to forecast the sensing probability of that secondary user [14]. A fusion center is proposed taking in inputs from the ANN-based secondary users on a cooperative fashion. It is shown that the chances of false alarm are greatly reduced. Artificial neural network-based digital classifier is developed in [14], which will aid in finding out various types of basic radio signals as feeble, sturdy, identified, or unidentified. Such a classifier will aid in various spectrum sensing methods. The key concept is developed on the reality that cognitive user does not have information regarding the primary user's signal category.

Cooperative spectrum sensing for Cognitive Radios is proposed using concept of perceptron learning concept in artificial neural networks [15]. Several scenarios have been simulated involving single and multiple secondary users, and performance and detection precision of the network have been evaluated. Artificial neural network-based double-threshold energy detection method is proposed toward improved spectrum sensing [16]. The noise element is drastically reduced by double threshold compared to single-threshold-based technique. The advantages of fuzzy logic and artificial neural networks are combined in adaptive neuro-fuzzy inference systems (ANFIS). ANFIS can be accomplished through grid partitioning and scatter partitioning. Fuzzy clustering means (FCM) is a type of scatter partitioning in ANFIS. The possibility of usage of FCM on energy detection-based cooperative spectrum sensing is explored in [17].

Two different combining methods of selection combining and optimal gain combining are done in the fuzzy cluster. The primary aim of the approaches is to reduce the energy consumption of secondary users. The results indicate improved performance compared to existing energy detection-based cooperative spectrum sensing methods. Cooperative spectrum sensing based on multi-objective evolutionary algorithm and fuzzy decision-making is investigated [18]. The chances of false alarm and detection are two contradicting objectives. This multi-objective optimization problem is solved using evolutionary algorithms. Among the different sets of solutions obtained, a middle solution is obtained using fuzzy logic-based method.

The impact of the overall performance is illustrated through various trials. Supervised soft computing technique of support vector machines used for analyzing data for classification also plays a role in spectrum sensing.

The classification of the sample data as a primary user or not is done using support vector machine which is used for spectrum sensing [19]. Presence of signal and noise indicates primary user, while only noise indicates absence of primary user. Support vector machines are also used for medium access control recognition [20]. The SVM prototype finds the MAC protocol types of current transmission and alters its transmission parameters accordingly. Support vector machines also aid in building an eigenvalue-based spectrum sensing technique for Cognitive Radio systems [21]. The training prototype is built using N data. For the N data taken, covariance matrix eigenvalues are generated. The fresh data is categorized depending on the existing data set. The essential part of SVM indicates the existence or nonexistence of primary user.

Bayesian approach of learning based on probability is used to solve various issues related to spectrum sensing. A multichannel cooperative spectrum sensing finding the state of the channel is explored using Bayesian approach [22]. The network is developed with and without restriction on the number of channels the secondary user is able to sense and access. A Bayesian learning method for spectrum sensing with an aim to exploit the frequency range to the fullest is developed [23]. Bayesian decision rule is being used to find the presence of modulated primary signal in this method. A cooperative spectrum sensing scheme based on Bayesian network is formulated in [24]. The characteristics of the secondary users are found, and its quality is established using Bayesian model. The usage of the secondary user is determined by the service quality of the user found. This system prevents the usage of secondary users of poor quality. Based on the concept of sparsity in cooperative spectrum sensing, a Bayesian hierarchical learning model has been developed [25]. Sparsity is generally present when the primary user occupies less space in bandwidth or when the number of users' present is less in number.

10.7.2 Cognitive Engine

The heart of the Cognitive Radio is the cognitive engine which contributes the main factor of intelligence to the Cognitive Radio. Soft computing approaches such as Bayesian networks, artificial neural networks, support vector machines, and genetic algorithms have been prevalently used in various levels for different functionalities in a cognitive engine. The cognitive module comprises of knowledge foundation, reasoning module, and learning module as shown in Fig. 10.4.

A Bayesian network-based Cognitive Radio decision module is formulated [26]. Junction tree algorithm is used to develop probabilistic model obtained from Bayesian networks. The radio parameters are altered in the Cognitive Radio toward maintaining quality of service of the users.

Fig. 10.4 Constituents of
the cognitive engine

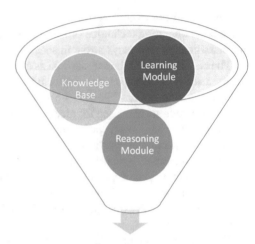

Cognitive Engine

Different algorithms and approaches of artificial neural networks have been used toward learning, reasoning, and decision-making toward improving the performance of communication in Cognitive Radio systems [14, 27]. Support vector machines have been used to include a learning plan to the cognitive module [28]. The model depends on parameters like data rate, error rate, and modulation type. The model is built for different types of fading. Once the prototype is built, it is being tested with different types of inputs. Genetic algorithms have also been explored toward providing awareness, processing, decision-making, and learning in a Cognitive Radio system [29, 30].

The design of the cognitive engine has also been approached using combination of genetic algorithm and radial basis function (RBF) networks to vary the parameters of the system, so that the system successfully adapts to the changes in the surroundings [31]. The decision-making module is developed using genetic algorithm, while the training is accomplished using RBF.

10.7.3 Dynamic Spectrum Allocation

After the functionalities associated with Spectrum Sensing and Cognitive Engine, the major module in any Cognitive Radio is Dynamic Spectrum Allocation. From the literature, it is evident that fuzzy logic has been used in spectrum allocation.

Various evolutionary algorithms like genetic algorithm, particle swarm optimization, and ant colony optimization have been popularly employed in performing dynamic spectrum allocation. A centralized fuzzy inference system has been proposed toward spectrum allocation. The bands of frequencies are allotted with deciding factors such as amount of traffic and kind and class of service importance

[32]. The cognitive users give their frequency requirements to the master cognitive user who uses fuzzy logic to give frequency admission. Based on the input parameters fed in, the master allots the frequency characterized in fuzzy as very large, large, intermediate, small, and very small.

The fuzzy inference rule has also been used in resource management [33] as well as in power management in wireless Cognitive Radio environment. Toward resource management, convergence of fuzzy happens at the local and global level. In power management schema, input parameters taken in are classified in terms of fuzzy as small, medium, large, and very large. These inputs are applied to fuzzy rules which in turn decide the power adjustments, again in terms of fuzzy.

Genetic algorithm has been commonly used toward solving problems related to spectrum allocation in Cognitive Radio systems. The system characteristics are improved using genetic algorithms while handling multi-objective problems that work toward minimization of error rate and power while maximization of throughput [34]. The various working parameters have been encoded in the chromosomes. The results indicate that the data from previous cognitive cycles can bring down the time taken toward convergence in GA, and the performance is improved. Spectrum optimization has also been explored using genetic algorithm [35]. Here the exclusiveness property of genetic algorithm is being used where the best chromosomes in the population are being passed on to the next generation before crossover or mutation is being done. Through this process, the most likely solution is not lost.

RF parameter optimization in Cognitive Radio is accomplished using genetic algorithms [36]. With different crucial parameters as inputs, fitness measure is found. The highest fitness assessment and its corresponding chromosomes are found as well as stored. Toward optimal solution for fixing the RF parameters, this best member is being used. Modified genetic algorithm approach is presented toward Cognitive Radio resource allocation [37]. In this approach, set crossover and mutation probabilities are changed toward achieving optimal performance. The major task of the process is to maximize the entire transfer pace of the Cognitive Radio systems taking into consideration the bare requirements.

A novel approach of spectrum allocation using genetic algorithms for Cognitive Radio networks have been proposed [38]. The utilization of the system and fairness are the major objectives toward which the system works. Because of the optimization, the space of the search is drastically reduced. A comprehensive analysis of adaptive optimization technique for Cognitive Radio network depending on multi-objective genetic algorithm is presented [39]. This method provides more opportunities in the spectrum for allocation. Genetic algorithm approach is being used to mitigate the problem of congestion in Cognitive Radio networks [40]. Route scheduling and best path selection method is being used to find the maximum range of frequencies in this approach.

Particle swarm optimization (PSO) is another popular evolutionary soft computing algorithm which is used for dynamic spectrum allocation. Other optimization techniques like ant colony optimization (ACO) and bee colony optimization (BCO) are also used for solving various issues related to spectrum allocation. It is also found in the literature that to build robust allocations, hybrid systems have been developed combining graph color problem (GCP) technique with genetic algorithm, PSO, or ACO. The comparison of six evolutionary algorithms for optimization of fitness functions in radio environment is done [41]. Genetic algorithm, differential evolution, PSO, bacterial foraging optimization, bee colony optimization, and cat swarm optimization are the techniques tested in single and multicarrier communication systems. The performance of these algorithms is compared based on their convergence and statistical metrics. Resource allocation is accomplished for Cognitive Radio systems based on quantum particle swarm optimization and RBF neural networks for Cognitive Radio systems [42]. The simulation results show that the total power consumption is greatly reduced in this technique.

An improved particle swarm optimization algorithm is explored to solve large-scale optimization problems [43]. The original PSO converges around the local optima. To overcome this drawback, an improved version of PSO combining chaos theory is proposed which helps toward searching solution beyond local optima in the global regions. Further, co evolutionary methodology is used where the problem is decomposed into number of small ones so that the convergence process is accelerated. Simulation results demonstrate reduced number of iterations and increased energy efficiency compared to other techniques.

Ant colony optimization (ACO) and graph color problem (GCP) is combined to perform spectrum allocation [44, 45]. The results are improved compared to classical PSO. GCP is also combined with PSO to investigate spectrum allocation satisfying user service demands and accomplish better utilization and fairness [46, 47]. These algorithms converge rapidly and provide good iteration efficiency.

10.7.4 Significance of Learning in Cognitive Engine

The cognitive engine is an important element contributing to intelligence in the Cognitive Radio. The cognitive engine consists of knowledge base, learning module, and reasoning module. Learning mechanisms or modules work on the information acquired from the surroundings and gather familiarity and intelligence. These learning mechanisms are responsible for storing of knowledge in knowledge base. Further these learning accomplished through learning mechanisms also guide in the decision-making and actions in the cognitive engine.

The learning mechanism are the prime contributor toward building the information and hence the knowledge base. The learnt knowledge is stored in the knowledge base from which it may be retrieved whenever necessary. The knowledge is of no use unless it is converted into useful action based on inferences. These inferences are done by the reasoning mechanisms. A reasoning module finds which actions can be implemented in a given radio is surrounding. Hence for the operation of reasoning mechanisms also, learning is extremely essential. These reasoning mechanisms could be computationally a time-consuming process, when the numbers of actions are large. In such cases, learning becomes all the more imperative such that heavy reasoning process is relaxed. The amalgamation of the learning engine can be essential for channel assessment and prediction stage toward improving steadiness and consistency of estimation of configuration capabilities. These learning mechanisms can be implemented and integrated in a cognitive engine through a variety of artificial intelligence and soft computing techniques.

10.7.5 Review on Learning Scheme for Cognitive Radio Using Soft Computing

The three main constituents in the cognitive engine are knowledge base, reasoning module, and learning module. This learning module greatly contributes to the addition of knowledge in the knowledge base. There is a requirement to understand what are the inputs and outputs to the learning module. The reasoning of sensed information is required to contribute as inputs to the learning engine [48]. Cognitive Radio performs reasoning on the sensed measurements and employs machine learning techniques. The complete overview of operation of cognitive engine can be understood by various solutions available.

The outputs of the learning modules also lead to inferences. The learning module can be formulated using different methods. Radio environment map (REM) approach has been initially developed to cognitive engine capable of awareness, learning, reasoning, and decision support [49]. Different learning algorithms have been implemented using REM. The simulations have been conducted in MATLAB and NS-2 environment, and it is observed that the hidden node problem is removed and secondary users exist along with primary users with minimum intrusion.

Learning and reasoning in a Cognitive Radio can be accomplished using different techniques. The learning module has been implemented using different artificial intelligence and soft computing techniques. A complete overview on different learning algorithms employed in cognitive engine is discussed [50]. Different learning and reasoning mechanisms can be employed in a Cognitive Radio [51]. The classification of reasoning type for cognitive networking is done. Also, different reasoning types and specific reasoning realizations are discussed.

A Bayesian network-based learning, inference, and decision-making engine has been formulated [26]. The simulation results indicate the possibility of modeling a cognitive learning module using Bayesian networks. Genetic algorithm has also been explored toward building cognitive engine for a Cognitive Radio. A genetic algorithm-based cognitive engine with the capacity to learn under supervised and unsupervised means has been developed [52]. In this work, the genes in the chromosomes characterize the varying parameters of the radio. These chromosomes are altered to result in a group of factors that optimize the radio according to the user's present requirements. The performance of such an engine has been tested with software simulations and in hardware domain.

The presence of genetic algorithm contributes for optimization of parameters, while neural networks can enhance learning capacity. Hence a cognitive engine combining genetic algorithms and radial basis function (RBF) neural networks has been proposed [31]. RBF learning scheme is trained using a decision-making table formulated using pre-processed initial information. Subsequently, genetic algorithms are used for adjusting the operating parameters and choose the suitable solution by the definition of fitness function. The cognitive engine developed is capable of learning and reconfiguration.

Game theory-based techniques have also been used toward enhancing decision-making and learning abilities in a Cognitive Radio. A comparative study of various decision-making solutions available through different techniques such as game theory, multi-armed bandit problem, Markovian process, and optimal stopping problem is done [53].

Stochastic learning methods are used in channel selection technique where the cognitive user learns based on their individual action-reward history and changes its characteristics accordingly [54]. This method doesn't require any previous data about the availability of channel or about the amount of cognitive users available. The simulation results obtained show that this stochastic learning technique achieves high system throughput with good fairness. Toward channel selection, game theory-based solutions are probed with an aim to achieve global solution with local information [54]. This solution proposed maximizes network throughput and minimizes the network collision level.

Artificial neural networks (ANN) is a very popular soft computing technique used to forecast outputs with known inputs based on the ability to generalize after training the network with known inputs and outputs. ANN is used in predictions both in sensing phase as well as learning phase. Adaptive resonance theory is a neural network model combining both supervised and unsupervised techniques. A channel sensing algorithm for cognitive radio based on ART has been proposed [55]. The simulation results indicate that the proposed methodology gives precise sensing results and improves the cognitive performance. Beyond sensing, neural networks majorly play a role in predictive modeling [56]. A new technique for spectrum prediction is proposed where the RF features are modeled as time series. This is given as input to ANN that forecast the time series to make a decision whether unlicensed secondary user can exploit the spectrum or not. This work acts as useful input toward inclusion of cognition in software-defined radio. A neural network-based channel prediction model is formulated which works with different types of traffic and learns and classifies traffic type of each channel [57]. This model chooses the prediction method based on the traffic type.

A novel learning technique has been developed which determines the details of the primary user channel usage over a period of time [58]. Through this technique the dynamic sensing capability of Cognitive Radio is greatly improved. In this work, it is also seen that secondary users apply learning algorithms to increase their data rate according to variation in primary users' activities. The interaction between the main primary radio and the secondary Cognitive Radio and a new active learning, supervised transmission for Cognitive Radio is explored [59]. From the feedback obtained from primary radio, the Cognitive Radio is able to find the channel gain and achievable rates.

Apart from learning contributing in spectrum sensing, learning module play a major role in cognitive engine. The initial review on learning ability of Cognitive Radio and role of genetic algorithm and neural network in learning module is discussed [60]. The supervised feedforward back propagation algorithm trained using Levenberg-Marquardt technique is used in the implementations.

The ability to learn and adapt in Cognitive Radio using artificial neural networks is explored [61]. The work uses multilayered feedforward networks for description of communication performance and displays some learning abilities. An intelligent scheme for Cognitive Radio involving many layer neural networks based learning scheme is designed to extort knowledge from cognitive users [62]. The work makes use of self-aware Cognitive Radio, where case-based learning mathematical replica has been formulated. The simulation results indicate that there is considerable increase in system performance with increase in the learning rate.

An intelligent elucidation involving many layers neural network model is proposed. This methodology proposed is applied for call admission control for VOIP communication in WLAN networks [63]. The performance is evaluated through simulations and test-bed experiments. The obtained solutions outperform existing methods for admission decision. A neural network-based learning technique is proposed to implement learning scheme in a cognitive engine [64]. The neural network model uses multilayer perceptron (MLP) resulting in quicker convergence and works on problem related to overfitting in neural networks.

Learning-based Cognitive Radio receiver capable of demodulating different types of modulated signal is formulated [65]. By making use of feature-based classification, the learning engine learns the signal features. For every demodulation bit pattern, a neural network structure is proposed. The proposed method is validated using different modulation signals with varying signal-to-noise ratio. It is observed that the proposed method provides higher flexibility than the traditional demodulation methods. A supervised multilayer feedforward network is presented for cognitive access point selection [66]. The past information and throughput performance is accumulated by the mobile station. This accumulated information is used to forecast the performance of the access point and find the most suitable one. The simulations are carried in a test bed using IEEE 802.11 technology. This method of forecasting performs much better than the conventional access point selection techniques.

Two different learning schemes based on radial basis function neural network (RBF-NN) are proposed [67]. The details of the modeling of the network along with preprocessing done are illustrated in detail. The simulation results indicate improvised performance of RBF-NN compared to back propagation neural networks. A perception-based learning scheme is proposed helping Cognitive Radio to decide their action taken based on the information gathered on the spectrum bands and aid in obtaining a band of spectrum [68]. In a scenario with bands of different types, the Cognitive Radio involuntarily chooses the best band greedily resulting in collision. A regret minimization mechanism is incorporated to resolve this problem and improve system performance.

A neural network-based Cognitive controller capable of learning and predicting the performance of different channels based on surrounding measurements is implemented [69]. This implementation of cognitive controller based on multilayered feedforward networks forms useful inputs for dynamic channel selection in IEEE 802.11 environment. It is observed from the controller that the model is capable of learning and generalizations. A cognitive engine dependent on neural network is developed, where the technique for assessing and learning paramount decision is formulated [70]. The drawbacks of cognitive model developed using genetic algorithm are detailed. Different important architectural features of the cognitive model for cognitive radio are elaborated with knowledge-based architecture and learning technique.

A learning technique to forecast spectrum behavior developed on many-layer perceptron-based artificial neural networks is initiated [71]. Supervised learning techniques are used to predict the state of various channels in future periods. The results indicate that the method is able to predict future spectrum behavior with minimum root-mean-square error. Learning schemes for Cognitive Radio that are dependent on artificial neural networks are evaluated [72] and are used for forecasting the abilities (e.g., data rate) that can be attained by a particular radio arrangement. Interesting settings are simulated and predictions of key operational parameters are forecasted. Time series is also added as an additional useful and meaningful parameter to observe if there is any improvement in prediction accuracy [73]. The simulations are repeated with time as an additional factor and important parameters are forecasted.

Adaptive Neuro Fuzzy Interference System dependent learning method for Cognitive Radio towards prediction of data rate is proposed in [74] which incorporates brainpower in Cognitive Radio. The performance of this technique is weighted against neural network-dependent learning technique. It is observed that this method has reduced complication compared to basic neural network-based learning method. Unsupervised neural networks schemes have also been used in cognitive engine of a Cognitive Radio. Self-organizing maps (SOM) and linear vector quantization (LVQ) have been used to reconfiguring transceivers based on the input parameters taken from the signal [75]. The reconfiguration aids in cognition and hence learning in the cognitive engine. The implementations of cognition have been incorporated in a software-defined radio on a multi-core, single-chip Xilinx Virtex-4 FPGA.

An incremental self-organizing map combined with hierarchical neural network (ISOM-HNN) has been formulated for signal classification as well as to improve learning and prediction accuracy in Cognitive Radio systems [76]. The details of the learning scheme and the simulation result are shown to illustrate the efficiency of the method. The intelligence in the Cognitive Radio is realized using self-organizing maps [77]. The implementations of self-organizing maps have been done using Gaussian and Mexican Hat neighborhood learning functions. Self-organizing map networks based on Gaussian learning functions perform better than the other networks. The proposed method also effectively works for improvement in intrusion detection rate compared to the existing methods.

Self-organizing map-based learning schemes are proposed for Cognitive Radio, with improved design flexibility. This method has prediction accuracy up to 78%, forecasting various key operating parameter of data rate [78].

10.7.6   Comparative Study and Summary

The different learning techniques adopted in the cognitive engine along with the role of the technique and the advantages and future scope of the technique are illustrated in Table 10.1.

It is observed from the table that different soft computing techniques can be applied toward implementation of the cognitive engine. Fuzzy logic, genetic algorithm, game theory, Bayesian networks, and neural networks are some of the techniques whose performances have been compared. It is summarized that neural network is the most popular and widely used learning technique. This is mainly because of neural network's ability to converge and give better network performance, thus contributing to intelligence, prediction, and learning ability of the cognitive engine.

Table 10.1 Comparative study of different techniques along with methods adopted and advantages and disadvantages of each method is discussed

Learning technique adopted	Role of the technique	Remarks
Fuzzy logic	Learning in transmission rate	Reduced complexity
		More accurate
	Prediction	Further parameters can be incorporated in the learning scheme resulting in finding of best radio arrangement
Genetic algorithms	Learning and optimization	Multi-objective performance, nonmathematical, non-closed form constraints could be extended for incorporation of learning machine to automatically update weights
Game theory	Learning in channel selection	Optimum usage of spectrum can result in high-end applications
		Results in innovative, futuristic products and services with improved abilities
Bayesian networks	Secondary system modeling	Ability to converge
		Better network-wide performance
	To implement cognitive cycle	Can be extended toward realizing intelligence of the cognitive engine
Neural networks	Intelligence	Ability to converge
	Learning engine	Better network-wide performance
	Learning in dynamic channel	Application of the model to different protocols and scenarios needs to be analyzed
	Selection	

10.7.7 Gap Identification

Based on the literature review, there has been learning models developed for forecasting and predicting key parameters. However, there is tremendous scope to improve prediction accuracy and design flexibility with improvisation in existing techniques. Based on the literature review, though learning schemes toward predictive modeling have been investigated in the literature, there is tremendous scope to improve prediction accuracy and design flexibility with improvisation in existing techniques and algorithms as well as alternate methods and algorithms.

Grid partition-based ANFIS learning schemes have been developed, giving room for exploration of subtractive clustering-based learning schemes. Further, self-organizing map networks based on sequential algorithms have been developed. The learning schemes using self-organizing maps batch algorithm have not been explored in the literature. All learning schemes developed primarily work toward prediction of only data rates. There is scope for prediction of other key functional parameters such as throughput. Further, the database in all existing work has been developed in IEEE 802.11 g WLAN environment, and the databases have not been developed using other WLAN standards. Primarily these learning algorithms are to be applied in Cognitive Radio networks. Hence development of database in IEEE 802.22 Cognitive Radio environment will be more relevant and suitable. This has also been an unexplored terrain in the literature.

Apart from soft computing techniques, alternate approaches like stochastic-based models could also be developed toward building learning scheme which is not investigated in the literature. The predictions from the learning schemes can form useful inputs for dynamic spectrum allocation which is the main attribute of a Cognitive Radio. This has also been an unseen territory in the literature.

10.7.8 Latest Contributions in Field of Learning in Cognitive Radio

There are many developments in the field of Cognitive Radio technologies. The various current improvements in the field of Cognitive Radio are listed below:

- In a Cognitive Radio model, spectrums can be dynamically and opportunistically shared benefitting licensed and unlicensed users. Exclusive Cognitive Radio-based models have been developed which uses approaches such as game theory, market equilibrium, and hybrid models [79].
- The operation in a Cognitive Radio could be either half duplex or full duplex. In half duplex sensing and transmission cannot happen simultaneously, while in full duplex both can happen together. This further increases the throughput and reduces the amount of collision in comparison to half-duplex Cognitive Radios. A survey of many spectrum sensing techniques, security requirements, control

protocols, and open issues of research in full-duplex Cognitive Radio is presented [80].

- Toward effective dynamic spectrum access in Cognitive Radio networks, there is a huge need to find the channel quality. Hence different metrics to assess the quality of channel have been formulated, and a survey on different criteria such as error rate, outage probability, etc. has been carried out [81].
- Most of the multimedia-based applications require more bandwidth. Cognitive Radio-based networks have been developed for multimedia-based applications. A survey of such CRNs which have been developed for multimedia-based applications working on improving the quality of experience and security is being presented [82].
- A complete study of different latest prototype independent learning techniques in Cognitive Networks is presented [83]. A complete understanding of learning techniques in the field of single-agent, multi-agent, and many-player games is being done. Further, a complete overview on how these solutions improve the performance of the system is also being done.
- The different learning problems arising in Cognitive Radio is surveyed, and the significance of artificial intelligence is analyzed in Cognitive Radio [84]. The different algorithms which can be applied to Cognitive Radio for learning are being discussed.
- There are several methods for obtaining and inferring information that are required for learning in Cognitive Radio system [51]. These techniques are being reviewed and detailed.

10.7.9   Comparison of Different Networks Which Can Be Used for Learning

In the case of supervised network, which covers feedforward network, focused time-delay network, and Elman network, the following are the major characteristics associated:

- Supervised networks perform well during validation.
- These networks are capable of generalization, where they are able to make predictions with unseen data.
- Supervised networks make predictions at good level.
- Time zone, when added as additional input to supervised networks, results in reduced values of root-mean-square error and improved prediction accuracy. This proves that adding meaningful inputs to the system improves the performance of networks.
- When large training sets are used, it takes longer time to train; however, it paves way to reduced MSE and improved prediction accuracy.

Three different ANFIS-based models have been presented in these theses which are grid partition method, fuzzy clustering means-based ANFIS method, and subtractive clustering-based ANFIS method. The following are the major inferences about ANFIS model:

- ANFIS-based networks combine advantage of neural network and fuzzy theory.
- These networks create rule as in fuzzy theory and are capable of adaptive learning as in neural network.
- ANFIS-based networks display better performance metrics compared to all other methods.
- ANFIS-based method has least MSE compared to all other methods.
- These networks are capable of generating huge rule when number inputs are increased.
- The numerical complexity related to this method is less than NN-based methods.
- Among the different techniques available in ANFIS, the disadvantage associated with grid partitioning is overcome by FCM and subtractive clustering methods.
- Overall, ANFIS methods are more effective than feedforward, FTDNN, and Elman networks

The following are the major inferences drawn based on unsupervised network of self-organizing map networks:

- SOM algorithms have higher MSE than that of the previous methods.
- SOM-based networks take less epochs (10) and take less training time for building knowledge.
- This characteristic of fewer epochs is very significant. This makes SOM networks most suitable for dynamic and online training.
- While neural network-based methods require more training data, SOM-based networks require less training data. SOM networks are especially helpful when less training data is available.
- In SOM networks, it is possible to use/add more variables without changing network design pattern.
- Summarizing the above inferences, even if there are cases where SOM demonstrates worst results, in terms of higher deviation (MSE) than other supervised NN-based schemes or hybrid networks, the former performs better in terms of needing less training data and epochs and also offering higher design flexibility (Table 10.2).

10.8   Conclusions

The feedforward networks, focused time delay networks, Elman networks, adaptive neuro-fuzzy inference system-based networks, and self-organizing map-based networks are the different networks used toward building learning techniques for Cognitive Radio. The prediction or forecasting of performance of key operating

Table 10.2 Comparison between supervised, hybrid, and unsupervised networks

Supervised networks (feedforward networks, FTDNN, Elman networks)	Hybrid networks (ANFIS-based networks)	Unsupervised networks (SOM-based networks)
More training time, more input required, high MSE	Minimum MSE	Least input required
	Reduced training time (epochs)	Least training time
		Less MSE
	Sufficient inputs required	Most suitable for online training
	More membership functions/rules	

parameters of throughput and maximum achievable data rate in Cognitive Radio based on learning done through the learning model can be done for various learning configurations leading to intelligent Cognitive Radio.

The adaptive neuro-fuzzy inference system – fuzzy clustering method-based learning schemes – has the highest prediction accuracy. Self-organized map-based learning schemes have adaptability and flexibility suitable for online dynamic systems. These learning outlines can be validated on Cognitive Radio networks built in real life like environment according to IEEE 802.22 standards. The development of such learning outlines paves way to its usage in complex situations.

The learning model built will aid to analyze the feasibility of implementing such learning models in larger scales toward improving efficiency of the Cognitive Radio in the highly varying radio environment. These models could be further expanded to look into real-life performances. Such approaches will bring out the unique dimensions of machine learning research for Cognitive Radios. It will also pave way toward designing new strategies toward accomplishing cognitive abilities of Cognitive Radios.

References

1. Doyle L (2009) Essentials of cognitive radio, Cambridge Wireless Essential Series. Cambridge University Press, Cambridge
2. Kwang-Cheng C, Prasad R (2009) Cognitive radio networks. Wiley Publications, Chichester
3. Mitola J, Maguire GQ (1999) Cognitive radio: making software radios more personal. IEEE Pers Commun 6:13
4. Haykins S (2005) Cognitive radio: brain-empowered wireless communications. IEEE J Sel Areas Commun 23(2):201–220
5. Akyildiz IF, Lee W-Y, Vuran MC, Mohanty S (2006) Next generation dynamic spectrum access cognitive radio wireless networks: a survey. Comput Netw 50:2127–2159
6. Haykin S (2006) Cognitive dynamic systems. Proc IEEE 94(11):1910–1911
7. Negnevitsky M (2011) Artificial intelligence: a guide to intelligent systems, 3rd edn. Pearson Education, Harlow/New York
8. Rich E, Knight K, Nair SSB (2011) Artificial intelligence, 3rd edn. TMH Publications, New Delhi

9. Matinmikko M, Ser Del J, Rauma T, Mustonen M (2013) Fuzzy-logic based framework for spectrum availability assessment in cognitive radio systems. IEEE J Sel Areas Commun 31 (11):1136–1159
10. Mohammadia A, RezaTaban M, JamshidAbouei HT (2013) Fuzzy likelihood ratio test for cooperative spectrum sensing in cognitive radio. Signal Process 93:1118–1125
11. Joshi GP, Acharya S, Kim SW (2015) Fuzzy-logic-based channel selection in IEEE 802.22 WRAN. Inf Syst 48:327–332
12. Jacoba J, Jose BR, Mathew J (2015) A fuzzy approach to decision fusion in cognitive radio. International conference on Information and Communication Technologies (ICICT 2014), Procedia Comput Sci 46:425–431
13. Khader AAH, Reja AH, Hussein AA, Beg MT, Mainuddin (2015) Cooperative Spectrum sensing improvement based on fuzzy logic system. Second International Symposium on Computer Vision and the Internet. Procedia Comput Sci 58:34–41
14. Zhang T, Wu M, Liu C (2012) Cooperative Spectrum sensing based on artificial neural network for cognitive radio systems. In: 8th international conference on Wireless Communications, Networking and Mobile Computing
15. Popoola J, van Olst R (2011) Application of neural network for sensing primary radio signals in a cognitive radio environment, IEEE Africon 11
16. Varatharajana B, Praveen E, Vinoth E (2012) Neural network aided enhanced Spectrum sensing in cognitive radio. In: International conference on Modeling Optimisation and Computing (ICMOC-2012), Procedia Eng 38:82–88
17. Maitya SP, Chatterjeea S, Acharya T (2016) On optimal fuzzy c-means clustering for energy efficient cooperative spectrum sensing in cognitive radio networks. Digital Signal Process 49:104–115
18. Pradhan PM, Panda G (2013) Cooperative spectrum sensing in cognitive radio network using multiobjective evolutionary algorithms and fuzzy decision making. Ad Hoc Netw 11:1022–1036
19. Zhang D, Zhai X (2011) SVM-based Spectrum sensing in cognitive radio. In: 7th international conference on Wireless Communications, Networking and Mobile Computing
20. Hu S, Yao Y-D, Yang Z (2014) MAC protocol identification using support vector machines for cognitive radio networks. IEEE Wirel Commun 21(1):52–60
21. Awe OP, Zhu Z, Lambotharan S (2013) Eigenvalue and support vector machine techniques for spectrum sensing in cognitive radio networks. In: 2013 conference on technologies and applications of artificial intelligence
22. Jiang C, Chen Y, Liu KJR (2014) Multi-channel sensing and access game: Bayesian social learning with negative network externality. IEEE Trans Wirel Commun 13(4):2176–2188
23. Zheng S, Kam P-Y, Fellow LY-C, Zeng Y (2013) Spectrum sensing for digital primary signals in cognitive radio: a Bayesian approach for maximizing spectrum utilization. IEEE Trans Wirel Commun 12(4):1774–1782
24. Zhou M, Shen J, Chen H, Xie L (2013) A cooperative spectrum sensing scheme based on the Bayesian reputation model in cognitive radio networks. In: IEEE Wireless Communications and Networking Conference (WCNC)
25. Li F, Xu Z (2014) Sparse Bayesian hierarchical prior modeling based cooperative spectrum sensing in wideband cognitive radio networks. IEEE Signal Process Lett 21(5):586–590
26. Huang Y, Wang J, Jiang H (2010) Modeling of learning inference and decision-making engine in cognitive radio. In: Second international conference on Networks Security, Wireless Communications and Trusted Computing
27. Tan X, Huang H, Ma L (2013) Frequency allocation with Artificial Neural Networks in cognitive radio system. IEEE 2013 Tencon–Spring
28. Huang Y, Jiang H, Hu H, Yao Y (2009) Design of learning engine based on support vector machine in cognitive radio. In: International conference on Computational Intelligence and Software Engineering

29. Kantardzic M (2003) Data mining: concepts, models, methods, and algorithms. Wiley, Hoboken
30. Lee K, El-Sharkawi M (2008) Modern heuristic optimization techniques: theory and applications to power systems. Wiley, Hoboken
31. Yang Y, Jiang H, Liu C, Lan Z (2012) Research on cognitive radio engine based on Genetic Algorithm and Radial Basis Function Neural Network. Spring congress on Engineering and Technology
32. Kaur P, Uddin M, Khosla A (2010) Fuzzy based adaptive bandwidth allocation scheme in cognitive radio networks. In: Eighth international conference on ICT and Knowledge Engineering
33. Qin H, Zhu L, Li D (2012) Artificial mapping for dynamic resource Management of Cognitive Radio Networks. In: 8th international conference on Wireless Communications, Networking and Mobile Computing
34. Chen S, Newman TR, Evans JB, Wyglinski AM (2010) Genetic algorithm-based optimization for cognitive radio networks. IEEE Sarno_ Symposium, Princeton, pp 12–14
35. Riaz Moghal M, Altaf Khan M, Bhatti HA (2010) Spectrum optimization in Cognitive Radios using elitism in genetic algorithms. In: 6th international conference on Emerging Technologies (ICET)
36. Hauris JF, He D, Michel G, Ozbay C (2007) Cognitive radio and RF communications design optimization using genetic algorithms. MILCOM 2007 – IEEE military communications conference
37. Zeng C, Zu Y (2011) Cognitive radio resource allocation based on niche adaptive genetic algorithm. In: IET international conference on Communication Technology and Application (ICCTA 2011)
38. Liu M, Zhang H, Fan R, Duan Z (2011) The GA solution of dynamic Spectrum allocation in cognitive radio based on collaboration and fairness. In: Third Pacific Asia conference on Circuits, Communications and System (PACCS)
39. Amudha V, Ramesh GP (2013) Dynamic spectrum allocation for cognitive radio using genetic algorithm. Int J Technol Eng Sci [IJTES] 1970:1092–1097
40. Balieiro A, Yoshioka P, Dias K, Cavalcanti D, Cordeiro C (2014) A multi-objective genetic optimization for spectrum sensing in cognitive radio. Expert Syst Appl 41:3640–3650
41. Pradhan PM, Panda G (2014) Comparative performance analysis of evolutionary algorithm based parameter optimization in cognitive radio engine: a survey. Ad Hoc Netw 17:129–146
42. Xu L, Qian F, Li Y, Li Q, Yang Y-w, Xu J (2016) Resource allocation based on quantum particle swarm optimization and RBF neural network for overlay cognitive OFDM system. Neurocomputing 173:1250–1256
43. Tang M, Xin Y (2016) Energy efficient power allocation in cognitive radio network using co evolution chaotic particle swarm optimization. Comput Netw 100:1–11
44. Koroupi F, Talebi S, Salehinejad H (2012) Cognitive radio networks spectrum allocation: an ACS perspective. Scientia Iranica 19(3):767–773
45. Miao Y, Jian-ping A (2011) An ant colony optimization algorithm for spectrum assignment in cognitive radio networks. J Electron Inform Technol 33(10):2306–2311
46. Liu Q, Niu H, Xu W, Zhang D (2014) A service oriented spectrum allocation algorithm using enhanced PSO for cognitive wireless networks. Comput Netw 74(2014):81–91
47. Zhang J, Qi Z, Zhou J (2009), Advanced graph-coloring spectrum allocation algorithm for cognitive radio. In: International conference on IEEE Wireless Communications Networking and Mobile Computing, pp 1–4
48. Adamopoulou E, Demestichas K, Theologou M (2008) Enhanced estimation of configuration capabilities in cognitive radio. IEEE Commun Mag 46(4):56–63
49. Zhao Y, Gaeddert J, Bae KK, Reed JH (2006) Radio environment map enabled situation-aware cognitive radio learning algorithms. In: Proceeding of the SDR 06 technical conference and product exposition

50. Sharma V, Bohara V (2014) Exploiting machine learning algorithms for cognitive radio. In: International conference on Advances in Computing, Communications and Informatics (ICACCI)
51. Gavrilovska L, Atanasovski V, Macaluso I, DaSilva LA (2013) Learning and reasoning in cognitive radio networks. IEEE Commun Surv Tutorials 15(4):1761–1777
52. Rondeau TW, Le B, Rieser CJ, Bostian CW (2004) Cognitive radios with genetic algorithm: intelligent control of software defined radios. In: Proceeding of the SDR 04 technical conference and product exposition
53. Xu Y, Anpalagan A, Wu Q, Liang S, Gao Z, Wang J (2013) Decision-theoretic distributed channel selection for opportunistic Spectrum access: strategies, challenges and solutions. IEEE Commun Surv Tutorials 15(4):1689–1713
54. Xu Y, Wang J, Wu Q, Anpalagan A, Yao Y-D (2012) Opportunistic spectrum access in cognitive radio networks: global optimization using local interaction games. IEEE J Sel Top Sig Proces 6(2):180–193
55. Zhu X-L, Liu Y-A, Weng W-W, Yuan D-M (2008) Channel sensing algorithm based on neural networks for cognitive wireless mesh networks. In: 4th international conference on Wireless Communications, Networking and Mobile Computing
56. Taj MI, Akil M (2011) Cognitive radio Spectrum evolution prediction using artificial neural networks based multivariate time series modelling. In: European wireless 2011, Vienna, Austria
57. Hoyhtya M, Sofie P, Mammela A (2011) Improving the performance of cognitive radios through classification, learning, and predictive channel selection. Adv Electron Telecommun 24:28–38
58. Tekin C, Hong S, Stark W (2009) Enhancing cognitive radio dynamic spectrum sensing through adaptive learning. In: MILCOM 2009 I.E. military communications conference. IEEE
59. Zhang R (2010) On active learning and supervised transmission of spectrum sharing based cognitive radios by exploiting hidden primary radio feedback. IEEE Trans Commun 58 (10):2960–2970
60. Ruslan R, Wan TC (2008) Learning ability in cognitive radio. In: International conference on Network Applications, Protocols and Services 2008 (NetApps 2008) Executive Development Center, Universiti Utara Malaysia
61. Baldo N, Zorzi M (2008) Learning and adaptation in cognitive radios using neural networks. In: 5th IEEE consumer communications and networking conference. IEEE
62. Al-Dulaimi A, Al-Saeed L (2010) An intelligent scheme for first run cognitive radios. In: 2010 fourth international conference on next generation mobile applications, Services and Technologies. IEEE
63. Baldo N, Dini P, Nin-Guerrero J (2010) User-driven call admission control for VoIP over WLAN with a neural network based cognitive engine. In: 2010 2nd international workshop on cognitive information processing. IEEE
64. Dong X et al. (2010) A learner based on neural network for cognitive radio. In: Communication Technology (ICCT), 2010 12th IEEE international conference on. IEEE
65. He F, Xu X, Zhou L, Man H (2011) A learning based cognitive radio receiver. In 2011-MILCOM 2011 military communications conference, pp 7–12. IEEE
66. Bojovic B, Baldo N, Nin-Guerrero J, Dini P (2011) A supervised learning approach to cognitive access point selection. In: 2011 I.E. GLOBECOM Workshops (GC Wkshps), pp 1100–1105. IEEE
67. Chun W, Xuedong C (2012) Neural network-based learning in cognitive radios. In: IEEE international conference on Oxide Materials for Electronic Engineering (OMEE)
68. Tosh DK, Sengupta S (2013) Self-coexistence in cognitive radio networks using multi-stage perception learning. In: Vehicular Technology Conference (VTC Fall), 2013 I.E. 78th. IEEE
69. Baldo N, Tamma BR, Manoj BS, Rao RR, Zorzi M (2009), A neural network based cognitive controller for dynamic channel selection. In: IEEE international conference on communications, vol. 9, pp 1–5

70. Zhang Z, Xie X. (2007) Intelligent cognitive radio: research on learning and evaluation of CR based on neural network. In: 2007 ITI 5th international conference on Information and Communications Technology
71. Yin L, Yin SX, Hong W, Li SF (2011) Spectrum behavior learning in cognitive radio based on artificial neural network. In: MILCOM 2011 Military Communications conference
72. Katidiotis A, Tsagkaris K, Demestichas P (2010) Performance evaluation of artificial neural network-based learning schemes for cognitive radio systems. Comput Electr Eng 36 (3):518–535
73. Tsagkaris K, Katidiotis A, Demestichas P (2008) Neural network-based learning schemes for cognitive radio systems. Int J Comput Commun 31:3394–3404
74. Hiremath S, Kumar Patra S (2010) Transmission rate prediction for cognitive radio using adaptive neural fuzzy inference system. In: 5th international conference on Industrial and Information Systems
75. Taj MI, Akil M, Hammami O (2010) Standard recognising self organizing map based cognitive radio transceiver. In: Proceedings of the 5th international ICST conference on Cognitive Radio Oriented Wireless Networks and Communications
76. Cai Q, Chen S, Li X, Hu N, He H, Yao Y-D, Mitola J (2010) An integrated incremental self-organizing map and hierarchical neural network approach for cognitive radio learning. In: The International Joint Conference on Neural Networks (IJCNN)
77. Sunilkumar G, Thriveni J, Venugopal KR, Patnaik LM (2011) Cognition based self-organizing maps (CSOM) for intrusion detection in wireless networks. In: Annual IEEE India conference
78. Tsagkaris K, Bantouna A, Demestichas P (2012) Self-organizing maps for advanced learning in cognitive radio systems. Elsevier, Int J Comput Electr Eng 38:862–881
79. Hassan MR, Karmakar GC, Kamruzzaman J, Srinivasan B (2017) Exclusive use Spectrum access trading models in cognitive radio networks: a survey. IEEE Commun Surv Tutorials 19 (4):2192–2231
80. Amjad M, Akhtar F, Rehmani MH, Reisslein M, Umer T (2017) Full-duplex communication in cognitive radio networks: a survey. IEEE Commun Surv Tutorials 19(4):2158–2191
81. Elderini T, Kaabouch N, Reyes H (2017) Channel quality estimation metrics in cognitive radio networks: a survey. IET Commun 11:1173
82. Amjad M, Rehmani MH, Mao S (2018) Wireless multimedia cognitive radio networks: a comprehensive survey. IEEE Commun Surv Tutorials
83. Wang W, Kwasinski A, Niyato D, Han Z (2016) A survey on applications of model-free strategy learning in cognitive wireless networks. IEEE Commun Surv Tutorials 18 (3):1717–1757
84. Bkassiny M, Li Y, Jayaweera SK (2013) A survey on machine-learning techniques in cognitive radios. IEEE Commun Surv Tutorials 15(3):1136–1159

Chapter 11
Multiuser MIMO Cognitive Radio Systems

Mostafa Hefnawi

11.1 MU-MIMO Cognitive System

Cognitive radio (CR) [1, 2] has been proposed as an effective dynamic spectrum allocation policy where secondary users (SUs) can utilize the spectrum of the primary users (PUs) as long as the interference caused by SUs at PUs is below a given interference threshold. This SUs' transmission power constraint imposed by PUs leads to a limitation in quality of service (QoS) and the wireless coverage of secondary networks. To extend the wireless coverage and guarantee the QoS of the secondary networks, the combination of MU-MIMO and CR (referred as MIMO CR) has been introduced as another promising method. This is achieved by multiplexing multiple users on the same time-frequency resources and allowing concurrent spectrum sharing instead of opportunistic sharing. MU-MIMO techniques have been successfully deployed in 4G cellular systems for traditional static spectrum assignment approaches [3–8], and a vast number of multiuser detection algorithms are presently being tailored toward solving the MU-MIMO processing in cognitive networks by imposing additional constraints to protect licensed users' QoS [9–18]. More specifically, capacity-aware MU-MIMO schemes have been proposed for both static spectrum [6, 7] and CR networks [14], using different multiuser detections schemes such as maximum ratio combining (MRC) and minimum mean-squared error (MMSE), and have shown the potential to exhibit better system capacity and provide better SER enhancement than traditional singular value decomposition (SVD)-based MU-MIMO systems. It is also well known that optimal performance can be achieved by using massive MIMO (large number of antenna), [19, 20, 22] with the simplest forms of user detection and beamforming techniques,

M. Hefnawi (✉)
Royal Military College of Canada, Kingston, ON, Canada
e-mail: hefnawi-m@rmc.ca

© Springer International Publishing AG, part of Springer Nature 2019 259
M. H. Rehmani, R. Dhaou (eds.), *Cognitive Radio, Mobile Communications and Wireless Networks*, EAI/Springer Innovations in Communication and Computing, https://doi.org/10.1007/978-3-319-91002-4_11

i.e., MRC and eigenbeamforming (EigBF). In this chapter, therefore, we will focus on large-scale MU-MIMO CR networks based on capacity-aware algorithms. Two applications of MU-MIMO CR will be explored: cognitive radio-based wireless sensor networks (CR-WSNs) [23, 29] and cognitive small cells in heterogeneous networks (HetNets) [30–32]. These two applications will be based on the uplink access scenario shown in Fig. 11.1, where L_s secondary users (SUs) and one secondary base station (SBS) coexist with L_p primary users (PUs) and one primary base station (PBS) via concurrent spectrum access. The users in both networks and the base stations are equipped with multiple antennas. It is also assumed that both the SBS and the PBS receivers detect independent OFDM data streams from multiple SUs and PUs simultaneously on the same time-frequency resources. Let $\mathbf{x}^s[k] = \left\{x_1^s, x_2^s, \cdots, x_{L_s}^s\right\}$ and $\mathbf{x}^p[k] = \left\{x_1^p, x_2^p, \cdots, x_{L_p}^p\right\}$ denote, respectively, the set of L_s SU signals and L_p PU signals transmitted on each subcarrier, $k = 1, \cdots, N_c$, where N_c denotes the number of subcarriers per OFDM symbol in the system. It is assumed that x_i^s and x_i^p are complex-valued random variables with unit power, i.e., $\mathrm{E}\left[\left\|x_i^s\right\|^2\right] = \mathrm{E}\left[\left\|x_i^p\right\|^2\right] = 1$.

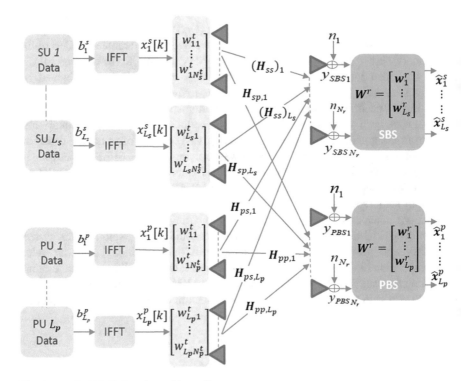

Fig. 11.1 MU-MIMO-based cognitive radio system

The expression for the array output of the SBS in Fig. 11.1 can be written for each subcarrier as

$$\mathbf{y}_{\text{SBS}}[k] = \sum_{l_s=1}^{L_s} \mathbf{H}_{\text{ss},l_s}[k]\mathbf{w}_{l_s}^t[k]\mathbf{x}_{l_s}^s[k] + \mathbf{n}[k] + \mathbf{I}_{\text{PU}}[k] \tag{11.1}$$

where $\mathbf{y}_{\text{SBS}}[k] = \left[y_1^s[k], y_2^s[k], \cdots, y_{N_s^r}^s[k] \right]^T$ is the $N_s^r \times 1$ vector containing the outputs of the N_s^r – element array at the SBS, with $(.)^T$ denoting the transpose operation, $\mathbf{H}_{\text{ss},l_s}[k]$ is the $N_s^r \times N_s^t$ frequency-domain channel matrix representing the transfer functions from secondary user l_s's N_s^t – element antenna array to the SBS's N_s^r – element antenna array, $\mathbf{w}_{l_s}^t[k] = \left[w_{l_s1}^t[k], w_{l_s2}^t[k] \quad \cdots \quad, w_{l_sN_s^t}^t[k] \right]^T$ is the $N_s^t \times 1$ complex transmit weight vector for SU l_s, $l_s = 1, \cdots, L_s$, $\mathbf{n}[k] = [n_1[k], \quad n_2[k] \ldots \quad, n_{l_s}[k]]^T$ is the $N_s^r \times 1$ complex additive white Gaussian noise vector, and $\mathbf{I}_{\text{PU}}[k]$ represents the interference introduced by PUs at the SBS, given by

$$\mathbf{I}_{\text{PU}}[k] = \sum_{l_p=1}^{L_p} \mathbf{H}_{\text{ps},l_p}[k]\mathbf{w}_{l_p}^t[k]\mathbf{x}_{l_p}^p[k] \tag{11.2}$$

where $\mathbf{H}_{\text{ps},l_p}[k]$ is the $N_s^r \times N_p^t$ channel matrix representing the fading coefficients from PUs to the SBS's N_s^r – element antenna array. On the other hand, the interference power seen by the primary base station due to secondary transmission is given by

$$J_{\text{sp}}[k] = \sum_{s=1}^{L_s} \mathbf{H}_{\text{sp},l_s}[k]\mathbf{w}_{l_s}^t[k]\mathbf{w}_{l_s}^{t,H}[k]\mathbf{H}_{\text{sp},l_s}^H[k] \tag{11.3}$$

where $(.)^H$ denotes the Hermitian transpose and $\mathbf{H}_{\text{sp},l_s}[k]$ is the $N_p^r \times N_s^t$ channel matrix representing the fading coefficients from the l_s-th SU to the PBS's N_p^r – element antenna array. It is also assumed that $\mathbf{H}_{\text{ss},l_s}[k] \in \mathbb{C}^{N_s^r \times N_s^t}$, $\mathbf{H}_{\text{sp},l_s}[k] \in \mathbb{C}^{N_p^r \times N_s^t}$, $\mathbf{H}_{\text{ps},l_p}[k] \in \mathbb{C}^{N_s^r \times N_p^t}$, and $\mathbf{H}_{\text{pp},l_s}[k] \in \mathbb{C}^{N_p^r \times N_p^t}$ consist of iid (independent and identically distributed) complex Gaussian entries, with zero mean and unit variance. The transfer functions from the l_s-th SU device to the SBS antenna array (the cascade of $\mathbf{H}_{\text{ss},l_s}[k]$ and $\mathbf{w}_{l_s}^t[k]$) result in a unique spatial signature for each SU, which can be exploited to effect the separation of the user data at the SBS using appropriate multiuser detection techniques. The SBS detects all L_s SUs, simultaneously at the multiuser detection module of the SDMA system, by multiplying the output of the array with the $N_s^r \times 1$ receiving weight vectors as follows:

$$\widehat{x}_{l_s}[\mathbf{k}] = \mathbf{w}_{l_s}^{r,H}[k]\, \mathbf{y}_{\text{SBS},}[k] \tag{11.4}$$

11.1.1 Gradient Search-Based Capacity-Aware Algorithm (GS-CA)

Capacity-aware algorithm seeks the optimal beamforming vector, $\left(\mathbf{w}_{l_s}^t[k]\right)_{\text{opt}}$, that maximizes the ergodic capacity of the cognitive MU-MIMO-OFDM channel for each SU l_s imposing the following two sets of constraints: (1) each secondary user l_s has a limited maximum transmission power equal to $P_{\text{max},l}^t$, and (2) the total maximum interference power at the PBS from the SUs does not exceed the maximum power constraint of $J_{\text{sp}}^{\text{max}}$.

The ergodic capacity of the cognitive MU-MIMO-OFDM is given by

$$C = E\left(\log_2\left\{\mathbf{I} + \frac{\rho_{l_s}}{N_s^t}\left|\mathbf{B}_{l_s}[k]^{-\frac{1}{2}}\tilde{\mathbf{H}}_{\text{ss},l_s}^t[k]\right|^2\right\}\right) \tag{11.5}$$

where

E [.] denotes the expectation operator,
ρ_{l_s} is the signal-to-noise ratio (SNR) of SU l_s,
$\tilde{\mathbf{H}}_{\text{ss},l_s}^t[k] = \mathbf{H}_{\text{ss},l_s}[k]\mathbf{w}_{l_s}^t[k]$, $\tilde{\mathbf{H}}_{\text{sp},l_s}^t[k] = \mathbf{H}_{\text{sp},l_s}[k]\mathbf{w}_{l_s}^t[k]$,
$\tilde{\mathbf{H}}_{\text{ps},l_p}^t[k] = \mathbf{H}_{\text{ps},l_p}[k]\mathbf{w}_{l_p}^t[k]$, $\mathbf{B}_{l_s}[k] = \mathbf{B}_{\text{ss}}[k] + \mathbf{B}_{\text{ps}}[k] + \sigma_n^2\mathbf{I}_{N_s^r}[k]$,
$\mathbf{B}_{\text{ss}}[k] = \sum_{i=1, i\neq l_s}^{L_s}[k]\tilde{\mathbf{H}}_{\text{ss},l_s}^{t,H}[k]$, and $\mathbf{B}_{\text{ps}}[k] = \sum_{l_p=1}^{L_p}\tilde{\mathbf{H}}_{\text{ps},l_s}^t[k]\tilde{\mathbf{H}}_{\text{ps},l_s}^{t,H}[k]$.

In mathematical terms, the two constraints are expressed as follows:

$$\max_{\mathbf{w}_{l_s}^t[k]}\left[E\left(\log_2\left\{\mathbf{I} + \frac{\rho_{l_s}}{N_s^t}\left|B_{l_s}^{-1/2}[k]\tilde{\mathbf{H}}_{\text{ss},l_s}^t[k]\right|^2\right\}\right)\right]$$

$$\text{Subject to}: \left\{ \begin{array}{c} \mathbf{w}_{l_s}^{t,H}[k]\mathbf{w}_{l_s}^t[k] \leq P_{\text{max},l_s}^t \\ J_{\text{sp}} = \sum_{i=1}^{L_s}\tilde{\mathbf{H}}_{\text{sp},l_s}^t[k]\tilde{\mathbf{H}}_{\text{sp},l_s}^{t,H}[k] \leq J_{\text{sp}}^{\text{max}} \end{array} \right\} \tag{11.6}$$

This problem is a constrained optimization problem, which is highly non-convex and complicated to solve. However, a suboptimal solution can be obtained by exploiting the method of Lagrange multipliers as follows:

$$
\mathcal{L}\left(\mathbf{w}_{l_s}^t, v_{l_s}, \lambda_{l_s}\right) = E\left(\log_2\left\{\mathbf{I} + \frac{\rho_{l_s}}{N_s^t}\left|\mathbf{B}_{l_s}[k]^{-\frac{1}{2}}\tilde{\mathbf{H}}_{\mathrm{ss},l_s}^t[k]\right|^2\right\}\right)
$$

$$
- v_{l_s}\left(\frac{\displaystyle\sum_{l_s=1}^{L_s}\tilde{\mathbf{H}}_{\mathrm{sp},l_s}^t[k]\tilde{\mathbf{H}}_{\mathrm{sp},l_s}^{t,H}[k]}{J_{\mathrm{sp}}^{\mathrm{max}}} - 1\right) \tag{11.7}
$$

$$
- \lambda_{l_s}\left(\frac{\mathbf{w}_{l_s}^{t,H}[k]\mathbf{w}_{l_s}^t[k]}{P_{\mathrm{max},l_s}^t} - 1\right)
$$

where v_{l_s} and λ_{l_s} are the Lagrange multipliers associated with the l_s-th SU transmission power and the PBS received interference, respectively.

In the gradient search-based cognitive capacity-aware algorithm (GS-CA), the weight vector for user l_s is updated at each iteration n, according to [13]

$$
\mathbf{w}_{l_s}^t(n+1) = \mathbf{w}_{l_s}^t(n) + \mu\,\nabla_{\mathbf{w}_{l_s}^t}\mathcal{L}\left(\mathbf{w}_{l_s}^t, v_{l_s}, \lambda_{l_s}\right) \tag{11.8}
$$

where $\nabla_{\mathbf{w}_{l_s}^t}$ is the gradient of $\mathcal{L}\left(\mathbf{w}_{l_s}^t, v_{l_s}, \lambda_{l_s}\right)$ w.r.t. to $\mathbf{w}_{l_s}^t$ and μ is an adaptation constant to be chosen relatively small in order to achieve convergence. Since the update is done separately on each subcarrier, we drop the frequency index $[k]$ and concentrate on the iteration index (n) in this recursion.

$$
\nabla_{\mathbf{w}_{l_s}^t}\mathcal{L}\left(\mathbf{w}_{l_s}^t, v_{l_s}, \lambda_{l_s}\right) = \frac{1}{\ln(2)}\left\{\frac{\rho_{l_s}\mathbf{B}_{l_s}(n)\mathbf{w}_{l_s}^t(n)}{1 + \rho_{l_s}\mathbf{w}_{l_s}^{t,H}(n)\mathbf{B}_{l_s}(n)\mathbf{w}_{l_s}^t(n)}\right\}
$$
$$
- \frac{v_{l_s}}{J_{\mathrm{sp}}^{\mathrm{max}}}\mathbf{H}_{\mathrm{ss},l_s}^H(n)\mathbf{H}_{\mathrm{ss},l_s}(n)\mathbf{w}_{l_s}^t(n) - \frac{\lambda_{l_s}}{P_{\mathrm{max},l_s}^t}\mathbf{w}_{l_s}^t(n) \tag{11.9}
$$

11.1.2   Performance Evaluation of CR-Based MU-MIMO

The system performance will be evaluated in terms of the symbol error rate and the ergodic channel capacity of PUs and SUs. The impact of the interference power constraints imposed to SUs and the number of antennas used at the base stations will be discussed when both capacity-aware MIMO and conventional MIMO-MRC are used.

11.1.2.1 Symbol Error Rate and Ergodic Channel Capacity

The SINR for SU l_s at iteration n, $\gamma_{l_s}(n)$, is given by

$$\gamma_{l_s}(n) = \frac{\mathbf{w}_{l_s}^{r,H}(n)\tilde{\mathbf{H}}_{ss,l_s}^{t}(n)\tilde{\mathbf{H}}_{ss,l_s}^{t,H}(n)\mathbf{w}_{l_s}^{r}(n)}{\mathbf{w}_{l_s}^{r,H}(n)\mathbf{B}_{l_s}\mathbf{w}_{l_s}^{r}(n)} \tag{11.10}$$

We observe that, in general, the off-diagonal elements of $\mathbf{B}_{l_s}$ are nonzero, reflecting the color of the interference. However in the asymptotic case of large N_s^r – element array, and given equal power transmitted by all users ($P_{l_s} = P_s$ and $P_{l_p} = P_p$), the central limit theorem can be invoked to show that

$$\mathbf{B}_{l_s} = \left(\frac{(L_s - 1)P_s + L_pP_p}{\sigma_n^2} + 1\right)\sigma_n^2\mathbf{I}_{N_s^r} \tag{11.11}$$

Thus, assuming MRC at the receiving SBS (i.e., $\mathbf{w}_{l_s}^{r}(n) = (\mathbf{B}_{l_s})^{-1}\mathbf{H}_{ss,1_s}(n)\mathbf{w}_{l_s}^{t}(n)$), we can express $\gamma_{l_s}(n)$ as

$$\gamma_{l_s}(n) = \left(\frac{\mathbf{w}_{l_s}^{t,H}(n)\mathbf{H}_{ss,l_s}^{H}\mathbf{H}_{ss,l_s}\mathbf{w}_{l_s}^{t}(n)}{(L_s - 1)P_s + L_pP_p + \sigma_n^2}\right) \tag{11.12}$$

For a large-scale MIMO, the channel vectors are nearly orthogonal, and hence $\mathbf{H}_{ss,1_s}^{H}\mathbf{H}_{ss,1_s}$ can be approximated by

$$\mathbf{H}_{ss,l_s}^{H}\mathbf{H}_{ss,l_s} = \mathbf{H}_{ss,l_s}\mathbf{H}_{ss,l_s}^{H} = \frac{\xi}{r}\mathbf{I}_r \tag{11.13}$$

where $r = \mathrm{N}_{\min} \triangleq \min(\mathrm{N}_T, \mathrm{N}_R)$ and $\xi = \sum_{i=1}^{r}\lambda_{i,l_s}$, with λ_{i,l_s} representing the eigenvalues of $\mathbf{H}_{ss,1_s}^{H}\mathbf{H}_{ss,1_s}$. Then (11.12) can be simplified as follows:

$$\gamma_{l_s}(n) = \left(\frac{\xi/r}{(L_s - 1)P_s + L_pP_p + \sigma_n^2}\right)\left|\mathbf{w}_{l_s}^{t}(n)\right|^2 \tag{11.14}$$

The symbol error rate, SER_{k,l_s}, associated with kth subcarrier of SU l_s, can be expressed as

$$\mathrm{SER}_{k,l_s} = E_{\gamma_{k,l_s}}\left[aQ\left(\sqrt{2b\gamma_{k,l_s}}\right)\right] \tag{11.15}$$

where $E[.]$ denotes the expectation operator, $Q(.)$ denotes the Gaussian Q-function, γ_{k,l_s} is the signal-to-interference-plus-noise ratio (SINR) associated with the kth subcarrier of cluster l_s, and a and b are modulation-specific constants. For binary phase-shift keying (BPSK), $a = 1$ and $b = 1$; for binary frequency-shift keying (BFSK) with orthogonal signaling, $a = 1$ and $b = 0.5$, while for M-ary phase-shift keying (M-PSK), $a = 2$ and $b = \sin^2(\pi/M)$.

The average SER$_{l_s}$ performance for user l_s can be estimated as

$$\text{SER}_{l_s} = \frac{1}{N_c} \sum_{k=0}^{N_c-1} \text{SER}_{k,l_s} \tag{11.16}$$

The Ergodic channel capacity, per subcarrier, for each cognitive user l_s is given by [4]

$$C(n) = E\left(\log_2 \left\{ \mathbf{I} + \frac{\rho_{l_s}}{\mathbf{B}_{l_s}} \frac{\mathbf{w}_{l_s}^{t,H}(n)\mathbf{H}_{\text{ss},l_s}^{H}(n)\mathbf{H}_{\text{ss},l_s}(n)\mathbf{w}_{l_s}^{t}(n)}{N_s^t} \right\} \right) \tag{11.17}$$

By noticing that for asymptotically large N_s^t (large cluster size), $\frac{1}{N_s^t}\mathbf{H}_{\text{ss},l_s}^{H}\mathbf{H}_{\text{ss},l_s} \to \mathbf{I}_{N_s^r}$ almost surely, and using (11.11), we can express the channel capacity asymptotically as

$$C(n) = E\left(\log_2 \left\{ 1 + \rho_{l_s}\mathbf{B}_{l_s}^{-1}\left|\mathbf{w}_{l_s}^{t}(n)\right|^2 \right\} \right)$$
$$= E\left(\log_2 \left\{ 1 + \frac{P_s}{(L_s-1)P_s + L_pP_p + \sigma_n^2}\rho_{l_s}\left|\mathbf{w}_{l_s}^{t}(n)\right|^2 \right\} \right) \tag{11.18}$$

and the gradient of $\mathcal{L}\left(\mathbf{w}_{l_s}^{t}, v_{l_s}, \lambda_{l_s}\right)$ as

$$\nabla\mathcal{L}\left(\mathbf{w}_{l_s}^{t}, v_{l_s}, \lambda_{l_s}\right) = \left(\propto (n) - \frac{v_{l_s}}{J_{\text{sp}}^{\text{max}}} \frac{\xi}{r} - \frac{\lambda_{l_s}}{P_{\text{max},l_s}^{t}} \right) \mathbf{w}_{l_s}^{t}(n) \tag{11.19}$$

where $\propto (n) = \dfrac{1}{\ln(2)}\left(\dfrac{\rho_{l_s}}{\frac{P_s}{(L_s-1)P_s+L_pP_p+\sigma_n^2} + \rho_{l_s}\left|\mathbf{w}_{l_s}^{t}(n)\right|^2} \right)$

Therefore, the iterative weight vector equation can be simplified as

$$\mathbf{w}_{l_s}^{t}(n+1) = \left(\mu + \propto (n) - \frac{v_{l_s}}{J_{\text{sp}}^{\text{max}}} \frac{\xi}{r} - \frac{\lambda_{l_s}}{P_{\text{max},l_s}^{t}} \right) \mathbf{w}_{l_s}^{t}(n) \tag{11.20}$$

11.1.2.2 Simulation Results

In our simulation setups, a CR-based MU-MIMO system with $N_s^t = N_p^t = 2$ transmit antennas is considered. The number of antennas at the PBS and at the SBS is the same, $N_p^r = N_s^r$, and varies from 16 up to 64. $L_s = 4$ SUs and $L_p = 4$ PUs. We assume BPSK modulation is assumed and the constraint imposed on the SUs are

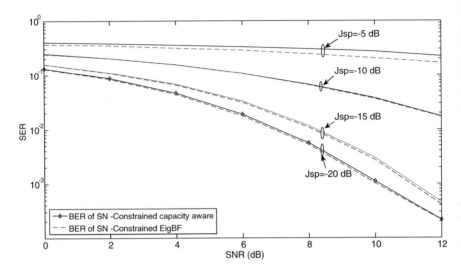

Fig. 11.2 Impact of interference power constraint, $J_{\text{sp}}^{\text{max}}$, on the SER of SUs. SUs using CCA and CEigBF schemes with $N_s^r = 32$

$P_{\text{max},l}^t = 0$ dB and $J_{\text{sp}}^{\text{max}} = -5, -10, -15$ *and* -20 dB. For the PN it assumed that a MU-MIMO system with non-constrained MIMO-MRC is used, i.e., EigBF at the transmitter and MRC at the receiving PBS.

Figure 11.2 shows the impact of $J_{\text{sp}}^{\text{max}}$ on the SER performance of SUs when using the constrained CA (CCA) and the constrained EigBF (CEigBF) schemes in MU-MIMO system with $N_s^r = 32$. As we can see, imposing a stronger $J_{\text{sp}}^{\text{max}}$ degrades significantly the SER of both schemes. It is also noted that CCA is slightly outperforming CEigBF.

Figure 11.3, on the other hand, shows the impact of the interference power constraints, $J_{\text{sp}}^{\text{max}}$, on the SER performance of PUs. For all cases of number of antennas at base stations, it is noted that as $J_{\text{sp}}^{\text{max}}$ becomes larger, the SER of PUs is improved due to the decreased and limited interferences from SUs.

It is also noted that the most significant interference at the PBS is caused when the SUs are using the non-constrained EigBF scheme, because it is designed without considering interference to PBS. It is also noted that as we increase N_p^r and N_s^r, the SER performance of PUs varies slightly with $J_{\text{sp}}^{\text{max}}$ variations. This means that when the number of base station antennas becomes large, the interference constrained imposed on SUs could be relaxed without impacting the SER performance of PUs.

Figure 11.4 shows the capacity of SN using CEigBF and CCA with various antenna configurations. For all cases it is noted that increasing the number of base station antennas results in an increase of system capacity. It is also seen that for the 2×64 and 2×32 cases, the CCA and CEigBF capacity curves are almost identical; however, for the lower number of base station antennas (2×24), CCA is outperforming CEigBF.

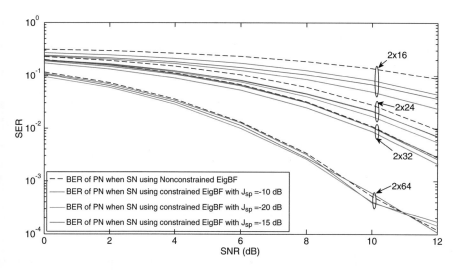

Fig. 11.3 Impact of interference power constraint, J_{sp}^{max}, on the SER of PUs

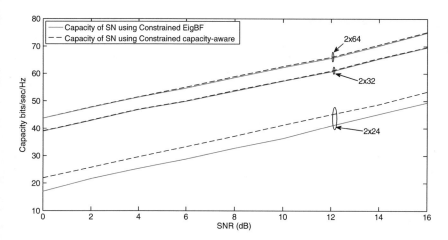

Fig. 11.4 Capacity of SN for different numbers of base station antennas

Figure 11.5 shows the impact of the interference power constraints, J_{sp}^{max}, on the capacity of SN. It is noted that, for both schemes, as J_{sp}^{max} becomes larger, the capacity of SUs is significantly reduced. On the other hand, it is noted that for the strongest interference constraint of $J_{sp}^{max} = -20$ dB the CCA is outperforming the CEigBF.

Figure 11.6 shows the impact of the interference power constraints, J_{sp}^{max}, on the capacity of PN. For all cases of number of antennas at base stations, it is noted that as J_{sp}^{max} becomes larger, the capacity of PUs is improved due to the decreased and limited interferences from SUs.

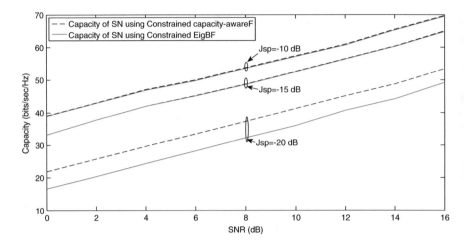

Fig. 11.5 Impact of interference power constraint, $J_{\text{sp}}^{\text{max}}$, on the capacity of SN when SUs use CCA and CEigBF

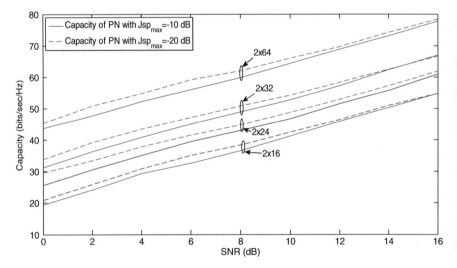

Fig. 11.6 Impact of interference power constraint, $J_{\text{sp}}^{\text{max}}$, on the capacity of PN

In summary, it was shown that in all cases, the CA MU-MIMO system is outperforming the well-known eigenbeamforming approach. It is also shown that the use of large number of antennas at the SBS could efficiently control the interference to PBS and could relax the constraints imposed on SUs. Thus, to achieve the QoS's needs for the SN without negatively affecting the PN, one should use a large-scale MIMO system.

11.2   MU-MIMO in Cognitive Radio Wireless Sensor Networks

In cognitive radio-based wireless sensor networks (CR-WSNs), sensor nodes are considered as SUs and can access the channels as long as they do not interfere with the PUs. Many recent works have proposed energy-efficient CR-WSNs where a large number of spatially distributed energy-constrained nodes with cognitive capabilities were deployed [27–29]. However, all these publications on CR-WSNs have assumed opportunistic spectrum sharing which may not be reliable and may limit the system capacity since it suffers from the interruptions imposed by the primary network (PN) on the SUs who must leave the licensed channel when PUs emerge. In cellular systems, one way to overcome these limitations is to incorporate MU-MIMO approach into CRNs to achieve higher spectral efficiency by multiplexing multiple SUs on the same time-frequency resources and protecting PNs from SUs' interferences. However, since it is difficult to integrate large antenna arrays in small sensors due to space limitations and circuit complexity, cooperative MIMO (CMIMO) can be used instead of real MIMO. In WSNs, CMIMO approach can be achieved using clustering of sensor networks; that is, the network is divided into groups of nodes called clusters, and all nodes within one cluster cooperate to form a virtual antenna array that can beamform a common signal to a base station or to distant relays. CMIMO approach is very attractive for WSNs because it can offer large increase in energy efficiency by allowing nodes with simple omnidirectional antennas to emulate a highly directional antenna array and focus their transmission in the direction of the intended receiver. CMIMO-based WSNs have been investigated in many works [23–27]. The focus in this chapter will be on CR-WSN with a very large number of WS nodes that can form a large-scale cooperative MIMO (LS-CMIMO). Such a system can significantly improve the system performance in terms of link reliability, spectral efficiency, and energy efficiency. It can also achieve optimal performances with the simplest forms of user detection techniques, i.e., MRC [4]. The capacity-aware LS-CMIMO schemes for CR-WSNs we will be based on the free-derivative particle swarm optimization (PSO) algorithm that is well known by its simple/fast hardware implementation and more suitable for the randomly distributed sensor nodes [34]. The PSO-based capacity-aware algorithm will be used at the cluster side to seek iteratively the transmit beamforming weights that maximize the uplink MIMO channel capacity for each cognitive cluster while controlling the interference levels to PUs without involving any gradient search.

11.2.1   PSO-Based Capacity-Aware Algorithm (GS-CA)

The gradient-based equation involves the derivative of the Lagrange multiplier given in (11.7), which might be difficult to achieve (highly nonlinear and non-differentiable). As an alternative, particle swarm optimization (PSO) algorithm can be used to optimize the transmit weight vector. PSO is a stochastic algorithm, where the birds or particles are mapped to the transmit beamforming weights and fly in the search space, aiming to optimize a given objective. In this context, the beamforming weights are optimized toward maximizing the constrained channel capacity given by (11.7). First, the PSO generates B random particles for each secondary user (i.e., random weight vector $\mathbf{w}_{l_s}^{t,(b)}$, $b = 1,\ldots, B$ of length $N_s^t \times 1$) to form an initial population set S (swarm). The algorithm computes the constrained channel capacity according to (6) for all particles $\mathbf{w}_{l_s}^{t,(b)}$ and then finds the particle that provides the global optimal channel capacity for this iteration, denoted $\mathbf{w}_{l_s}^{t,(b,\mathrm{gbest})}$. In addition, each particle b memorizes the position of its previous best performance, denoted $\mathbf{w}_{l_s}^{t,(b,\mathrm{pbest})}$. After finding these two best values, PSO updates its velocity $\mathbf{v}_{l_s}^{t,(b)}$ and its particle positions $\mathbf{w}_{l_s}^{t,(b)}$, respectively, at each iteration n as follows:

$$
\begin{aligned}
\mathbf{v}_{l_s}^{t,(b)}(n+1) &= \omega \mathbf{v}_{l_s}^{t,(b)}(n) + c_1\varphi_1\left(\mathbf{w}_{l_s}^{t,(b,\mathrm{pbest})}(n) - \mathbf{w}_{l_s}^{t,(b)}(n)\right) \\
&\quad + c_2\varphi_2\left(\mathbf{w}_{l_s}^{t,(b,\mathrm{gbest})}(n) - \mathbf{w}_{l_s}^{t,(b)}(n)\right)
\end{aligned}
\tag{11.21}
$$

$$
\mathbf{w}_{l_s}^{t,(b)}(n+1) = \mathbf{w}_{l_s}^{t,(b)}(n) + \mathbf{v}_{l_s}^{t,(b)}(n+1)
\tag{11.22}
$$

where c_1 and c_2 are acceleration coefficients toward the personal best position (pbest) and/or global best position (gbest), respectively, φ_1 and φ_2 are two random positive numbers in the range of $[0, 1]$, and ω is the inertia weight which is employed to control the exploration abilities of the swarm. Large inertia weights will allow the algorithm to explore the design space globally. Similarly, small inertia values will force the algorithms to concentrate in the nearby regions of the design space. This procedure is repeated until convergence (i.e., channel capacity remains constant for a several number of iterations or reaching maximum number of iterations). An optimum number of iterations are tuned and refined iteratively by evaluating the average number of iterations required for PSO convergence as a function of the target minimum square error for algorithm termination and as a function of the population size. Since random initialization does not guarantee a fast convergence, in our optimization procedure, we consider that the initial value of $\mathbf{w}_{l_s}^{t,(b)}(n)$ at iteration index $n = 0$ is given by the eigenbeamforming (EBF) weight, i.e., $\mathbf{w}_{l_s}^{t,(b)}(0) = \mathbf{w}_{l_s}^{t,\mathrm{EBF}} = \sqrt{P_{\max,l_s}^t}\,\mathbf{u}_{\max,l_s}$, where $\mathbf{u}_{\max,l_s}$ denotes the eigenvector corresponding to $\lambda_{\max,l_s}$, the maximum eigenvalue of $\mathbf{H}_{\mathrm{ss},l_s}^H \mathbf{H}_{\mathrm{ss},l_s}$. This initial guess enables the algorithm to reach a more refined solution iteratively by ensuring fast convergence and allows to compute the initial value of the received beamforming vector at iteration index $n = 0$. In our case we assume MRC at the receiving SBS; that is, $\mathbf{w}_{l_s}^r(0) = (\mathbf{B}_{l_s}(0))^{-1}\tilde{\mathbf{H}}_{\mathrm{ss},l_s}^t(0)$.

11.2.2   Performance Evaluation of CR-WSNs

We consider the uplink multicluster access scenario shown in Fig. 11.7 where a CR-WSN is divided into L_s clusters each having a leader, called the cluster head (CH). The cluster heads can be selected randomly or based on one or more criteria such as having the highest residual energy, the maximum number of neighbor nodes, and the smallest distance from base station.

In our analysis and for simplicity, we have assumed that each CH is connected to uniformly distributed sensor nodes. Each cluster consists of N_s^t cooperative nodes that collect the data broadcasted by its CH and relay it to a sink node (also known as a base station) using a beamforming scheme. It is also assumed that the capacity-aware beamforming weights are computed by each cluster head then transmitted to the cooperative nodes in the cluster. During data transmission, each collaborating node receives the signal broadcasted by the CH and retransmits a weighted version of it to the base station. In each cluster, the CH and its N_s^t cooperative nodes act as one secondary user (SU) with N_s^t– element antenna array. The L_s clusters (SUs) and their secondary base station (SBS) coexist with L_p PUs and their primary base station (PBS) via concurrent spectrum access. It is also assumed that both the sink node and

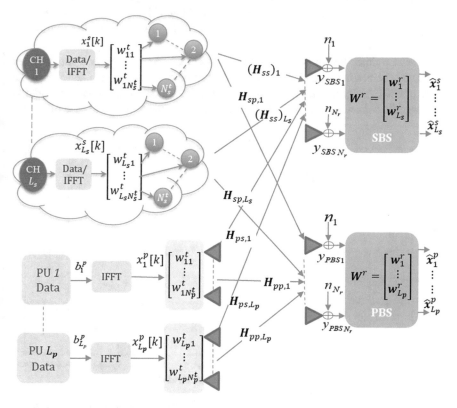

Fig. 11.7 Cognitive radio-based wireless sensor network

the PBS detect independent OFDM data streams from multiple clusters and PUs simultaneously on the same time-frequency resources.

In our simulation setups, a CR-WSN is organized into five clusters ($L_s = 5$), each with N_s^t cooperative nodes (N_s^t varying from 8 to 50) and one cluster head per each cluster. The number of antennas at the PBS and at the sink node (SBS) is the same, $N_p^r = N_s^r$, and is varying from 8 to 200. The number of PUs is $L_p = 5$ PUs, each transmitting with a single antenna ($N_p^t = 1$). We assume QPSK modulation. We impose $P_{max,l}^t = 0$ dB and $J_{sp}^{max} - 15$ dB on each cluster. For the OFDM configurations, we assume the 256-OFDM system ($N_c = 256$), which is widely deployed in broadband wireless access services. For the primary network, we assume a MU-MIMO system with non-constrained MIMO-MRC, i.e., eigenbeamforming at the transmitter and MRC at the receiving PBS. For the PSO parameters, the swarm size is 30, the maximum iteration number is 25, and the acceleration coefficients are $c_1 = c_2 = 2$. The inertia weight ω ranges from 0.9 to 0.4 and varies as the iteration goes on. The swarm size and the number of iterations were chosen by evaluating the average number of iterations required for PSO convergence as a function of the target minimum square error (MSE) threshold for algorithm termination (Figs. 11.8 and 11.9).

Figures 11.10 and 11.11 show, respectively, the SER performance and system capacity of the proposed PSO-CA and the traditional GS-CA schemes for different cluster size ($N_s^t = 8, 25$ *and* 50) with $N_s^r = 200$. It is observed from the results that for all cases, PSO-CA is outperforming GS-CA.

It is also noted that as we increase the cluster size, the performance gap between the two schemes is reduced. This difference can be explained by the fact that in the case of GS-CA algorithm, the weight vector Eq. (11.20) was based on the gradient of the ergodic capacity of (11.19) which was derived in the asymptotic case of a large

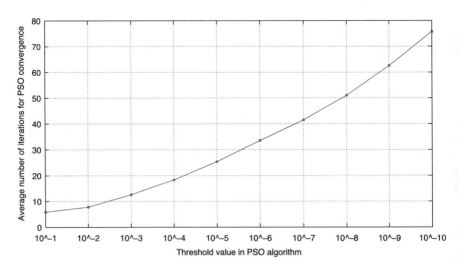

Fig. 11.8 Average number of iteration required for PSO convergence versus MSE threshold

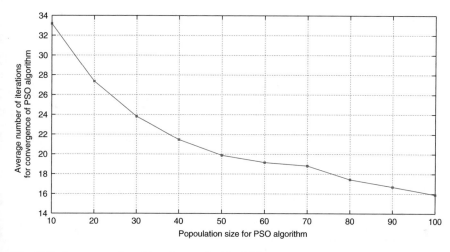

Fig. 11.9 Average number of iteration required for PSO convergence versus population size

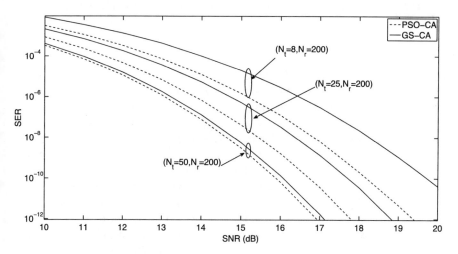

Fig. 11.10 SER performance of LS CR-WSN using PSO-CA and GS-CA schemes with different cluster size

number of transmitting antennas (large cluster size), whereas in the case of the free-derivative PSO-CA (11.17), the vector weights were updated using the exact expression of the ergodic capacity.

11.2.3 Energy Efficiency

Numerous studies have shown that multiuser massive MIMO can achieve a substantial reduction in the transmit power of each user because of their very high array gain. In [21] it was shown that under perfect channel state information (CSI)

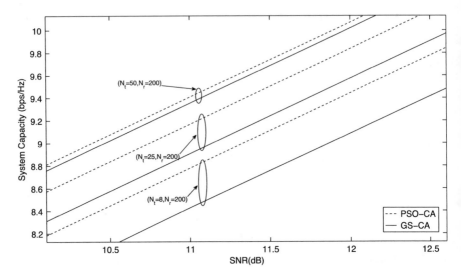

Fig. 11.11 Ergodic channel capacity of LS CR-WSN using PSO-CA and GS-CA schemes with different cluster size

and considering only intra-cell interferences, when the number of antennas at the base station, M, grows without bound, the transmit power of each user reduces proportionally to 1/M. However, this optimal energy efficiency (EE) value will be affected by different factors such as the imperfect CSI, the number of antennas, the order of the modulation scheme, the intercell interference, and the beamforming and multiuser detection schemes. In this chapter, where massive MIMO concept was exploited in CR-WSNs using virtual MIMO techniques, the optimal EE value will be reduced due to the inter-cluster interference and the primary user interference. For example, from Fig. 11.10, we can see that for the same SER performance of 10^{-5}, the difference in SNR when the number of sensor nodes per cluster goes from 8 to 50 is about 3dB. Consequently, when using 50 sensor nodes per cluster, the power can be scaled down by half compared to 8 sensor nodes per cluster.

11.3 Capacity-Aware Multiuser Massive MIMO for Heterogeneous Cellular Network

Deploying small cognitive cell networks over existing macro-cellular networks, also known as heterogeneous networks (HetNets), has emerged as a promising solution to deal with the increasing wireless traffic demands in next-generation 5G cellular networks [30–33]. The users in these HetNets are offloaded from the congested macro-cell base stations (MBSs) to the small-cell base stations (SBSs), which

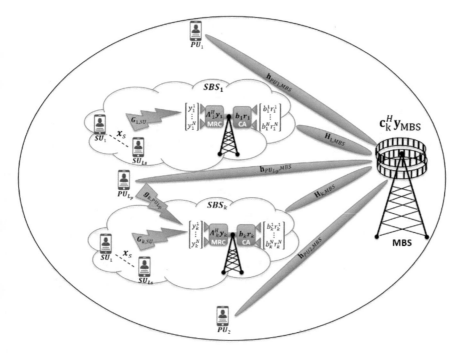

Fig. 11.12 HetNet consisting of one macro-cell and K small cells and their corresponding users

enhanced their quality of service (QoS) and increase the overall system capacity. These HetNets are supported by gigahertz bandwidth backhaul links that connect MBSs and the associated SBSs. Such gigahertz bandwidth can be achieved by conventional optical fiber- or millimeter wave (mmWave)-based wireless backhauls. Optical fiber backhauls, while reliable, might be expensive and difficult to deploy in HetNets where several small cells are unplanned and installed quite arbitrarily. Wireless backhauls, on the other hand, are more attractive to overcome the restriction of deployment and installation and can provide a cheap and scalable solution. The concept of massive MU-MIMO is applied to the HetNet of Fig. 11.12, where K small cells and one macro-cell are sharing the same frequency band. Each small cell includes one SBS equipped with massive N-element antenna array and L_s single-antenna secondary users (SUs). Each SBS and its users act as a cognitive network that coexist, via concurrent spectrum access, with L_p macro-cell primary users (PUs) and their primary MBS, which are also equipped with massive M-element antenna array. It is also assumed that both the SBS and the MBS detect independent OFDM data streams from their mobile users simultaneously on the same time-frequency resources.

The N × 1 received signal vector at the kth SBS is given by

$$y_k = \sqrt{p_u}G_{k,\text{SU}}x_s + n_{k,\text{SBS}} + \mathbf{I}_{\text{PU,SBS}} \qquad (11.23)$$

where $G_{k,\text{SU}} \in \mathbb{C}^{N \times L_s}$ is the channel matrix between the kth SBS and its L_s users, $x_s \in \mathbb{C}^{L_s \times 1}$ is the transmitted signal vector of L_s users in the kth small cell, p_u is the average power transmitted by each user (here we assume equal power allocation for all users), $n_{k,\text{SBS}} \in \mathbb{C}^{N \times 1}$ is the received AWGN vector at the SBS, and $\mathbf{I}_{\text{PU,SBS}}$ represents the interference introduced by macro-cell users (PUs) at the SBS and is given by

$$\mathbf{I}_{\text{PU,SBS}} = \sqrt{p_p}G_{k,\text{PU}}x_p \qquad (11.24)$$

where $G_{k,\text{PU}} \in \mathbb{C}^{N \times L_p}$ is the channel matrix between the kth SBS and L_p users, p_p is the average power transmitted by each PU, and $x_p \in \mathbb{C}^{L_p \times 1}$ is the transmitted signal vector of L_p users in the HetNet.

We consider MRC detection scheme at each SBS. The kth SBS processes its received signal y_k by multiplying it by the N × L_s receive beamforming weight matrix A_k^H as follows:

$$r_k = A_k^H y_k = \sqrt{p_u}A_k^H G_{k,\text{SU}}x_s + A_k^H n_{k,\text{SBS}} + A_k^H \mathbf{I}_{\text{PU,SBS}} \qquad (11.25)$$

The detection of user l_s by its kth SBS can then be expressed as

$$r_{k,l_s} = a_{k,l_s}^H y_k = \sqrt{p_u}a_{k,l_s}^H G_{k,\text{SU}}x_s + a_{k,l_s}^H n_{k,\text{SBS}} + a_{k,l_s}^H \mathbf{I}_{\text{PU,SBS}} \qquad (11.26)$$

where r_{k,l_s} is the l_sth element of r_k and a_{k,l_s} is the l_sth column of A_k.

The kth SBS transmits the l_sth user signal r_{k,l_s} by multiplying it by the N × 1 transmit beamforming weight vector b_{k,l_s} as follows:

$$s_{k,l_s} = r_{k,l_s}b_{k,l_s} = a_{k,l_s}^H y_k \, b_{k,l_s} \qquad (11.27)$$

The expression for the array output of the MBS in Fig. 11.1 can be written for each subcarrier as

$$y_{\textbf{MBS}} = \sum_{k=1}^{K} \sum_{l_s=1}^{L_s} \mathbf{H}_{k,\text{MBS}}s_{k,l_s} + \mathbf{n}_{\text{MBS}} + \mathbf{I}_{\text{PU,MBS}} \qquad (11.28)$$

where $y_{\textbf{MBS}}$ is the M × 1 vector containing the outputs of the M-element array at the MBS and $\mathbf{H}_{k,\text{MBS}}$ is the M × N frequency-domain channel matrix representing the transfer functions from the N-element antenna array of the kth SBS to the M,-element antenna array of the MBS, $b_k = [b_1, b_2 \quad \ldots \quad , b_N]^T$ is the N × 1 complex transmit weight vector of the kth SBS, $\mathbf{n}_{\text{MBS}}$ is the received M × 1 complex additive white Gaussian noise vector at the MBS, and $\mathbf{I}_{\text{PU, MBS}}[k]$ represents the interference introduced by PUs to SUs at the MBS and is given by

$$I_{\text{PU,MBS}} = \sqrt{P_p} H_{\text{PU,MBS}} x_p \tag{11.29}$$

where $H_{\text{PU, MBS}}$ is the $M \times L_p$ channel matrix from the L_p PUs to the MBS's M – element antenna array.

The MBS detects the l_sth user signal of the kth SBS by multiplying the output of the array y_{MBS} with the $M \times 1$ receiving weight vector, c_{k,l_s}^H as follows:

$$\widehat{x}_{k,l_s} = c_{k,l_s}^H y_{\text{MBS}} \tag{11.30}$$

where $S_{k,l_s} = c_{k,l_s}^H H_{k,\text{MBS}} s_{k,l_s}$ is the l_sth user signal of the kth SBS, $S_{I_s} = c_{k,l_s}^H \sum_{i=1, i \neq k}^{K} \sum_{l_s=1}^{L_s} H_{k,\text{MBS}} s_{k,l_s}$ is the multiple-access interference (MAI) from the $K - 1$ other SBSs, $S_{I_p} = \sqrt{P_p} c_{k,l_s}^H H_{\text{PU,MBS}} x_p$ is the MAI from L_p PUs, and $N = c_{k,l_s}^H n_{\text{MBS}}$ is the noise signal at the array output of the MBS. The signal detected from the l_sth user of the kth SBS can be expressed by (11.31), and the signal detected by the MBS from the l_sth user of the kth SBS can be expressed by (11.32).

$$
\begin{aligned}
S_{k,l_s} &= c_k^H H_{k,\text{MBS}} b_k r_{k,l_s} = c_k^H H_{k,\text{MBS}} b_k a_{k,l_s}^H y_k \\
&= c_k^H H_{k,\text{MBS}} b_k a_{k,l_s}^H G_{k,\text{SU}} x_s + c_k^H H_{k,\text{MBS}} b_k a_{k,l_s}^H n_{k,\text{SBS}} \\
&\quad + c_k^H H_{k,\text{MBS}} b_k a_{k,l_s}^H I_{\text{PU,SBS}}
\end{aligned}
\tag{11.31}
$$

$$\widehat{x}_{k,l_s} = c_k^H y_{\text{MBS}} = S_{k,l_s} + S_{I_s} + S_{I_p} + N \tag{11.32}$$

The SINR at the MBS for user l_s of the kth SBS can thus be depicted as

$$\gamma_{k,l_s} = \frac{c_{k,l_s}^H H_{k,\text{MBS}} s_{k,l_s} s_{k,l_s}^H H_{k,\text{MBS}}^H c_{k,l_s}}{c_k^H B_k c_k} \tag{11.33}$$

where B_k is the covariance matrix of the interference-plus-noise and is given by

$$B_k = B_{\text{SBS}} + B_{\text{PU,MBS}} + B_{\text{PU,SBS}} + B_n \tag{11.34}$$

where

$$B_{\text{SBS}} = \sum_{i=1, i \neq k}^{K} \sum_{l_s=1}^{L_s} H_{k,\text{MBS}} s_{k,l_s} s_{k,l_s}^H H_{k,\text{MBS}}^H$$

$$B_{\text{PU,MBS}} = p_p H_{\text{MBS},l_p} H_{\text{MBS},l_p}^H$$

$$B_{\text{PU,SBS}} = p_p H_{k,\text{MBS}} b_k a_{k,l_s}^H G_{k,\text{PU}} G_{k,\text{PU}}^H a_{k,l_s} b_k^H H_{k,\text{MBS}}^H$$

$$B_n = c_k^H c_k + H_{k,\text{MBS}} b_k a_{k,l_s}^H a_{k,l_s} b_k^H H_{k,\text{MBS}}^H$$

The ergodic backhaul channel capacity is given by

$$C(\mathbf{H}_{k,\text{MBS}}, s_{k,l_s}) = E\left(\log_2\left\{+\frac{1}{N}\frac{\mathbf{H}_{k,\text{MBS}}s_{k,l_s}\mathbf{s}_{k,l_s}^H\mathbf{H}_{k,\text{MBS}}^H}{\mathbf{B}_k}\right\}\right) \qquad (11.35)$$

11.3.1 Performance Evaluation

In the simulation setups, a HetNet organized into K SBSs (K = 10) and one macro-cell are considered. The number of antennas at the SBSs and at the MBS is the same, $N = M = 100$. Each SBS is serving $L_s = 10$ users, and the macro-cell is serving $L_p = 10$ users, each transmitting with a single antenna. For the backhaul link, we assume an MU-MIMO system with capacity-aware beamforming at each SBS and MRC detection at the MBS. For the access link, we assume MRC detection at each SBS. For the PSO parameters, the swarm size is 30, the maximum iteration number is 25, and the acceleration coefficients are $c_1 = c_2 = 2$. The inertia weight ω ranges from 0.9 to 0.4 and varies as the iteration goes on.

Figure 11.13 compares the system capacity of the CA MU-MIMO-based HetNet and the traditional macro-cell network. We observe that CA MU-MIMO based HetNet is outperforming the macro-cell network, especially at high SNR. Figure 11.14, on the other hand, compares the SER performance of both networks. It is again observed the HetNet is outperforming the traditional macro-cellular network.

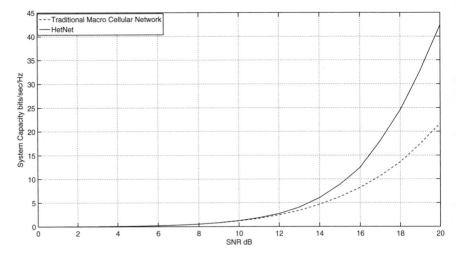

Fig. 11.13 Ergodic channel capacity for K = 20 SBSs and M = N = 100 antennas: HetNet vs traditional macro-cellular network

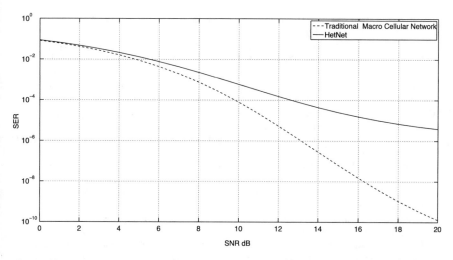

Fig. 11.14 SER performance for K = 20 SBSs and M = N = 100 antennas: HetNet vs traditional macro-cellular network

11.4   Summary

In this chapter we have considered the combination of MU-MIMO and CR to improve the spectral and energy efficiency of the secondary networks and guarantee their QoS. Two applications were presented: MU-MIMO in CR-WSNs where cooperative MIMO and clustering were exploited to increase the range and spectral efficiency of CR-WSNs and MU-MIMO in CR-HetNets where both the access and backhaul links of cognitive cells were improved.

References

1. Mitola J, Maguire GQ (1999) Cognitive radio: making software radios more personal. IEEE Pers Commun 6(6):13–18
2. Haykin S (2005) Cognitive radio: brain-empowered wireless communications. IEEE J Select Areas Commun 23(2):201–220
3. Munster M, Hanzo L (2001) Performance of SDMA multi-user detection techniques for Walsh-Hadamard-spread OFDM schemes. IEEE-VTC 4:2319–2323
4. Kang M (2004) A comparative study on the performance of MIMO MRC systems with and without cochannel interference. IEEE Trans Commun 52(8):1417–1425
5. Sulyman I, Hefnawi M (2008) Adaptive MIMO beamforming algorithm based on gradient search of the channel capacity in OFDMSDMA system. IEEE Commun Lett 12(9):642–644
6. Sulyman I, Hefnawi M (2010a) Performance evaluation of capacity-aware MIMO beamforming schemes in OFDM-SDMA systems. IEEE Trans Commun 58(1)

7. Sulyman I, Hefnawi M (2010b) Capacity-aware linear MMSE detector for OFDM-SDMA systems. IET Commun 4(9)
8. Yang LL, Wang LC (2008) Zero-forcing and minimum Mean-Square error multiuser detection in generalized multicarrier DS-CDMA Systems for Cognitive Radio. EURASIP J Wirel Commun Netw:1–13
9. Zhang R, Liang YC (2008) Exploiting multi-antennas for opportunistic spectrum sharing in cognitive radio networks. IEEE J Sel Top Sign Proces 2(1):88–102
10. Vu M, Yiu S, Tarokh V (2008) Interference reduction by beamforming in cognitive networks. In: IEEE GLOBECOM Telecom. Conference
11. Hamdi K, Zhang W, Letaief KB (2009) Opportunistic spectrum sharing in cognitive MIMO wireless networks. IEEE Trans Wirel Commun 8(8):4098–4109
12. Hefnawi M (2012) SDMA aided cognitive radio networks. In: IEEE 26th Biennial symposium on communications
13. Hefnawi M (2014) SER performance of large scale OFDM-SDMA based cognitive radio networks. Int J Antennas Propag:1–8
14. Hefnawi M, Abubaker A (2014) Channel capacity maximization in multiuser large scale MIMO-based cognitive networks. Int J Microwave Opt Technol 9(6):437–444
15. Bixio L, Oliveri G, Ottonello M et al (2010) Cognitive radios with multiple antennas exploiting spatial opportunities. IEEE Trans Signal Process 58(8):4453–4459
16. Wang L, Quoc NH, Elkashlan M et al (2017) Massive MIMO in Spectrum sharing networks: achievable rate and power efficiency. IEEE Syst 11(1):20–31
17. Nguyen VD, Tran LN, Duong TQ et al (2017) An efficient precoder design for multiuser MIMO cognitive radio networks with interference constraints. IEEE Trans Veh Technol 66 (5):3991–4004
18. Deng HJ, Chen SH, KU ML (2017) Multiuser MIMO Precoders with proactive primary interference cancelation and link quality enhancement for cognitive radio relay systems. IEEE Access 5:17701–17712
19. Marzetta L (2010) Noncooperative cellular wireless with unlimited numbers of base station antennas. IEEE Tran Wirel Comm 9(11):3590–3600
20. Rusek F, Persson D, Lau B, Larsson E, Marzetta et al (2013) Scaling up MIMO: opportunities and challenges with very large arrays. IEEE Signal Process Mag 30(1):40–60
21. Ngo HQ, Larsson EG, Marzetta TL (2013) Energy and spectral efficiency of very large multiuser MIMO systems. IEEE Trans Commun 61(4):1436–1449
22. Hoydis J, ten Brink S, Debbah M (2013) Massive MIMO in the UL/DL of cellular networks: how many antennas do we need? IEEE J Sel Areas Commun 31(2):160–171
23. Hefnawi M (2016) Large-scale multi-cluster MIMO approach for cognitive radio sensor networks. IEEE Sensors J 16(11):4418–4424
24. Coso D, Spagnolini U, Ibars C (2007) Cooperative distributed MIMO channels in wireless sensor networks. IEEE J Sel Areas Commun 25(2):402–414
25. Cui S, Goldsmith AJ, Bahai A (2004) Energy-efficiency of MIMO and cooperative MIMO techniques in sensor networks. IEEE J Sel Areas Commun 22(6):1089–1098
26. Jayaweera SK (2006) Virtual MIMO-based cooperative communication for energy-constrained wireless sensor networks. IEEE Trans Wirel Commun 5(5):984–989
27. Yuan Y, He Z (2006) Virtual MIMO-based cross-layer design for wireless sensor networks. Vehicular Technol IEEE Trans 55(3):856–864
28. Vijay G, Bdira BA, Ibnkahla M (2011) Cognition in wireless sensor networks: a perspective. IEEE Sensor J 11:582–592
29. Akan OB, Karli OB, Ergul O (2009) Cognitive radio sensor networks. IEEE Netw 23:34–40
30. Hefnawi M (2017) Channel maximization in wireless backhaul based HetNets. Int J Wirel Mobile Netw 9(3):51–60
31. Siddique U, Tabassum H, Hossain E, Kim DI (2015) Wireless backhauling of 5G small cells: challenges and solution approaches. IEEE Wirel Commun 22(5):22–31

32. Gao Z, Dai L, Mi D et al (2015) MmWave massive MIMO based wireless backhaul for 5G ultra-dense network. IEEE Wirel Commun 22(5):13–21
33. Tabassum H, Hamdi SA, Hossain E (2016) Analysis of massive MIMO-enabled downlink wireless backhauling for full-duplex small cells. IEEE Trans Commun 64(6):2354–2369
34. Kennedy J, Eberhart RC (1995) Particle swarm optimization. In: Proceedings of the IEEE conference on neural networks

Index

A

Ad hoc architecture, 67
AdapCode, 108
Adaptive neuro-fuzzy inference systems
    (ANFIS), 240, 252, 253
Adaptive resonance theory, 246
Adaptive turbo trellis modulation scheme
    (ATTCM), 107
Additive white Gaussian noise (AWGN), 55, 57
Ant colony optimization (ACO), 243, 244
Artificial intelligence (AI), 238
Artificial neural networks (ANN), 39, 239
    240, 246
Australian Communications and Media
    Authority (ACMA), 11
Authorized shared access (ASA), 5–6, 12–13

B

Band allocation technique, 54
Base transceiver stations (BTSs), 34
Bayesian networks, 239
Bee colony optimization (BCO), 243
Binary frequency-shift keying (BFSK), 264
Binary phase-shift keying (BPSK), 264
Bit error rate (BER), 227
Broadcasting
    agile nature, channel availability, 85
    CCC, 80–81, 85
    channel diversity and heterogeneity, 85
    characteristics
        effective spectrum utilization, 72
        efficient data transmission, 72
        minimum time consumption, 73

        PR user occupancy, 73
        PU constraint modelling, 73
        reduced channel switching, 73
        reduced overhead, 73
    collison avoidance, 86–87
    complete, 83
    group-based, 82–83
    metric-based, 81
    multi-hop, 71
    multiple-channel broadcast, 79
    neighbour channel selection, 86
    neighbour discovery, 86
    networking tasks, 72
    protocols, 77–78
    randomly selected channel, 79–80
    rapid channel switching, 87
    route selection, 87
    schemes, 79
    set of channels, 83–84
    single-channel broadcast, 79
    single-hop, 71
    SUs, 72
    transmitted signal, 71
    wireless transmission, 71

C

Capital expenditure (CAPEX), 3
Cellular architecture, 6
Centralized cooperative spectrum sensing
    (CCSS), 31–32
Channel handover process (CHP), 51, 52
Channel state information (CSI), 273
Channel-hopping sequence, 80

© Springer International Publishing AG, part of Springer Nature 2019
M. H. Rehmani, R. Dhaou (eds.), *Cognitive Radio, Mobile Communications and
Wireless Networks*, EAI/Springer Innovations in Communication and Computing,
https://doi.org/10.1007/978-3-319-91002-4

Printed in the United States
By Bookmasters